Innovative Renewable Energy

Series Editor

Ali Sayigh, World Renewable Energy Congress, Brighton, UK

The primary objective of this book series is to highlight the best-implemented worldwide policies, projects and research dealing with renewable energy and the environment. The books are developed in partnership with the World Renewable Energy Network (WREN). WREN is one of the most effective organizations in supporting and enhancing the utilisation and implementation of renewable energy sources that are both environmentally safe and economically sustainable. Contributors to books in this series come from a worldwide network of agencies, laboratories, institutions, companies and individuals, all working together towards an international diffusion of renewable energy technologies and applications. With contributions from most countries in the world, books in this series promote the communication and technical education of scientists, engineers, technicians and managers in this field and address the energy needs of both developing and developed countries.

Each book in the series contains contributions from WREN members and cover the most-up-to-date research developments, government policies, business models, best practices, and innovations from countries all over the globe. Additionally, the series publishes a collection of best papers presented during the annual and bi-annual World Renewable Energy Congress and Forum each year.

More information about this series at http://www.springer.com/series/15925

Ali Sayigh • Antonella Trombadore
Editors

The Importance of Greenery in Sustainable Buildings

Springer

Editors
Ali Sayigh
World Renewable Energy Congress
Brighton, United Kingdom

Antonella Trombadore
DIDA Department of Architecture
University of Florence
Firenze, Firenze, Italy

ISSN 2522-8927 ISSN 2522-8935 (electronic)
Innovative Renewable Energy
ISBN 978-3-030-68555-3 ISBN 978-3-030-68556-0 (eBook)
https://doi.org/10.1007/978-3-030-68556-0

This Springer imprint is published by the registered company Springer Nature Switzerland AG
The registered company address is: Gewerbestrasse 11, 6330 Cham, Switzerland

Introduction

Greenery is as important to buildings as its walls are because it requires natural lighting and gives beauty to naked walls and yards. Greenery reduces energy consumption in buildings. Moreover, shrubs, plants, and green grass make the inhabitants cheerful and indirectly reduce stress. Greenery acts as an absorbent of toxins and a provider of oxygen.

The old saying in the Middle East is "The three most beautiful things in life are: Water, Greenery and a Beautiful Face."

Whenever architects and builders have integrated greenery within the facades of their buildings or in the various spaces or on the roofs, plants have contributed to the physical and mental well-being making indoor spaces healthier.

This book is part of the World Renewable Energy Network's drive to encourage architects and builders to use greenery as much as possible in their design to reduce energy consumption and provide a pleasant appearance and pleasing aspect to their buildings. It shows and demonstrates how widespread the use of greenery is in buildings. The 12 chapters were chosen from 12 different countries representing the use and benefit of using greenery in buildings.

Chapter 2 consists of two parts: Part I deals with green facades and vertical walls while Part II shows how urban farming can create a circular economy and benefits to the occupants based on their experience in Italy.

Chapter 3 provides useful guidelines for the use of greenery in the Municipality of Bahrain, while Chap. 4 describes the thermal aspects of green walls in reducing energy consumption in buildings of Bahrain.

Chapter 5 to delves into vernacular architecture of courtyard housing with greenery considering it as a natural resource specifically in hot arid climates such as that of Iran where it provides cooling.

Chapter 6 has been written by three eminent architects, two from Mexico and one from Spain, outlining the importance of vegetation in buildings.

Although most chapters are related to a hot arid climate where cooling is a priority, Chap. 7 demonstrates the use of greenery to create a micro-cool climate in buildings.

Professor Aboulnaga in Chap. 8 describes the importance of urban farming in improving lifestyle, viability, and buildings in the cities. The chapter mostly describes the Egyptian scenario.

Chapter 9 discusses the use of green-roof housing design and its importance.

Chapter 10 again outlines the complementarity between vernacular and urban green architecture and patios in achieving sustainable contemporary architecture in Argentina.

Professor El Adli and his co-author concentrate on an integrated approach to planning sustainable green cities in Egypt.

Professor Scudo and his two associates, in Chap. 12, write about multiscale greening of buildings with a discussion on tradition and innovation of the built environment.

Professor Battisti in Chap. 13 envisages a green dream in regenerating cities through nature. In Chap. 14 the author is looking forward to the day when photosynthesis powers buildings.

Professor Zeiler writes, in Chap. 15, about the new development in the Netherlands towards the implementation of greenery in sustainable buildings in the country.

Chapter 16 is devoted to the importance and advantages of using greenbelts in a major complex such as the International Islamic University in Malaysia.

Finally Chap. 17 addresses the importance of using trees and shrubs in the Jordanian environment.

Greenery makes buildings alive instead of being a lump of dead construction of steel, concrete, and glass. The more greenery is introduced in buildings the more they become attractive, healthier, and alive.

This book is aimed at all architects, building construction authorities, planners, and policymakers to encourage the use of greenery in their future buildings and explain why it is important to do so.

Brighton, UK Ali Sayigh

Contents

The Added Value of Greenery for Sustainable Building: The Perspective from the Netherlands

Wim Zeiler

Introduction

Human activity is the dominant influence on climate and the environment. Risks to health from high temperatures are increasing with climate change, and new ways to mitigate urban heat island (UHI) effects are needed as nearly three-quarters of the global population will be living in towns and cities by 2050. We have to act responsibly for the future of humanity. In the early 1990s researchers began to discuss the contribution that design could make in the development towards sustainable development (Bhamra 2019). Design for sustainability over the last 30 years has developed and let to good examples of nature-inspired architecture and modern plant breeding by understanding the complexity of sustainable development (Bhamra 2019). Designers started looking for integrated techniques from plant science, building physics and architecture to use plants in the built environment (e.g. green infrastructure: green walls and roofs). This to search for energy reduction and comfort improvement and to identify opportunities for food production within cities. Important is the right balance in the sustainability triangle between environment, economy and society (Bhamra 2019); see Fig. 1; this forms the basis for responsible design.

Responsible design requires the actions of individual designers and design teams, focusing on economical responsible behaviour, environmental responsibility and social responsibility. It comprises three aspects (Eppinger and Maier 2019):

- People: Socially responsible design that enables better work conditions, healthy social interactions, and improvement of the human condition

W. Zeiler (✉)
Technische Universiteit Eindhoven, Eindhoven, Netherlands
e-mail: w.zeiler@bwk.tue.nl

A. Sayigh, A. Trombadore (eds.), *The Importance of Greenery in Sustainable Buildings*, Innovative Renewable Energy,
https://doi.org/10.1007/978-3-030-68556-0_1

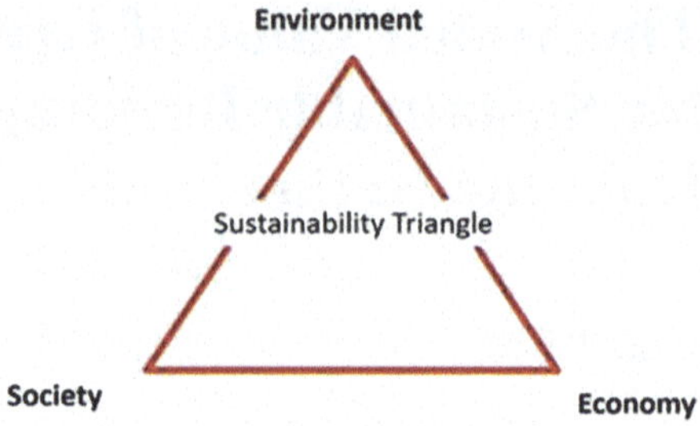

Fig. 1 The sustainability triangle (Bhamra 2019)

- Planet: Environmentally responsible design that meets the needs of the present without compromising the ability of future generations to meet their own needs
- Profit: Economical responsible behaviour design that encourages people and organizations to interact with others in a respectful and sustainable manner

One of the pioneers and promotors looking at the added value of greenery in sustainable building design was Ken Yang (1972). In his book The Green Skyscraper (1999) he laid the basis for designing sustainable intensive buildings and clearly demonstrated the added value of greenery. The Netherlands as densely populated country combined with high standards of living had always to (and knows how to) mould the natural environment to suit its needs (MVDRV 2019). A good example is the Netherlands' specific contribution to the ecological spectrum of the World Fair in Hannover 2000. Demonstrating that technology and nature need not be mutually exclusive, they can perfectly well reinforce one another. Nature arranged on many levels provided both an extension to existing nature and an outstanding symbol of its artificiality (MVDRV 2019).

By working with nature, cities can become more resilient to the changing climate. Embracing green roofs and facades helps to achieve liveable, sustainable built spaces (Young 2014). Furthermore, green roofs retain a high amount of rainwater, so they are perfect for harvesting, thus reducing the amount of water reaching urban sewage systems. With green roofs, water roofs, vertical farming and even high-rise greening with tree like the Bosco Verticale by Stefano Boeri in Milan (Smith 2015)—from large-scale interventions to smaller scale enrichments—cities in the future will need to look vastly different to cities now (Arup 2015; MVDRV 2019). The Nieuw Bergen project of MVRDV architects is such an example (Malone 2017). It is a residential development combining urban and green qualities in the neighbourhood of Bergen, Eindhoven; see Fig. 2.

Vertical farming techniques and urban agricultural systems, such as hydroponics, can potentially be utilized to help address the local food production as well as contribute to the environmental conditions within the cities. As a result of the economic, environmental and social developments, urban agriculture will become part of the urban culture in the twenty-first century (Zoellner 2013). New technology will allow roofs, walls and building facades to be 'greened', creating a filter for pollution, absorbing carbon dioxide by acting as a carbon 'sink' as well as providing

Fig. 2 Nieuw Bergen project MVRDV Eindhoven, Netherlands. (Source: https://mvrdv.nl/projects/290/nieuw-bergen?photo=17070)

natural cooling and insulation to enhance air quality within cities. Bio-inspiration from plants for the future built environment through green infrastructure and bio-based materials might lead to new architectural designs with benefits for (Wootton--Beard 2018):

- Thermal comfort—cooling the atmosphere (indoor and outdoor) to reduce the impact of heat waves
- Energy use/building thermoregulation—applying an external layer to regulate internal temperature and energy use
- Indoor environment—improving air quality, improving hygrothermal conditions and regulating noise
- Outdoor air pollution—trapping pollutants in canopy or substrate, removal by absorption or engineered solutions
- Mental health—visual stimulation, aesthetic quality and evolutionary preferences

All resulting design actions should emphasize the benefits of improved energy efficiency of buildings; new sustainable building technologies; low-carbon, renewable energy production; and reduction of energy demand. Building greenery, plants and architecture and bio-inspiration from plants all offer new opportunities for the future built environment. However, it is important to get a true understanding of its values as well.

The Added Value of Greenery: An Integrated Characterization Framework

The study by Zhao et al. (2019) presents a summary of the added value and benefits of green buildings through a bibliometric research approach. They reviewed in total 2980 articles published in 2000–2016 and the following key research topics were identified: vertical greening systems, green and cool roof, water efficiency, occupants' satisfaction and financial benefits of green building. This study forms the basis for the discussion of the different greenery aspects combined with the overview by Wootton-Beard (2018) and completed with additional recent Dutch research.

Green and Cool Roof

Cool roofs are the roofing systems using the cool materials that present a high albedo. In this cluster, Romeo and Zinzi (2013) found that the cool-roof application on a non-residential building decreased the roof surface temperature by 20 °C, and that it was effective in reducing cooling and total net energy demand. Green roofs, which are the roofs partially or completely covered with vegetation, are also common in green buildings. Santamouris (2014) reviewed the technologies to mitigate heat island phenomenon and analysed the effect of increasing the albedo of cities by cool roofs and using vegetative green roofs on heat island mitigation. Gill et al. (2007) indicated that greening roofs in areas with a high proportion of buildings can significantly lower down the surface temperature and reduce the rainwater runoff.

Akbari et al. (2001, 2012) indicated that urban trees, cool roofs and cool pavements had significant influence on urban air temperature, thereby reducing cooling energy consumption. Similarly, Ng et al. (2012) investigated the cooling effect of urban greening in Hong Kong, and revealed that roof greening was not effective for human thermal comfort near the ground and that grass was less effective than trees in lowering the temperature of pedestrian areas.

Sailor (2008) developed a model that provides energy modeller with vegetated green-roof design options. Additionally, Takebayashi and Moriyama (2007) found that the sensible heat flux was small on the surfaces of green roofs and the roofs with highly reflective white paint.

The representative research on 'land surface temperature' was undertaken by Wong et al. (2003), who performed a field measurement of the thermal impacts of roof gardens in Singapore and confirmed that roof garden provided thermal benefits for both buildings and surrounding environments. Additionally, Santamouris et al. (2007) calculated both the cooling and heating load for a whole building with a green roof and for its top floor, and found the great reduction in the building's cooling load in summer as well as the insignificant effect on heating load in winter.

Substrate can neutralize localized acid rain (Wootton-Beard 2018). Depending on atmospheric deposition rates, green roofs can become a source of pollution in the longer term, particularly metals, e.g. Pb, Cr, Cd, Cu and Zn. Fertilization will create a source of water contamination for N, P and K. It is highly dependent on soil depth, type, age and weather conditions (e.g. wind direction) as well as on urban geometry (Wootton-Beard 2018).

The Netherlands has an estimated 200 million m^2 of flat roofs. Most of these roofs could be made green. Ever more municipalities adopted a policy to actively promote green roofs. Not yet on such a wide scale as in some other countries, but it is a start. Most municipalities opt for a grant scheme for both residents and the business sector which contributes to the cost of installing a green roof, usually 50% of the total cost.

Vertical Greening Systems

These are usually classified into green facades and living walls in accordance with their growing method (Ottelé et al. 2011; Perini and Rosasco 2013). Living walls can support the vegetation that is rooted on the walls or in the substrates attached to the walls (Chen et al. 2013) while green facades are made by the climbing plants attached directly to the wall surface (Perini et al. 2011). Vertical greening systems can be used as passive systems to save energy, with the consideration into the following mechanisms: blockage of solar radiation because of the shadow provided by vegetation, evaporative cooling produced by blocking the wind and by evapotranspiration from substrate and plants, and thermal insulation brought by the substrate and vegetation (Perez et al. 2011). Additionally, Wong et al. (2010) found that vertical greenery systems lowered down the surface temperature of building facades in the tropical area, thereby reducing the cooling load and energy expense. The principal cooling effect is evapotranspiration (evaporation + transpiration) (Wootton--Beard 2018) which is fundamentally determined by the availability of moisture (and wind flow), vegetation cover, precipitation, irrigation and humidity. The vegetation replaces sensible heating with latent heating and reduces the Bowen ratio, the ratio of heat flux to moisture flux near the surface. The vegetation modifies surface roughness and wind flow, thus altering the convective heat exchange. Heterogeneity of the greenery is necessary and leads to best effects on heat loss (Wootton-Beard 2018). There is a minimum size of 50,000 m^2 to have significant cooling effect, so it is only relevant at town/city scale or for large green spaces. There is not yet any evidence for the network of smaller spaces owing to necessary fetch (Wootton-Beard 2018).

Incorporation of vegetation has a potential impact on the building design process and on building energy use. The principal effect of greenery on building thermoregulation is that insulation/shading is provided to building envelope which results in more stable internal temperatures (Wootton-Beard 2018). Plant albedo values are higher than grey surfaces creating more stable roof temperature. Also the shading potential is considerable, and can be a cost-effective mitigation measure (Gupta and Gregg 2012). Rural vegetation reflects 15–25% of short-wave radiation (Armson et al. 2012); however, effectiveness is driven by leaf size, crown area and leaf area index (Santamouris 2014). Mangone and van der Linden (2014) studied the effect of a vegetation shade canopy of an 11,000 m^2 office building on the energy consumption and carbon emissions. The thermal effects of the vegetation canopy were found to have a slightly greater effectiveness than the original shading solution (Mangone 2015).

A special class of vertical greening systems are the vertically integrated greenhouses (VIG) which combine a double-skin building facade (DSF) with a hydroponic greenhouse (Levenston 2019). The DSF can reduce the energy used for space conditioning in modern high-rise buildings considerately up to 20%. A DSF consists of a vertically continuous void space enclosed by a second curtain of glazing over the entire facade. A DSF provides solar heat in winter and buoyancy-driven cooling flows in summer, and allows opening of windows year-round. Already in 2003 van Passen (Stec et al. 2005; Stec and van Passen 2005) started their research aimed at defining the thermal performance of the double-skin facade with plants. Further research by Fang et al. (2011) and among others Larsen et al. (2013) proved that the use of plants in building facades is a useful bioclimatic strategy providing many benefits to the thermal, acoustic and psychological comfort of the inhabitants. The VIG combines a DSF with a novel system of hydroponic food production, for installation on new high-rise buildings and as a potential retrofit on existing buildings. In addition to producing food, plants can reduce building maintenance costs by providing shade, air treatment and evaporative cooling to building occupants.

Occupants' Satisfaction

One of the main disturbances in open-space office buildings is noise. This is something where green partitioning walls can help. The principal positive effect of green living walls inside buildings for sound insulation is that it provides a physical noise barrier and/or sound insulation (Wootton-Beard 2018). It is effective at reducing sound reduction index (15 db) when used as a physical barrier. A sound absorption co-efficient of 0.4 is reachable (Azkorra et al. 2015). It provides some acoustic insulation for buildings (Pérez et al. 2016) but it can also be used to provide a source of pleasant sounds, e.g. bird song and wind through leaves (Irvine et al. 2009).

Indoor thermal comfort is to a large extent determined by the availability of moisture (and airflow) (Wootton-Beard 2018). Important influencing factors of the applied greenery are vegetation cover, irrigation and humidity. They result in positive effects on thermal comfort as proven in controlled studies throughout a range of seasons and temperatures leading to reduced energy consumption (Wootton-Beard 2018). However, large number of plants are required to have significant cooling effect which is then also highly influenced by building characteristics and use patterns. Nevertheless the relatively lower air temperature and higher relative humidity levels proximate to the living wall (Gunawardena and Steemers 2019a, b) could be used as a kind of green air conditioning using indoor green wall systems as a climate control method (Armijos Moya et al. 2017, 2018). Not only passive green air conditioning systems can be applied but also active systems incorporating fans have been designed, e.g. a smart wall (Naava 2019).

Several short-term studies have found evidence that plants may improve occupant thermal comfort, yet this phenomena had not yet been rigorously evaluated (Mangone et al. 2014). Therefore they evaluated in a quasi-experiment the effect of indoor plants on the thermal comfort of 67 office workers within an office building for four separate months, 1 month in each season. The presence of a substantial quantity of plants in the work environment had a significant effect on the perceived thermal comfort of the participants. On average the occupants of the rooms in which the presence of plants was alternated were 12.0% more thermally comfortable when plants were present in the room. This would allow the temperature set point to be raised in the summer and lowered in the winter leading to reduced energy consumption and carbon emission rates (Mangone et al. 2014).

The principal effect of plants on indoor air quality is that they absorb/trap harmful air pollutants and 'clean' the air (Wootton-Beard 2018; Armijos Moya et al. 2018). Already in the 1980s, the NASA Clean Air Study presented some studies about the behaviour of plants regarding IAQ (Wolverton et al. 1984, 1989). Plants trap particulate matters PM2.5 and PM10 which are harmful to health, e.g. the tree cover of West Midlands estimated to reduce PM10 by 4% (McDonald et al. 2007). Its performance is related to canopy density and leaf traits, e.g. waxiness and hairiness (Wootton-Beard 2018). In addition plants can absorb some pollutants, e.g. NO*x* and SO2, and the substrate can be a sink for heavy metal deposition and pollutants, and is preferable to surface dust for water quality.

There are active green wall products available that include either absorbents or a rhizosphere, the vicinity of plant roots together with the rhizobacteria (Torpy et al. 2016). House plants (Kapoor 2017) can remove some harmful chemicals, e.g. formaldehyde, benzene, xylene, ammonia, trichloroethylene, acetone and some VOCs. Leaves trap PM10/PM2.5 to varying degrees based on physiology. However, the effectiveness is strongly dependent on building users and behaviour (i.e. opening a window). Furthermore, a large number of plants are needed in the average-sized home (10–20 per room).

Wolverton (2010) compiled a list with 50 of the best, most useful, easiest-to-maintain plants after his research, plus information regarding exactly what you need to do with them, and combined all this knowledge (Wolverton 2008). The most practically applicable part of the book is the assessment of a total of 50 houseplants. Wolverton (2008) has done this based on the following points:

- The degree to which hazardous substances are filtered
- Ease of maintenance
- Sensitivity to diseases
- The degree of water evaporation

The golden cane palm, also known as the areca palm or in Latin *Chrysalidocarpus lutescens*, is one of the best houseplants for removing all air pollutants in a house (Wolverton 2008). The broadleaf lady palm, or the *Rhapis excelsa*, is just slightly less effective in purifying, but is more user friendly and more resistant to diseases. Every plant has its own strength. Studies to the effects are mainly based on small-scale, controlled-chamber experiments, not real living scenarios (Wootton-Beard 2018). Plants absorb CO2 during the day. Improving the air quality in school classrooms, where the pupils produce high levels of CO2, using plants proved to be a practically executable solution. Research by van Duijn et al. (2011) showed that the classrooms from the third group, with plants and plenty of light, achieved some excellent scores (IntoGreen 2019): 10–20% CO2 reduction and degradation of

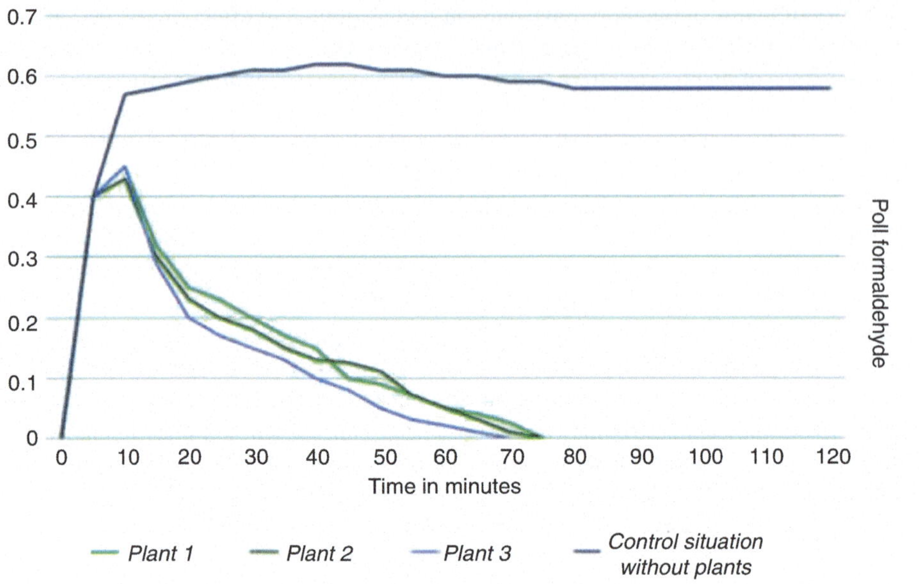

Fig. 3 Degradation of pollutant formaldehyde within 45 min to 0 level, as a result of the plants' air-purifying properties (Velzeboer et al. 2019)

formaldehyde within 45 min to 0 level, as a result of the plants' air-purifying properties. Figure 3 shows the breakdown of formaldehyde in a classroom without plants, the blue line; in classrooms with plenty of light and plants, the pink/magenta line; and in classrooms with plants with little light, the blue and purple lines. So plants can even significantly contribute to the breakdown of this substance in darker classrooms.

Velzeboer et al. (2019) studied the potential of air-purifying plant systems (plants with ventilation systems) to remove particulate matter (PM) and volatile organic compounds (VOC) from air. A new experimental set-up (a climatic chamber) was used with high-end reference equipment to measure air quality and where specific concentrations of PM and VOC can be generated. They selected six air-purifying plant systems and two mechanic air-purifying systems that use classical filters. For the plant systems the removal ranged between 41% and 50%. The mechanical systems showed a much higher aerosol removal ranging from 71% to 87%. The study nevertheless showed how plant systems can contribute to an improvement of the indoor air quality.

Visual/aesthetic qualities of transforming grey-green have positive psychological effects (Wootton-Beard 2018). There is a clear link between well-being and green space (Takano et al. 2002; de Vries et al. 2003). Well-managed greenery increases the sense of community attachment (Kuo et al. 1998). Viewing greenery ameliorates attention fatigue and stress (Ulrich 1983; Ulrich et al. 1991); it is naturally superior to urban for attention function and emotional state (Hartig et al. 2003). Effects have been replicated in hospital patients, school children, office workers, etc. Leaman and Bordass (2007) examined the occupants' opinions on British green buildings, and found that green buildings were difficult to manage and usually repeated past mistakes in conventional buildings. Additionally, Paul and Taylor (2008) compared occupants' perceptions on the aesthetics, serenity, lighting, acoustics, ventilation, temperature, humidity and satisfaction of a green building with those of two conventional buildings, but did not find that green buildings were more comfortable. Still the positive effects of greenery in urban working environments have been becoming more clear in the last years as shown by Hiemstra et al. (2019). Greenery in and around offices and other working environments is good for both the indoor and outdoor climate. It has a positive effect on the health and general well-being of employees. It aids concentration, reduces stress and increases staff productivity (Hiemstra et al. 2019). Hermans et al. (2019) developed a conceptual model, describing the short-term, medium and long-term effects of plants on the indoor climate and the health and well-being of people. They tested the model by means of intervention research at three companies and eight homes for the elderly. The effects of plants on the physical indoor climate were measured with sensors, and the effects on the employees concerning health and well-being were measured with questionnaires. At the companies, significant effects were found on relative humidity (up), attractiveness of the workplace (more attractive), state of mind (more positive), satisfaction with own functioning (higher) and sickness absence reporting (less). The need for recovery after a working day showed a reversed effect (rising).

Although the Netherlands is a very small country it is big when it comes to green. The Dutch sector associations VHG (horticulturists and interior landscapers) and VGB (trade in plants and flowers) promoted that plants are so much more than just pretty. The conclusive insights were published on the online platform Into Green; see Fig. 4.

Financial Benefits of Green Building

Kat (2003) investigated the financial and cost performance of green building projects, revealing that the premiums for green buildings in the USA were below 2% and that energy consumption was on average 28% less than conventional buildings (Wootton-Beard 2018). However, more upfront cost is usually needed for green sustainable office buildings than non-green buildings (Davis Langdon 2007). In addition, Newsham et al. (2009) surveyed 100 institutional and commercial buildings with LEED certification, and indicated that 28–35% of these buildings had higher energy consumption than non-green buildings, despite an average energy saving of 18–39% per floor area. As for the residential developers, Deng and Wu (2014) indicated that the Singaporean developers paid for nearly all the extra expenses for higher energy efficiency occurring in the construction process but cannot receive significant financial benefits from the development of green residential buildings. However, Zhang et al. (2011) argued that green strategies can offer developers competitive advantages, and Eichholtz et al. (2010) revealed that investment

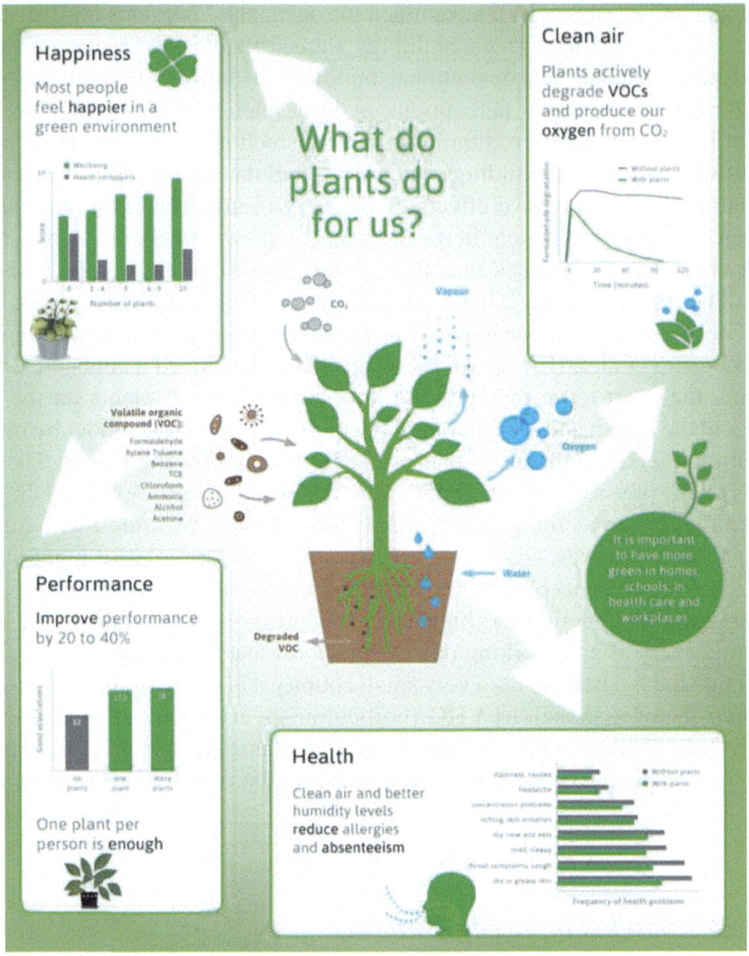

Fig. 4 The benefits of plants for occupants (IntoGreen 2019)

in green building could provide developers with economic benefits. Organizations revolve around returns and profit, and in most organizations this is generated by the employees (Wootton-Beard 2018). The potential of greenery for buildings to influence the level of sick leave and to affect productivity positively has been scientifically proven for decades (Wootton-Beard 2018). Potential effect of designs on revenues is not easy to calculate and 'prove', but the financial payback is much shorter than for example energy for energy reduction. For example: measures that improve productivity of office employees typically create 25 times more gain than measures that save energy costs! While numerous studies have been conducted on its potential benefits and its environmental impacts, there has been little focus on its life cycle cost (LCC) which is a crucial factor for building owners to adopt greenery such as a VGS as well as to compare among various types of system (Huang et al. 2019). To fill this gap, the study by Huang et al. (2019) aims to identify and calculate the LCC of main types of VGS available in Singapore. The cost incurred during each life cycle stage including initialization, installation, operation and demolition, is recorded based on eight completed and ongoing cases selected in Singapore. The framework provides a systematic perspective for building stakeholders to understand the life cycle costing of VGS by including the calculative processes, cost components and earnings from greenery benefits (Huang et al. 2019).

Water Efficiency

Mentens et al. (2006) found that the annual rainfall-runoff relationships for green roofs were significantly influenced by the depth of the surface layer, and indicated that the use of green roofs can significantly reduce the rainfall-runoff for a region and individual buildings. Berndtsson (2010) discussed various factors influencing runoff dynamics from green roofs and the influence of a green roof on runoff water quality. Green walls, installed on the side of the buildings, mainly for their aesthetics and microclimate benefits, could become effective on-site grey water treatment solutions (Fowdar et al. 2017; Prodanovic et al. 2019). A first step in the development of grey water-treating green walls is to examine how variation in plant spaces and operational conditions (hydraulic loading rate, inflow concentrations and intermittent drying) influences nutrient removal from light grey water. Prodanovic et al. (2019) conducted an experiment over 12 months on a large-scale pilot green wall located in a laboratory. They showed that ornamental plant species can play an important role in nitrogen and phosphorus uptake from grey water (Prodanovic et al. 2019). However, more research is needed to convert these nature-based systems into truly robust and effective low-cost water purification technologies.

Summary of the Benefits of Greenery

To summarize the benefits of greenery Xing et al. (2017) describe an integrated characterization framework (to aid the mainstreaming of urban green planning, design and retrofit). To make greening design decisions in relation to the importance of greenery in sustainable buildings (plants, substrate/supporting elements, irrigation and maintenance) is complex. He identifies ten categories of potential benefits (Fig. 5, left) and four key design decisions (Fig. 5, right).

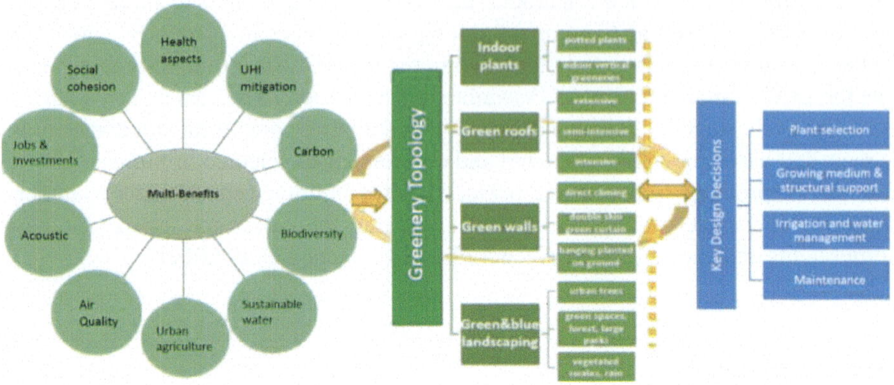

Fig. 5 A generic characterization framework for the multi-benefits of greenery and the related key design decisions (after Xing and Bosch 2017)

Examples of Greenery in Sustainable Buildings in the Netherlands

Green House Technology

The Netherlands is a small, densely populated country, lacking almost every resource necessary for large-scale agriculture. Despite this, the small country of the Netherlands is the world's number two exporter of food, second only to the USA, which is incomparable in size! It is this lack of resources and not-so-optimal climate conditions that has let the Dutch greenhouse industry grow into what it has become today. They took control over the climate by growing in a protected environment and by applying smart technology and improved yields significantly on the same growing area (Raus 2018). By using the world's most efficient agricultural technologies, the Netherlands is able to do far more with less. This important aspect of the Dutch culture was also transferred to the built environment where greenhouse technology was applied.

Lumen: Wageningen—1998

One of the Dutch pioneer buildings was the Alterra building, now called Lumen, in Wageningen, designed by Behnisch and Behnisch (Behnisch 2019) in 1993 and finished in 1998; see Fig. 6. The design brief was for a functional, user-friendly research facility working in harmony with nature, versatile and ecologically sound. Its design is centred around the building's users and surroundings. The building facilitates researchers at the Agricultural University in Wageningen in a centralized, collaborative work environment. The institute's scientists are engaged in work on

Fig. 6 Lumen—Wageningen (Behnisch 2019)

forestry, urban landscapes, soil microbiomes, biodiversity and climate adaptation through the world. Lumen is characterized by light, natural materials, greenery and water, on the inside as well as on the outside. The building was designed not to dominate its rural setting, but to embrace the landscape, with all workplaces in direct contact with indoor and outdoor gardens. Two indoor gardens provide the focus for daily activities and function as informal meeting areas. Beyond this, they are an integral component of the building's energy concept, improving the performance of the external envelope. The office building has two gardens (2000 m^2) inside the building. The gardens help with the building's climate control and create a better environment to the people working in this building. All the rooms are connected to the garden for optimal benefits.

The regional landscape structure of 'polders'—small rectangular fields divided by wind breaks and canals—served as an organizing principle for the building's form. On either side of the building, the gardens reflect distinct native Dutch landscape forms, while between the office wings of the building the gardens are roofed, and the more temperate microclimate allows for subtropical species, external circulation, outdoor meetings and chance encounters between colleagues. The lab wing to the north is the only portion of the 12,000 m^2 building with mechanical cooling, while the water features and vegetation within the greenhouses (designed in collaboration with the artist Michael Singer) play a key role in the environmental control.

Villa Flora: Venlo—2011

One other interesting example is the Villa Flora by Kristinsson (2011) that was built in Venlo in 2011. It is a combination of a greenhouse and an office building, designed to reach for the sustainable synergy between both building elements and building service systems. In 2009 it became part of the Dutch Government UKP NESK programme to stimulate innovation for energy-neutral buildings. UKP means unique chance projects and NESK means 'towards energy-neutral schools and offices' (Naar Energieneutrale Scholen en Kantoren). This programme of the Dutch Government gave in 2010 funding to projects which show exceptional innovation in the area of energy conservation, sustainability or organization within the building industry. Characteristics of the Villa Flora project are very-low-temperature heating at 25 °C, very-high-temperature cooling at 15 °C and seasonal storage of heat and cold in the groundwater. By the sloping roof and many glasses the transparent building looks very much like a greenhouse; see Fig. 7. There is, therefore, the use of smart energy-saving applications from the greenhouse horticulture and an optimal use of solar energy.

The concrete skeleton of the building is demon table, flexible and reusable. That makes a fast construction time as possible and also provides for a simple integration of all to apply techniques. The building can therefore keep pace with the development in the field of sustainability.

The seasonal surplus of heat and cold from the solar power system is stored in the aquifer approximately 50 m beneath the building. The building is equipped with a heating system based on very low temperatures (24–25 °C), enabled by fine-wire heat exchangers recently invented. The cooling of the indoor climate in summer is based mainly on the same system.

The vacuum sewer system and organic waste collection are aspects of ecological innovation. By means of an anaerobic bioreactor and a micro-turbine, electricity can be produced on location with addition of organic waste. The filtered exhaust gases are used as nutrients up till 1000 ppm, in order to take care of the CO2 balance. Pure water can be retained within this process, which takes 2 or 3 days.

Hogeschool Dronten: Dronten—2012

Hogeschool Dronten is a college focusing on education, research and application of agricultural knowledge. Its new premises are encased in a glass structure with a height of just over 16 m: the tallest glass casing in Europe; see Fig. 8. Definitely

Fig. 7 Villa Flora—Venlo—Netherlands (Kristinsson)

worthy of the name 'BrightBox'! Hogeschool Dronten wanted its college building to have a unique look befitting its agricultural character. And it was to incorporate a smart, open area suitable for large events, to compensate for the lack of such a venue in the city centre of Dronten. It was decided not to make use of traditional curtain walls and roof systems for the new build of this large-scale daylight structure. Instead, an entirely new daylight solution was introduced for utility facilities inspired by greenhouses which design concept offers Hogeschool Dronten many advantages:

- Great appeal—the building's look and the possibility of organizing events in the large meeting area are very attractive for students.
- Lower exploitation costs—the new building implies a saving of more than a 100,000 euros a year in energy costs for the Hogeschool relative to its last premises.
- Greater welfare—the pleasant indoor climate and large amount of daylight make the employees and students feel fitter and more energetic and alert.

The solution was realized in a short space of time and for the same budget as a traditional college building. The enormous glass structure is a highly distinctive

Fig. 8 CAH in Dronten, the Netherlands (BDG Architecten Zwolle)

eye-catcher in the world of education. Many people still visit Hogeschool Dronten to find out precisely how the architecture and the technology behind it work.

The entire building for advanced education is placed within a high glasshouse, which refers to the kind of education. This has great advantages on the field of energy consumption and exploitation and results in extra space for a more informal way of education. The two buildings for education are placed in the glasshouse and are connected to each other on the ground floor. Two big stairs on each side of the building lead you to the inner plaza on the first floor. The classrooms, auditoria and offices are placed in the two buildings; the stairs and the plaza are meant as places where student and teacher can meet and discuss with each other. The space between the buildings and the glasshouse functions like an air duct, which is regulated by a smart climate system. By using proved technologies from the greenhouse sector, like the solar blinds and the printed glass panels, the need for active cooling in the summer is limited. The rainwater is collected and being reused for the toilets and cleaning of the building.

Green Walls

SportPlaza Mercator: Amsterdam—2006

Sustainable architecture has to do with optimizing the quality of life, energy efficiency and use of materials in the design. The cultural context and the direct surroundings of the building play an important part in this, because a significant part of the quality of life and the life expectancy is determined by the relationship between a building and its environment. It is also possible to achieve gains in terms of energy and materials in relation to the larger network environment (Venhoeven 2012). An integrated approach reveals that a building is not a static, isolated object, but interacts with its environment as if it is a living organism. Designers draw inspiration from the natural world and use this to improve technology in all areas. This biomimicry offers the foundation for a new type of functionalism. Eco-functionalism, the new style that is emerging right now all over the world, connects a diversity of functions and use, decentralized intelligence energy generation, exchange of waste flows and maximization of use options with optimizing relationships with the environment; see Fig. 9. This approach makes it possible to combine affordable, low-tech measures with the advantages of profitable advanced technology. The search for smart opportunities to coordinate production and use of multiple buildings, building sections or building clusters, not only functionally and energetically, but also socially, economically and ecologically, is the foundation for this new eco--functionalism. This leisure centre SportPlaza Mercator in Amsterdam by Dutch architects VenhoevenCS (Venhoeven 2009) was designed as a fortress covered in plants. SportPlaza Mercator is positioned at the entrance to a park. The architects

Fig. 9 SportPlaza Mercator, Amsterdam (photography by Luuk Kramer)

wanted it to fit in with its surroundings, so they added a camouflaging facade of bushy plants and flowers. Windows nestle in among the planted exterior but feature tinted glass to reduce visibility into the swimming-pool halls. Skylights were also added to bring in more natural light (Frearson 2013). From a distance, it seems like an overgrown fortress flanking and protecting the entryway to the nineteenth--century city. The building was designed as a city—a society in miniature—inside a cave. The building is full of lines of sight and keyholes that offer perspectives on the various visitors, activities and cultures in the building. Sunlight penetrates deep into the building's interior through all sorts of openings in the roof. Low windows frame the view of the street and the sun terrace (Frearson 2013).

City Hall: Venlo—2017

The City Hall of Venlo was designed by Kraaijvanger (2019) on the basis of cradle--to-cradle (C2C) principles (C2C center 2019); see Figs. 10, 11 and 12. With 630 flexible workplaces and public functions, the building embodies the ambition to have the entire city and region function. Well-being of the occupants was the starting point: a good building makes people happier and boosts productivity. The spatial design is based on three goals:

- To bring as much daylight and greenery into the interior as possible
- To create routes through the building that stimulate people to move around
- Encounter others, and to use only healthy materials

Around a patio with a helophyte filter for water purification, people can enjoy the greenery, the water and the views. The building refers to the agricultural tradition of this city and its top floor features a greenhouse with seasonal workplaces and room to grow regional products that also heats and humidifies the air that enters the

Fig. 10 City Hall—Venlo (Kraaijvanger 2019)

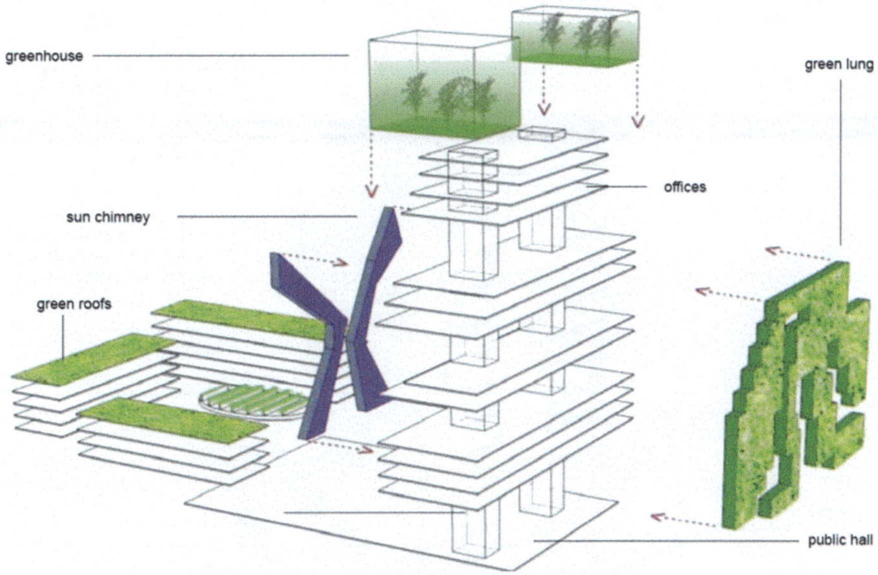

Fig. 11 Climate concept, City Hall, Venlo (Kraaijvanger 2019)

building. The green air-purifying facade is the largest green building facade in the world and forms a protective shell against traffic and railway pollution. The air in Venlo's new city hall is purer than the air outside and purifies air in a 500 m radius around the building. Cradle-to-cradle principles and holistic thinking should be in action in a city's buildings.

Green Roofs

Library Technical University of Delft: Delft—1997

Technical University of Delft originally designed their library with its intensive green roof from 1993 to 1995, resulting in construction from 1996 to 1997; see Fig. 13. Although nearly 15 years old, the visionary TU Delft Library is still an excellent example of forward-thinking sustainable design. The 15,000 m^2 library contains extensive underground book archives, reading rooms, university publisher, offices, study spaces and a book binder and book shop. It won the National Steel Award 1998 and Corus Construction Award 2000. Mecanoo Architecten's (Mecanoo 2019) vision was to develop a light-filled landmark that would serve as 'a gateway to the digital highway'. The library is designed as a sloped plane, extending the grass from the ground to the very edge of the roof allowing people to walk to the top. What they achieved is a public place that not only is a core for information, but

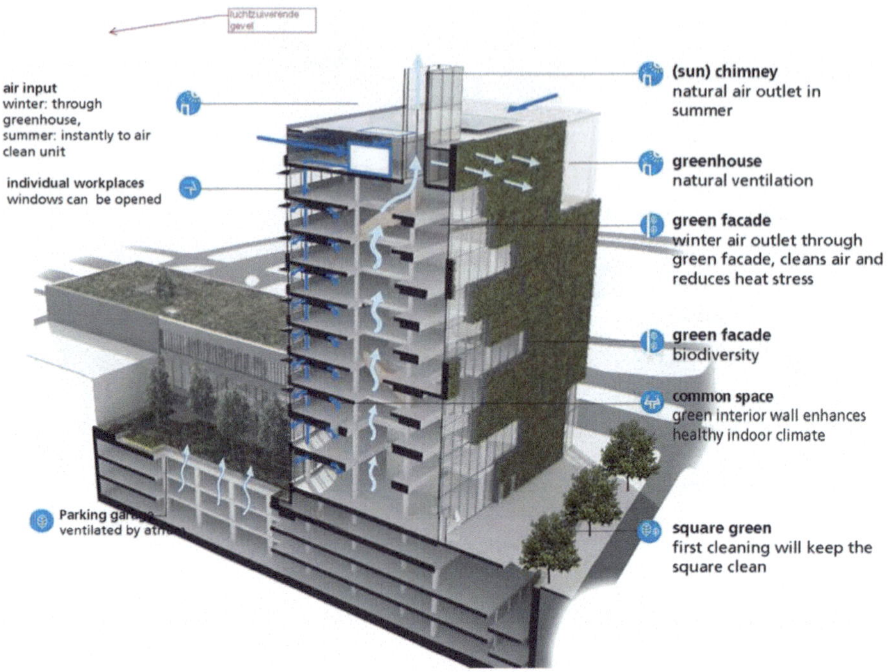

Fig. 12 Greenery aspects, City Hall, Venlo (Kraaijvanger 2019)

also features a prodigious public green roof for people to explore, relax and play (Mecanoo 2015). Enjoyed throughout the year, the green roof is converted into a sledding hill in the winter for further recreation. Visitors can catch some sun on the inclined lawn during the warmer months, and when it snows the roof makes for the best sledding in town—after all, where else are you going to find a hill in the Netherlands? The green roof also provides the interior with a much-reduced need for cooling and helps the Delft University of Technology campus control storm water runoff (Michler 2011). With a renovation in 2009, the green roof has continued to benefit the university, continuing to meet expectations more than two decades after its opening. The roof is supported by slender, splayed steel columns in a huge hall enclosed with canted, fully glazed walls. The base of the slope to the west is marked by a broad flight of steps leading up to a recessed entrance. The library is topped by a huge steel cone which houses four levels of traditional study spaces connected by a helical stair. Within the cone, a central void provides daylight from a glazed roof to the internal reading spaces. Extending 40 m above grade and floodlit at night, the cone acts as a beacon on the campus day and night.

Fig. 13 The library is a building that does not really want to be a building, but a landscape (Mecanoo 2019)

Urban Rooftop Farms Dakakker: Rotterdam—2017

In Rotterdam, an architecture collective ZUS has reclaimed an old building in the centre of the city and built an urban farm on top of the roof; see Fig. 14. The Dakakker is a 1000 m^2 rooftop farm on top of the Schieblock in Rotterdam in the Netherlands (Dakakker 2019). The Dakakker is the largest open-air roof farm in the Netherlands and one of the largest in Europe. The Smartroof is located on the roof of the rooftop pavilion. It is a test site for smart water storage and management as it has a smart flow control that is driven by the weather forecast. When predicting extreme rainfall, the smart flow control responds by making extra water storage capacity available 24 h in advance. The rooftop system is made up of different layers: a protective absorption layer (not roof resistant), a drainage buffer layer and a filter layer with the rooftop vegetable garden substrate on top. The conditions on the Dakakker are comparable to a Mediterranean climate: dry rocky soil and a lot of wind sometimes quite warm. That is why herbs such as mint, lemon verbena and lavender are grown on the Dakakker, as well as vegetables such as Jerusalem artichoke, beetroot, carrot, radish and garlic varieties such as onion, leek, garlic and strong plants such as raspberry, rhubarb and pumpkin. The edible flowers of the roof are the top product and are delivered to six restaurants in the immediate vicinity of the rooftop farm.

Fig. 14 Dakakker—Rotterdam (Dakakker 2019)

Based on the gained experience, with a total potential of 18.5 km^2 of flat rooftops in Rotterdam, the city has set a target of putting 1 km^2 of that space to good use by 2030, by greening it with plants and trees, adding solar panels and utilizing it for recreation or even homes.

Hotel Roof Garden: Zoku, Amsterdam—2016

The existing ‘Metropoolgebouw’ is a new hotel that was built in 2016: Zoku, Amsterdam. It provides a home base for travelling professionals who are living and working in a city for periods from a few days to a few months; see Fig. 15. Zoku created a new category in the hotel industry: a re-invented apartment hotel, a home--office hybrid, which is a relaxed place to live, work and socialize with like-minded people while getting wired into the city (Zoku 2019). With a beautiful green-roof garden, it has an outdoor area on both sides of the hexagon-shaped greenhouse on the roof. ‘Hexagons’ are the theme of the hotel and this was implemented in the design of the roof park as well.

Roof Park Vierhavenstrip: Rotterdam—2011

On the roof of the shopping centre, an area of 1 km by 80 m and with a soil depth of 1.5 m has been used to create good growing conditions for trees and other plants; see Fig. 16. Steel mesh was implemented as an extra anchoring for the trees. The majority of the trees planted are multi-stems, including a number of monumental specimens. A varied range of trees was chosen, to ensure striking colours in every season. The eastern edge of the park is characterized by a high-rise urban design, with tall trees and hedges acting as structural elements. The western edge is oriented towards the district and has a very green character with themed gardens at the main entrances. Large trees, lawns, a playground, a community garden and a Mediterranean

Fig. 15 Hotel Roof Garden, Zoku, Amsterdam (De Dakdokters 2019)

garden with an orangery offer local residents a great place to relax and meet, and of course to enjoy a spectacular view of the city and the harbour. The main paths are bordered by water and there is a 70 m-by-7 m step waterfall on the roof. The Vierhavensstrip rooftop park connects living, shopping, working, study, culture and recreation in a unique way. The roof park has a drainage system with a perforated core for podium decks on an inverted roof. The drainage system prevents the build-up of hydrostatic pressure on the waterproofing membrane. The vapour-permeable filter geotextile glued (not thermally bonded) to the dimpled core on the back protects the XPS insulation from mechanical loads and ensures that vapour is able to evaporate freely (Nophadrain 2019).

Discussion and Conclusion

Nature-inspired architecture and modern plant breeding both seek to increase tolerance to climate extremes and try to make the built environment more resilient and also allow efficient use of energy, materials, water and light (Wootton-Beard 2018). Besides green roofs and green walls, plant factories with artificial lighting (PFAL), also known as indoor plant production systems, are expected to play a vital role in greenery within urban development (Kozai et al. 2016). Urban agriculture and vertical farming will grow enormously in the forthcoming decades as it has a tremendous potential. Therefor PFAL is attracting much interest among researchers,

Fig. 16 Roof park Vierhavenstrip—Rotterdam

community developers, architects, designers and entrepreneurs. In the next few decades, urban agriculture and vertical farming will not be just a modified form of existing agriculture and farming practices. It will provide entirely new social services in urban areas which are facing ecological, economic and social constraints.

The relationships between indoor building conditions (thermal aspects, ventilation, lighting, moisture and noise) and well-being (health and comfort) of occupants of office buildings, schools and homes are complex (Bluyssen 2019). Therefore it is also difficult to determine the positive effects of greenery in sustainable buildings as effects occur often in settings with a lot of changing circumstance. Nevertheless there is an overwhelming number of research outcomes that show that greenery has a positive effect within buildings as well as on city scale.

This chapter reviews the importance of greenery in sustainable buildings and their effect on the indoor and outside environmental quality (IEQ), based on scientific studies from the past 30 years. Indoor greenery can be used to reduce sound levels as a passive acoustic insulation system. Living wall systems in combination with biofiltration are emerging technologies to provide beneficial effects on the improvement of indoor comfort. Quite a number of studies have shown that biophilic workspaces and interaction with plants may change human behaviours and improve productivity and the overall well-being. Evapotranspiration from plants helps lowering the temperature around the planting environment and this can be utilized for air cooling and humidification on city scale as well. In addition to literature overview several recent applications of greenery in projects in the Netherlands are presented to illustrate the architectural added value as well.

References

Akbari, H., Pomerantz, M., & Taha, H. (2001). Cool surfaces and shade trees to reduce energy use and improve air quality in urban areas. *Solar Energy, 70*, 295–310.

Akbari, H., Damon Matthews, H., & Seto, D. (2012). The long-term effect of increasing albedo of urban areas. *Environmental Research Letters, 7*(2), 024004.

Armijos Moya, T., van den Dobbelsteen, A., Ottelé, M., & Bluyssen, P. M. (2017). Green Air Conditioning, using indoor living wall systems as a climate control method. *REHVA Journal, 2017*, 27–31.

Armijos Moya, T., van den Dobbelsteen, A., Ottelé, M., & Bluyssen, P. M. (2018). A review of green systems within the indoor environment. *Indoor and Built Environment, 28*(3), 298–309.

Armson, D., Stringer, P., & Ennos, A. R. (2012). The effect of tree shade and grass on surface and globe temperatures in an urban area. *Urban Forestry & Urban Greening, 11*, 245–255.

Arup. (2015). Cities Alive. Retrieved from http://www.arup.com/Homepage_Cities_Alive.aspxJournal.

Azkorra, Z., Pérez, G., Coma, J., Cabeza, L. F., Bures, S., Álvaro, J. E., Erkoreka, A., & Urrestarazu, M. (2015). Evaluation of green walls as a passive acoustic insulation system for buildings. *Applied Acoustics, 89*, 46–56.

Behnisch. (2019). Retrieved from https://behnisch.com/work/projects/0022, https://behnisch.com/content/2-work/1-projects/0-0022/022-e-ibn-wageningen.pdf.

Berndtsson, J. C. (2010). Roof performance towards management of runoff water quantity and quality: A review. *Ecological Engineering, 36*, 351–360.

Bhamra, T. (2019). *Back to the future ... of design for sustainability, Proceedings ICED 19*. Netherlands: Delft.

Bluyssen, P. M. (2019). *The need for understanding the indoor environmental factors and its effects on occupants through an integrated analysis, 10th International Conference on Indoor Air Quality, Ventilation and Energy Conservation in Buildings (IAQVEC 2019), 5–7 September*. Italy: Bari.

C2C Center. (2019). Retrieved from http://www.c2ccentre.com/sites/default/files/Case%20Study%20City%20Hall%20Venlo_Final.pdf.

Chen Q., Li B., Liu X. (2013). An experimental evaluation of the Living Wall system in hot and humid climate, Energy and Buildings *61*, 298–307.

Dakakker. (2019). https://dakakker.nl/site/?lang=en.

Davis Langdon. (2007). *The cost & benefit of achieving green building*. Sydney, NSW: Davis Langdon.

De Dakdokters. (2019). Retrieved from https://dakdokters.nl/portfolio-items/zoku/.

de Vries, S., Verheij, R. A., Groenewegen, P. P., & Spreeuwenberg, P. (2003). Natural environments—Healthy environments? An exploratory analysis of the relationship between greenspace and health. *Environment and Planning A: Economy and Space, 35*(10), 1717–1731.

Deng, Y., & Wu, J. (2014). Economic returns to residential green building investment: The developers' perspective. *Regional Science and Urban Economics, 47*, 35–44.

Eichholtz, P., Kok, N., & Quigley, J. M. (2010). Doing well by doing good? Green office buildings. *American Economic Review, 100*, 2492–2509.
Eppinger, S., & Maier, A. (2019). *Responsible design: Recognising the impact of how we design, Proceedings ICED 19*. Netherlands: Delft.
Fang, W., Xiaosong, Z., Junjie, T., & Xiuwei, L. (2011). The thermal performance of double skin façade with Tillandsia usneoides plant curtain. *Energy and Buildings, 43*, 2127–2133.
Fowdar, H. S., Hatt, B. E., Breen, P., Cook, P. L. M., & Deletic, A. (2017). Designing living walls for greywater treatment. *Water Research, 110*, 218–232.
Frearson A. (2013). *SportPlaza Mercator by VenhoevenCS*. Retrieved from https://www.dezeen.com/2013/03/13/sportplaza-mercator-by-venhoevencs/.
Gill, S. E., Handley, J. F., Ennos, A. R., & Pauleit, S. (2007). Adapting cities for climate change: The role of the green infrastructure. *Built Environment, 33*, 115–133.
Gunawardena, K. R., & Steemers, K. (2019a). Living wall influence on microclimates: An indoor case study. *Journal of Physics: Conference Series, 1343*, 012188.
Gunawardena, K. R., & Steemers, K. (2019b). Living walls in indoor environments. *Building and Environment, 48*, 478–487.
Gupta, R., & Gregg, M. (2012). Using UK climate change projections to adapt existing English homes for a warming climate. *Building and Environment, 55*, 20–42.
Hartig, T., Evans, G. W., Jamner, L. D., Davies, D. S., & Gärling, T. (2003). Tracking restoration in natural and urban field settings. *Journal of Environmental Psychology, 23*, 109–123.
Hermans, T., de Vries, S., Jeurissen, L., Kraan, J., Oppedijk, B., & van Duijn, B. (2019). *Planten vooreen prima binnenklimaat: Industriële spoor*. Wageningen, Netherlands: Regional Development and Spatial Use, WASS, Wageningen Environmental Research rapport No. 2952. Retrieved from https://library.wur.nl/WebQuery/wurpubs/551856.
Hiemstra, J. A., de Vries, S., & Spijker, J. H. (2019). *Greenery and Work: The positive effects of greenery in urban environments*. Wageningen, Netherlands.: WASS Nature and Society Wageningen University & Research. Retrieved from https://edepot.wur.nl/418847.
Huang, Z., Lu, Y., Wong, N. H., & Poh, C. H. (2019). The true cost of "greening" a building: Life cycle cost analysis of vertical greenery systems (VGS) in tropical climate. *Journal of Cleaner Production, 228*, 437–454.
IntoGreen. (2019). *Air quality in classrooms drastically improves with plants*. Retrieved from https://intogreen.eu/air-quality-in-classrooms-drastically-improves-with-plants/.
Irvine, K. N., Devine-Wright, P., Payne, S. R., Fuller, R. A., Painter, B., & Gaston, K. J. (2009). Green space, soundscape and urban sustainability: An interdisciplinary, empirical study. *Local Environment, 14*(2), 155–172.
Kapoor, M. (2017). Managing ambient air quality using ornamental plants-an alternative approach. *Universal Journal of Plant Science, 5*, 1–9.
Kat, G. H. (2003). *Green building costs and financial benefits*. Westborough, MA: Massachusetts Technology Collaborative.
Kozai, T., Niu, G., & Takagaki, M. (Eds.). (2016). *Plant factory, an indoor vertical farming system for efficient quality food production*. Amsterdam: Elsevier.
Kraaijvanger. (2019). Retrieved from https://www.kraaijvanger.nl/en/projects/city-hall-venlo/.
Kristinsson, J. (2011). Retrieved from http://www.kristinsson.nl/.
Kuo, F. E., Sullivan, W. C., & Coley, R. L. (1998). Fertile ground for community: Inner-city neighborhood common spaces. *American Journal of Community Psychology, 26*, 823.
Larsen, S. F., Filippin, C., & Lesino, G. (2013). Thermal simulation of a double skin façade with plants, Proceedings ISES Solar World Congress. *Energy Procedia, 57*, 1763–1772.
Leaman, A., & Bordass, B. (2007). Are users more tolerant of 'green' buildings? *Building Research and Information, 35*, 662–673.
Levenston M. (2019). *Vertically integrated greenhouse, city farmer news*. Retrieved from https://cityfarmer.info/vertically-integrated-greenhouse/.

Malone, D. (2017). *A sustainable mixed-use development springs from a Dutch city center like a green-fringed crystal formation*. Retrieved from https://www.bdcnetwork.com/sustainable-mixed-use-development-springs-dutch-city-center-green-fringed-crystal-formation.

Mangone, G. (2015). *Performative microforests*. PhD Thesis TU Delft: Delft, Netherlands.

Mangone, G., & van der Linden, K. (2014). Forest microclimates: Investigating the performance potential of vegetation at the building space scale. *Building and Environment, 73*, 12–23.

Mangone, G., Kurvers, S. R., & Luscuere, P. G. (2014). Constructing thermal comfort: Investigating the effect of vegetation on indoor thermal comfort through a four season thermal comfort quasi-experiment. *Building and Environment, 81*, 10–426.

McDonald, A. G., Bealey, W. J., Fowler, D., Dragosits, U., Skiba, U., Smith, R. I., Donovan, R. G., Brett, H. E., Hewitt, C. N., & Nemitz, E. (2007). Quantifying the effect of urban tree planting on concentrations and depositions of PM10 in two UK conurbations. *Atmospheric Environment, 41*(38), 8455–8467.

Mecanoo. (2015). *Delft University of Technology among greenest in the world*. Retrieved from https://www.mecanoo.nl/News/ID/180/Delft-University-of-Technology-among-greenest-in-the-world.

Mecanoo. (2019). *Library Delft University of Technology*. Retrieved from https://www.mecanoo.nl/Projects/project/27/Library-Delft-University-of-Technology?t=0.

Mentens, J., Raes, D., & Hermy, M. (2006). Green roofs as a tool for solving the rainwater runoff problem in the urbanized 21st century? *Landscape and Urban Planning, 77*, 217–226.

Michler, A. (2011). *Mecanoo's TU Delft Library Crowned with a Massive Green Roof, Inhabit*. Retrieved from https://inhabitat.com/mecanoos-tu-delft-library-crowned-with-a-massive-green-roof/.

MVDRV. (2019). *EXPO 2000*. Retrieved from http://www.mvrdv.nl/projects/EXPO/#.

Naava. (2019). *Not a normal green wall*. Retrieved from https://www.naava.io/.

Newsham, G. R., Mancini, S., & Birt, B. J. (2009). Do LEED-certified buildings save energy? Yes, but …. *Energy and Buildings, 41*, 897–905.

Ng, E., Chen, L., Wang, Y., & Yuan, C. (2012). A study on the cooling effects of greening in a high-density city: An experience from Hong Kong. *Building and Environment, 47*, 256–271.

Nophadrain. (2019). Retrieved from https://www.nophadrain.com/inspiration-projects/vierhavenstrip-roof-park/.

Ottelé, M., Perini, K., Fraaij, A., Haas, E., & Raiteri, R. (2011). Comparative life cycle analysis for green façades and living wall systems. *Energy and Buildings, 43*, 3419–3429.

Paul, W. L., & Taylor, P. A. (2008). A comparison of occupant comfort and satisfaction between a green building and a conventional building. *Building and Environment, 43*, 1858–1870.

Perez, G., Rincon, L., Vila, A., Gonzalez, J. M., & Cabeza, L. F. (2011). Green vertical systems for buildings as passive systems for energy savings. *Applied Energy, 88*, 4854–4859.

Pérez, l. G., Coma, J., Barreneche, C., de, G. A., Urrestarazu, M., Burés, S., & Cabeza, L. F. (2016). Acoustic insulation capacity of vertical greenery systems for buildings. *Applied Acoustics, 110*, 218–226.

Perini, K., & Rosasco, P. (2013). Cost–benefit analysis for green façades and living wall systems. *Building and Environment, 70*, 110–121.

Perini, K., Ottelé, M., Fraaij, A., Haas, E., & Raiteri, R. (2011). Vertical greening systems and the effect on air flow and temperature on the building envelope. *Building and Environment, 46*, 2287–2294.

Prodanovic, V., McCarthy, D., Hatt, B., & Deletic, A. (2019). Designing green walls for greywater treatment: The role of plants and operational factors on nutrient removal. *Ecological Engineering, 130*, 184–195.

Raus, T. (2018). *How on earth have the Dutch done it?* Retrieved from https://dutchgreenhouses.com/blog/how-on-earth-have-the-dutch-done-it.

Romeo, C., & Zinzi, M. (2013). Impact of a cool roof application on the energy and comfort performance in an existing non-residential building. A Sicilian case study. *Energy and Buildings, 67*, 647–657.

Sailor, D. J. (2008). A green roof model for building energy simulation programs. *Energy and Buildings, 40*, 1466–1478.

Santamouris, M. (2014). Cooling the cities—A review of reflective and green roof mitigation technologies to fight heat island and improve comfort in urban environments. *Solar Energy, 103*, 682–703.

Santamouris, M., Pavlou, C., Doukas, P., Mihalakakou, G., Synnefa, A., Hatzibiros, A., & Patargias, P. (2007). Investigating and analysing the energy and environmental performance of an experimental green roof system installed in a nursery school building in Athens, Greece. *Energy, 32*, 1781–1788.

Smith, A. (2015). Garden city. *CIBSE Journal, 2015*, 46–50.

Stec, W. J., & van Passen, A. H. C. (2005). Symbiosis of the double skin façade with the HVAC system. *Energy and Buildings, 37*, 461–469.

Stec, W. J., van Passen, A. H. C., & Maziarz, A. (2005). Modelling the double skin facade with plants. *Energy and Buildings, 37*, 419–427.

Takano, T., Nakamura, K., & Watanabe, M. (2002). Urban residential environments and senior citizens' longevity in megacity areas: The importance of walkable green spaces. *Journal of Epidemiology & Community Health, 56*, 913–918.

Takebayashi, H., & Moriyama, M. (2007). Surface heat budget on green roof and high reflection roof for mitigation of urban heat Island. *Building and Environment, 42*, 2971–2979.

Torpy, F. R., Zavattaro, M., & Irga, P. J. (2016). Green wall technology for the phytoremediation of indoor air: A system for the reduction of high CO2 concentrations. *Air Quality, Atmosphere and Health, 5*, 151–167.

Ulrich, R. S. (1983). Aesthetic and affective responses to natural environment in: I. Altman, J.F. Wohlwill (Eds.), Behavior and the natural environment, human behavior and environment. *Advances in Theory and Research, 6*, 85–125.

Ulrich, R. S., Simons, R. F., Losito, B. D., Fiorito, E., Miles, M. A., & Zelson, M. (1991). Stress recovery during exposure to natural and urban environments. *Journal of Environmental Psychology, 11*, 201–230.

van Duijn, B., Klein Hesselink, J., Kester, M., Jansen, J., & Splitters, H. (2011). *Rapport Project Plant in de klas*. Productschap Tuinbouw: Projectnummer 13908.

Velzeboer, I., Blom, M. J., & Duyzer, J. H. (2019). *Effects of plants on air quality, TNO report R10896*. Netherlands: Petten.

Venhoeven, T. (2009). *The wonderful world of venhoeven*. CS: Episode Publishers.

Venhoeven, T. (2012). Retrieved from https://venhoevencs.nl/vision-and-research/sustainable-architecture/?project=357.

Wolverton, B. C. (2008). *How to grow fresh air: 50 house plants that purify your home or office*. London: Weidenfeld & Nicolson.

Wolverton, B. C. (2010). *Plants: Why you can't live without them*. New Delhi: Roli Books.

Wolverton, B. C., Mcdonald, R. C., & Watkins, E. A. (1984). Foliage plants for removing indoor air-pollutants from energy efficient homes. *Economic Botany, 38*, 224–228.

Wolverton, B. C., Johnson, A., & Bounds, K. (1989). *Interior landscape plants for indoor air pollution abatement. Final report NASA (NASA-TM-101766, NAS 1.15:101766)* (pp. 1–22). Mississippi: National Aeronautics and Space Administration.

Wong, N. H., Chen, Y., Ong, C. L., & Sia, A. (2003). Investigation of thermal benefits of rooftop garden in the tropical environment. *Building and Environment, 38*, 261–270.

Wong, N. H., Tan, A. Y. K., Chen, Y., Sekar, K., Tan, P. Y., Chan, D., Chiang, K., & Wong, N. C. (2010). Thermal evaluation of vertical greenery systems for building walls. *Building and Environment, 45*, 663–672.

Wootton-Beard, P. C. (2018). *Building greening for urban wellbeing, What to think about when specifying on-building green infrastructure, Presentation ASBP Healthy Buildings Conference & Expo*. London, UK: UCL Roberts Building, Torrington Place. Retrieved February 14, 2018, from https://asbp.org.uk/wp-content/uploads/2018/04/Peter-Wootton-Beard-Aberystwyth-University-HealthyBuildings18.pdf.

Xing, Y., & Bosch, M. (2017). Exploring design principles of biological and living building envelopes: What can we learn from plant cell walls? *Intelligent Building International, 10*(2), 78–102.

Xing, Y., Jones, P., & Donnison, I. (2017). Characterization of nature based solutions for the built environment. *Sustainability, 9*(1), 149.

Yang, K. (1972). Bases for ecosystem design. *Architectural Design, 1972*, 434–436.

Yang, K. (1999). *The green skyscraper: The basis for designing sustainable intensive buildings*. Munich: Prestel Publishing.

Young, L. (2014). Into the blue. *CIBSE Journal, 4*(6), 24–27.

Zhang, X., Shen, L., & Wu, Y. (2011). Green strategy for gaining competitive advantage in housing development: A China study. *Journal of Cleaner Production, 19*, 157–167.

Zhao, X., Zuo, J., Wu, G., & Huang, C. (2019). A bibliometric review of green building research 2000–2016. *Architectural Science Review, 62*(1), 74–88.

Zoellner, T. (2013). *Sexy fish and rocking berries, presentation international strawberry congress, Antwerp*. Retrieved from http://hoogstraten.eu/presentations/thursday/9_TomZollner_UrbanFarmers_Stra wberryAntwerpEnTZ.pdf.

ZOKU. (2019). Retrieved from https://livezoku.com/.

Green Dreams: Regenerating Cities Through Nature

Alessandra Battisti

Land Consumption and Climate Change

Land is a limited natural resource, yet it plays a key role in combating climate change and is a necessary reserve of biodiversity, essential for our survival.

Almost 1000 km^2 of farmland or wilderness currently disappears every year in the EU following the construction of new urban areas and infrastructure (EEA 2017). An increasing proportion of the EU's territory is subject to degradation and, according to the estimates published in the latest IPBES[1] report, over 75% of the planet's land is extremely degraded, affecting the well-being of 3.2 billion people (IPBS 2017).

Such estimates signify the desertification of large swathes of global territory due to pollution, deforestation and intensive farming, thus becoming the main reason why many species become extinct. The IPBES's global assessment warns us that though we can intervene, we must do so immediately because if we continue at this rate, 95% of the world's land could be degraded by 2050, forcing hundreds of millions of people to migrate, while food production could collapse. Land degradation, loss of biodiversity and climate change are three sides to one single, crucial challenge: the growing impact of our choices on the health of our natural environment (IPBS 2017).

[1] https://www.ipbes.net/system/tdf/2018_ldr_full_report_book_v4_pages.pdf?file=1&type=node&id=29395.

A. Battisti (✉)
PDTA Planning Design Technology of Architecture Department, University La Sapienza, Rome, Italy
e-mail: alessandra.battisti@uniroma1.it

A. Sayigh, A. Trombadore (eds.), *The Importance of Greenery in Sustainable Buildings*, Innovative Renewable Energy,
https://doi.org/10.1007/978-3-030-68556-0_2

According to the US-based Natural Resource Conservation Service,[2] desertification is already affecting 33% of global soil, together with erosion, salinisation and other degenerative processes that gradually reduce the soil's ability to carry out its role as a substratum for biological communities. This phenomenon has led to a 20% reduction in the productivity of soils even in Asia, India and China, and up to a 40% reduction in Africa and in some parts of the United States, such as Ohio (Natural Resource Conservation Service NRCS 2013).

Already in 2008 researchers Erle Ellis and Navin Ramankutty, of the *Anthromes*[3] project, published a new world map. The conventional view classifying the earth's ecosystems into biomes organised on the basis of vegetation and climate—defined as tundras, temperate forests, grasslands, rainforests, etc.—was combined with research on the transformations produced by human intervention, revealing biomes that have been profoundly altered compared to the original evolutions. The result is that, today, more than ¾ of the earth's surface appears to have been redesigned by human activities.

The map of the planet's major natural environments between the years 1700 and 2000 elaborated by the two American researchers shows that over half the biosphere was in a wild state in the year 1700, whereas 5% was in a semi-natural state, i.e. with modest soil transformations due to farming activities and settlements; on the other hand, in the year 2000, most of the planet seemed to be covered by agricultural areas and other anthropic biomes, with less than 20% of the surface in a semi--natural state and less than ¼ in a wild state (Ellis and Ramankutty 2008). As a result of the profound changes listed above, Germany became the first European member state to recognise the need to stop land consumption as far back as 1985 and, since the late 1990s, economic development in this country has been no longer coinciding with the occupation of land. The maximum limit for land consumption for housing construction and infrastructure was set at 30 ha/day up until 2020, compared to a daily rate of consumption of 120 ha at that time. In Italy, 750,000 ha was consumed from 1995 to 2006, 390,000 of which was earmarked for housing, 210,000 for manufacturing and 150,000 for infrastructure. Every year, 68,000 ha is urbanised, at an average rate of 187 ha/day: the equivalent of 250 football pitches. The lowest demographic growth in Europe walks hand in hand with one of the highest rates of land consumption (ISPRA-SNPA 2018).

If the aim we need to focus on when developing our built environment is to save land, then we also need to ask ourselves how we should act in this day and age. Both the European Community and the United Nations are issuing serious directives on protecting land, our environmental heritage and the landscape, and we come across

[2] NRCS Natural Resource Conservation Service (2013). *Soil Survey Data*. Available from http://websoilsurvey.nrcs.usda.gov.

[3] This is how the project is described on the *Anthromes* website by researchers from the University of Maryland: 'The Anthromes Working Group aims to investigate, understand and model human transformation and management of the terrestrial biosphere based on the concept of Anthromes (Anthropogenic Biomes) as a new paradigm for incorporating human systems into global ecology and earth science research and education for the Anthropocene'.

two driving forces in these documents: the recognition of the value of natural capital and the urgent demand for a total stop to net land consumption by 2050 (Commissione Europea 2014), in line with demographic growth, and for a reduction in land degradation by 2030 (UN 2015).

At an international level, these public policies concerning urban areas and natural capital have stimulated two factors:

- Regulation and maintenance (climate regulation, carbon capture and storage, control of erosion, regulation of elements essential for soil fertility, regulation of water quality, protection and mitigation of extreme hydrological phenomena, genetic reserves, conservation of biodiversity, etc.)
- Guaranteed provision of cultural services (cultural and recreational services, ethical and spiritual functions, landscape, natural heritage, etc.)

These changes have affected the economy as a whole in various ways and, as regards their portrayal in cultural media, many of them have been suitably represented in documentaries and films. As regards academia, they have been accompanied by discussion regarding the ecological functions that good-quality land should ensure, which should guarantee not only intrinsic value but economic and social value as well by supplying a number of different ecosystem services. Ecosystem services can be considered the indirect contribution of 'natural capital', according to the World Bank's definition: 'the stock of natural resources that provides flows of valuable goods and services' (World Bank 2012). A landscape and its parkland can be considered to be of good quality if the human activities carried out are compatible with its ecosystems, and the conservation of natural capital is essential if we wish to maintain a high quantitative and qualitative level of services essential for making ecosystems work and for ensuring human well-being.

Ecosystems are constantly dependent on each other and, in such circumstances of reciprocal relationships, urban areas become heavily dependent on the functions of natural capital. In the study entitled The Economics of Ecosystems and Biodiversity, which attempts to combine our current scientific, economic and political knowledge, the benchmark measures adopted by local and regional authorities are given an official rendition so as to support them when implementing territorial management policies that can have a beneficial effect on biodiversity and ecosystem services and can propose a green economy-influenced vision, a vision that has made a great deal of headway in recent years. Such measures consider the environmental sector a producer of goods and services as well as innovative projects, which take their cue from people and their needs and then move on towards renaturalisation of the landscape and protection and development of green resources and water resources.

Urban Regeneration and the Renaturalisation of Farmland

In 2015, the United Nations' Global Agenda for Sustainable Development (UN 2015) established its Sustainable Development Goals (or SDGs) and indicated, amongst others, a number of targets that particularly concern the territory and the soil, which were to be integrated in national short- and medium-term programme and that had to be achieved by 2030:

- To ensure that land consumption does not exceed demographic growth
- To ensure universal access to safe, inclusive and accessible green and public space
- To achieve a land degradation-neutral world, as an essential basis for maintaining ecosystem functions and services

By signing the Agenda, all member states of the European Union chose to participate in a process that would monitor these objectives, a process handled by the United Nations Statistical Commission, using a system based on indicators, including a number of specific indicators concerning land consumption, land use and percentage of territory subject to degradation (World Bank 2012).

Despite this, Europe continues to be the continent with the highest level of urbanisation, second only to Latin America: over two-thirds of the population live in urban areas (three out of four citizens) and, of these, almost 25% are on the brink of poverty and social exclusion while only 70% have a job. Moreover, the way European cities have evolved has led to the abolition of legally established administrative borders: approximately 68% of the European population is concentrated in what are termed 'metropolitan' areas where 67% of the EU's GDP is produced (UNCCD 2017).

The data presented here is a wake-up call and places cities in a position of greater environmental responsibility, with dangerous consequences. In such circumstances, the increasingly complex challenges faced by European cities—managing the immigration emergency from the point of view of reception and integration; fighting urban poverty; handling demographic change; supporting environmentally sustainable growth and low emissions; improving air quality and living standards; developing policies that can lead to zero land consumption, energy efficiency, design and administration of sustainable city development—are all opportunities that have become real and tangible thanks to urban regeneration strategies and programme that foster an improvement in the quality of space, forms and practices and that make cities productive again (Battisti 2014).

In productive cities, the issue of public space as a place for social interaction and relationships, for localisation, its use and its size comes to the fore, an issue that must also include and tackle design projects with a focus on inclusiveness and diversity. From this point of view, the territory itself becomes the matrix of the human environment, where we must analyse the relationship between natural and artificial, identifying its environmental quality in terms of the potential and sensitivity of a place and its identity (Ricci 2018).

The vision of a productive city involves a reconnection of places, a careful approach to designing environmental and infrastructural networks (green infrastructure); to the exchange of material, energy and information; to empty spaces in urban areas; and to open ground in marginal areas and waste ground, collective services and facilities; productive cities build meaningful, stimulating relationships that influence how cities are organised and formed (Stevens 2011); they establish a backbone that links urban places and links their inhabitants.

The restoration and reuse of abandoned sites undoubtedly play a vital role in regenerating the productiveness of cities. This can be achieved by adopting practices where farmland is renaturalised; permitting an increase in a site's level of naturalness; reducing harmful emissions produced by the presence of static or dynamic polluting agents in the urban fabric; and a widespread process where environmental 'performance' is improved in all senses (de la Salle and Holland 2010), including in terms of its administrative profile, its elements and systems, and a gradual attempt is made to restore the balance of conditions of liveability in the public urban space as a whole. The territorial approach based on urban farming models allows differences (of an urban, geomorphological, social and cultural nature) to play an active role and significantly contributes to the restoration of the dialogue between man and his environment and the improvement of the relationships between people (Kaufman and Bailkey 2000). Wherever the urban form does not allow the assertion of, and dialogue between, different demands (whether they be ecosystem based, socio-economic or cultural), the institutionalisation of urban farming systems often results in the restoration of a sense of place and of community, restoring the network of internal morphological, historical, cultural and socio-economic relationships that clear the way for possible future communication and development within a community (Marsden et al. 1999).

Urban Farming: A Brief Historical Overview

Peri- and intra-urban agriculture was an intensive and highly productive sector before the mass transportation of food appeared with the nineteenth-century industrialisation. As far back as the Middle Ages, kitchen gardens were at the height of their splendour in Europe. In medieval cities, urban vegetable gardens were created within the walls of convents and monasteries for use in times of danger or of siege so as to provide the city's inhabitants with sustenance and guarantee its self-sufficiency when necessary.

Sometime later in seventeenth-century France, Murs à pêches appeared in Montreuil. These were urban vegetable gardens that supported a flourishing system of peach cultivation,[4] a type of farming that has left behind a particular architectural

[4]The peaches of Montreuil became famous amongst members of France's high society, ultimately reaching the court at Versailles. This fame made its way to the Queen of England, the Prince of Wales and the Russian Tsars.

system known by the same name, Murs à pêches (Carrière 1890), and which in 1907 occupied almost 300 ha of the 900 ha of a city that was mostly agricultural at that time, with over 600 ha dedicated to farming the land. What happened in Montreuil allowed the development of tree training techniques that are still in use today or are being rediscovered. In Montreuil, labyrinthine walls and farmland managed to create the perfect microclimate for growing peaches, which are notoriously suited to more temperate climates. The walls were approximately 3 m high and over a metre and a half thick, covered in locally sourced limestone plaster, thus creating elements that boasted high levels of thermal inertia and an excellent ability to store heat. Their orientation allowed the accumulation of daytime heat that would then be released during the night, a process that ensured the ripening of the fruit. Indeed, the temperature inside these walls was roughly 8°–12° higher than the surface, and the exploitation of the walls' height using wooden trellises allowed farmers to make the most of the space available: the fruit was grown along the walls and on wooden frames, while the central parts, between the walls, were earmarked for growing cut flowers, herbs and vegetables (Savard and Brunet 2005).

Again in France, an intensive farming technique was developed during the Haussmann era in Paris, in the Marais district (Jordan 2004). This involved growing crops in heated greenhouses that used horse manure as fertiliser and managed to produce 3–6 harvests a year, mostly growing vegetables that were subsequently sold throughout France and exported as far afield as London.

With the advent of the industrial revolution and the availability of more efficient transportation methods, the cost of exporting products fell and, with the application of industrial methods to agriculture, urban land was earmarked for other, more profitable purposes than farming.

In early twentieth-century England, a type of urban farming still existed that was practised for social reasons so as to improve the quality of life of the poorer classes living in big cities. At the same time in Germany, many working-class villages and social housing projects envisaged small communal or individual kitchen gardens set up for the same reasons as their English counterparts. Over the course of the twentieth century, there was a renewed interest in the dietary self-sufficiency of cities in the work and utopian visions of a number of architects and town planners, particularly intent on reconciling the contrast between town and country, seeking innovative forms of dialogue between these two dialectic extremes.

In the USA, town planners as early as the 1930s had identified ruralism as a catalyst between subsistence farming and increasing vulnerability of the urban workforce, partially in response to the Great Depression of 1929, as an element that could mediate between farming areas and industrial areas using a utopian labour system where hours spent working in factories would alternate with farm shifts. When it comes to this line of research, Frank Lloyd Wright's Broadacre City (1934–1935) is particularly significant: here, many of the implications of this combination of farm production and urban form were made obvious (Nelson 1995). Wright's work tackles the status of farmers and uses it as a springboard for launching the social and formal revival of cities,[5] which reflected the late-nineteenth--century intellectual debate between those who espoused the agricultural ideology of Thomas Jefferson on the one hand and controversy regarding the urban decay noted by writers such as Henry James and John Dewey on the other.

Nevertheless, as Serge Latouche declared in 2011: '… honourable attempts by architects and town planners to find a solution to the urban and social crisis by proposing ingenious schemes urban regions, garden cities, total cities, urban networks, conurbations [Geddes], Broadacre City [Wright], compact cities, extended cities, etc.—that seek a new relationship between city and countryside are bound to be defeated due to the lack of a comprehensive analysis of the failure of a society based on growth (Latouche 2011)'.

The dynamic vision of 'Agronica' proposed by Andrea Branzi follows Latouche's footsteps. It stands out due to the combination of forces and flows, linking social and environmental aspirations and the inexorable horizontal urban expansion created by the creeping occupation of land, dialectically engaging with the potential relationships between agricultural production and energy production and boasting the new 'culture' of consumption that these could generate: '… in the context of a post-industrial economy, the factory has lost its central role, and other systems of production, linked to weak and diffuse technologies, are developing new benchmark models that overturn the strong construction basis of modern architecture. These sophisticated serial production models focus on revisiting agriculture as a universe of complex natural technologies, as an environmental development system that can provide a range of different edible products, able to adapt to reversible programmes, fed by weak, seasonal and eco-compatible genetic energy; a type of agriculture that can offer a self-regulated model of industrial production based on controllable natural energy, and thus a kind of agriculture that no longer represents the world of pre-industrial technologies but, on the contrary, occupies a new position of extreme administrative and productive sophistication (Branzi 2000)'. History's utopian visions demonstrate how urban agriculture, in its various forms, is always welcomed and seen as an opportunity to heighten the social, cultural and environmental values of the territories involved, as well as their economic aspects.

Development Strategies

At the moment in its different forms, therefore, urban agriculture represents an opportunity to increase the social, cultural and environmental value of an area while improving its economic factors as well. According to René van Veenhuizen, there

[5] Wright compares modern cities to a 'great mouth' fed by the toil of farmers. Thanks to the formal redevelopment of agricultural structures, Broadacre City enables the social emancipation of farmers, finally setting up an equal relationship between them and citizens, between the city and the countryside.

are five main strategies that should be followed to support the development of urban agriculture as a way of achieving green, productive cities (van Veenhuizen 2009):

– Encouragement of a favourable environmental policy, along with formal acceptance of the zoning of certain urban areas for agricultural within the historical, established form of the city
– Optimisation of access to empty outdoor urban areas and of safe use of the land for agricultural purposes
– Optimisation of the productivity and economic profitability of urban agriculture through training, technical consulting and credit
– Public support for the creation and reinforcement of urban farmer organisations
– Enactment of measures to prevent/reduce risks to health and the environment tied to urban agriculture

In attempting to highlight the potential of urban agriculture in terms of bettering the quality of the environment and standards of sustainable life, the primary factors and indicators involving opportunities and problems tied to urban farming can be divided into:

Those tied to the quality of the environment: improving the ecological performance of cities; shortening the transportation routes of agricultural products; minimising food packaging; eliminating artificial fertilisers and pesticides; producing oxygen; reducing harmful emissions; the possibility of monitoring the full production chain; the occasion for regenerating residual spaces and optimising their use; and the gathering of environmental indexes (the possibility of monitoring air quality based on plants and animals)

Those tied to economic benefits: increased property values; yields of agricultural production; 0-km food production; lower food production costs; and creation of new jobs

Those tied to social benefits: tool for supervising the territory; means of sustenance for the weaker sectors of society and families; curative effect on physical and mental health; awareness of cycles of food production; reinforcement of the community and safeguarding of cultural identity; social aggregation of different generations; physical and therapeutic exercise; an educational experience at various levels of depth; and scholastic instruction

Principles and Objectives

In keeping with this outlook, the FAO formulated in 2014 (FAO 2014) a shared vision and integrated approach to agricultural sustainability, as summarised in these five principles of sustainable food and agriculture: increasing productivity and employment while adding value to food systems; protecting and improving natural resources; improving means of sustenance while favouring inclusive economic growth; improving the resilience of individuals, communities and ecosystems; and adapting methods of governance to meet new challenges. This unified outlook—valid in all sectors of agriculture, with consideration given to social, economic and environmental factors as well—ensures effective action in the field, being supported by the best available science and adapted to both the European Community and national levels, so as to ensure its pertinence and applicability.

Consistent with these principles, the trend, at least in countries with more advanced economies, is towards technological advances that not only increase production per unit of cultivated surface area, but also ensure that agricultural production in an urban setting is an economically and environmentally sustainable activity by pursuing the following objectives:

- Reduction and optimisation of land consumption
- Reduction and optimisation of water consumption
- Development of systems to supply plants with carbon dioxide, in order to increase their growth
- Experimentation with supports and materials representing alternatives to soil, and in the specific case of growing in closed spaces
- Computerised control of internal microclimatic conditions in terms of air quality, temperature, lighting, humidity, irrigation, fertilisation and presence of CO_2
- Optimisation of transparent materials designed to transmit natural light
- Development and optimisation of integrated systems of artificial lighting
- Integration and optimisation of mechanised systems of labour
- Integration and optimisation of systems for the production of renewable energy
- Biological control of illnesses and harmful insects

Two Examples of Urban Regeneration Tied to Agriculture

Observation of what is actually being done is the only way to fully appreciate the significance of these initiatives developed ‘from above’, thanks to commissions—as in the case of Thanet:

- From multinational corporations of the size of Asda, Sainsbury’s, Tesco, M&S and the HRGO agency, as well as ‘from below’, through well-informed communities that, in selecting the city sites best suited to bringing forth a productive city, make it possible to judge the potential for such practices in the future

Of particular interest, in terms of gauging the wealth of potential offered by urban agriculture when it comes to processes of urban regeneration, are two especially meaningful experiences: the Expo Porte de Versailles in Paris and the MUFI Urban Ag Campus in Detroit, both of which feature highly distinctive, original forms that are well integrated into their respective urban contexts and supported by the resident communities.

In the spring of 2020, an urban oasis was expected to be unveiled in Paris. Covering a surface area of approximately 14,000 m^2, it will be the world’s largest rooftop farm, built to renew this area of southwest Paris. Plans call for more than 30 different species of plants to be grown, with daily production of approximately 1000 kg of fruits and vegetables, thanks to the attentive labours of roughly 20 gardeners who use nothing but biological methods. 'The goal is to make the agricultural enterprise a model of sustainable production recognised throughout the world', states Pascal Hardy, the founder of Acropolis, the urban agricultural concern behind the project. 'We will use top-quality products, grown in accordance with the timing of natural cycles, all in the heart of Paris'.[6]

Found atop of a large exposition complex currently being restored in the 15th arrondissement, the Expo Porte de Versailles Paris, this farm will also feature an on-site restaurant and a café, with seating for approximately 300. Run by Le Perchoir, a renowned chain of Parisian rooftop establishments, the restaurant on the roof will offer panoramic views of the French capital plus, it goes without saying, a menu that includes seasonal products grown at the site (Fig. 1 and Table 1).

Also available at the farm will be a series of services related to urban farming, including educational tours, team-building seminars and special events. Last, but no less important, will be the opportunity for local residents to rent small plots of land in which to grow produce thanks to specially designed wood crates.

A number of crops will be grown in the open air, but use will also be made of the 'vertical' hydroponic technique, which is done without soil, nourishing the plants with a mist filled with nutrients, an approach that satisfies organic standards while using little water, thanks to towers designed for vertical growth, in a similar fashion to that used by covered farms, such as that run by Plenty.[7] With the plants stacked along the height of the towers, food can be grown on different levels without

Fig. 1 Expo Porte de Versailles hydroponic rooftop, Paris | © Acropolis

[6]The Guardian, 13 August 2019: 'World's largest urban farm to open—on a Paris rooftop'.

[7]A farm run by robots in the Silicon Valley.

Table 1 Matrix of Expo Porte de Versailles, Paris

EXPO PORTE DE VERSAILLES

CREDITS

Designer Team
Agripolis
Client
Viparis & co
Year
2017-2020
Place
Paris, France
Crops area
14'000 m²
Production
300 tons/year

48°51'24" N | 2°21'07" E | 78 m a.s.l

CLIMATE DATA

Temperate Oceanic

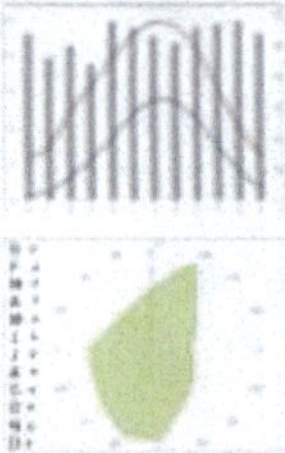

1°C (JANUARY)
25°C (JULY)
3 m/s (NNE)
5 m/s (S)

CHALLENGES

- ○ Heat Island
- ● Water management
- ● Pollution
- ○ Biodiversity defence
- ● Urban quality
- ● Food problems
- ○ Survival
- ● Economic income
- ○ Hobby

SUSTAINABILITY OF THE SITE

- ○ Empty space use
- ● Public transport access
- ○ Cycle path
- ○ Low emission vehicles
- ● Parking areas
- ● Area accesibility
- ● Enhancement of spaces
- ● Permeable surfaces
- ● Green integration

FORMS OF URBAN AGRICULTURE

- ● ALLOTMENT GARDEN
- ○ BACKYARD GARDEN
- ○ COMMUNITY GARDEN
- ○ EDIBLE GARDEN
- ● EDUCATIONAL GARDEN
- ○ NURSERY FARM
- ● ROOFTOP GARDEN
- ○ THERAPEUTIC GARDEN
- ● URBAN FARM

FOOD PRODUCTION

- ○ Food self-sufficiency
- ● Food security
- ● Zero kilomiter
- ○ Low production cost
- ● Educational activities
- ● Social participation

PRODUCTS

- ● VEGETABLES
- ○ FLOWERS
- ● FRUITS
- ○ MEDICAL HERBS

PRODUCTION PROCESSING

- ● BIOLOGICAL
- ○ PERMACULTURE
- ○ HYDROPONIC
- ● AEREOPONIC
- ○ AQUAPONIC
- ○ ORGANOPONIC

Fig. 2 The Michigan Urban Farming Initiative | © the Michigan Urban Farming Initiative

increasing the base area. Furthermore, by giving hiring preference to local residents, the company has assured itself an exceptionally low carbon footprint.

Based on a different set of premises, but ones that still ensure excellent efficiency and effectiveness, the Michigan Urban Farming Initiative is a non-profit effort that has been underway for years now in Detroit.

The mission of the Michigan Urban Farming Initiative, found in Detroit's North End, is to utilise urban farming as a platform for promoting education, sustainability and community, all in the interest of strengthening urban communities while resolving Detroit's many social problems and potentially developing a more extensive model for the renewal of other urban areas.

Of particular importance is the commitment to restoring this area of approximately 12,500 m^2 in Detroit's North End, under an approach inspired to a great extent by the 'retooling of constricted environments through adaptation', featuring experimentation with the best practices for sustainable urban agriculture, plus effective strategies for increasing food safety and economically competitive models whose scale can be adjusted, reducing downy mildew, along with innovative blue and green infrastructure.

The former residential complex holding the project consists of six units of three storeys each. Built in 1915, it was inhabited without interruption until 2009, after it was abandoned and began to deteriorate, only to be purchased by MUFI in October of 2011. In collaboration with BASF, the association is undertaking an environmentally sustainable restoration of the complex, with the goal of obtaining the LEED Platinum designation. Inside the project's campus, a series of activities have reached various stages of development. Approximately 1/3 of the campus is used for open--air agricultural production, another 1/3 for interactive farming and the remaining 1/3 serves a space for closed structures (Fig. 2).

Table 2 Matrix of MUFI Community Center, Detroit

MUFI URBAN AG CAMPUS

CREDITS

Designer Team
Integrity Building Group
Client
MUFI
Year
In progress
Place
Detroit, Michigan,USA
Crops area
12'140 m²
Production
23 tons/year

42° 59' 53" N
83° 02' 45" W
183 m a.s.l.

CLIMATE DATA

Humid Continental

–8°C (JANUARY)
28°C (JULY)
7 m/s (SW)
10 m/s (WNW)

CHALLENGES

- Heat Island
- Water management
- Pollution
- Biodiversity defence
- Urban quality
- Food problems
- Survival
- Economic income
- Hobby

SUSTAINABILITY OF THE SITE

- Empty space use
- Public transport access
- Cycle path
- Low emission vehicles
- Parking areas
- Area accesibility
- Enhancement of spaces
- Permeable surfaces
- Green integration

FORMS OF URBAN AGRICULTURE

- ALLOTMENT GARDEN
- BACKYARD GARDEN
- COMMUNITY GARDEN
- EDIBLE GARDEN
- EDUCATIONAL GARDEN
- NURSERY FARM
- ROOFTOP GARDEN
- THERAPEUTIC GARDEN
- URBAN FARM

FOOD PRODUCTION

- Food self-sufficiency
- Food security
- Zero kilometer
- Low production cost
- Educational activities
- Social participation

PRODUCTS

- VEGETABLES
- FLOWERS
- FRUITS
- MEDICAL HERBS

PRODUCTION PROCESSING

- BIOLOGICAL
- PERMACULTURE
- HYDROPONIC
- AEREOPONIC
- AQUAPONIC
- ORGANOPONIC

MUFI operates the campus with the help of more than 10,000 volunteers who have donated over 100,000 h of work, growing and distributing more than 22,500 kg of products (with biological methods) that are supplied free of charge to more than 2000 families in need. Over the years, the campus has become an international tourist destination in its own right, being visited every year by thousands of tourists from throughout the world. In addition to the agricultural facilities, there is a community resource centre, an administrative office and additional office space, a portion of which is used by the MUFI staff and interns, while other space is kept available for leasing, in order to serve the administrative needs of other non-profit organisations engaged in similar activities.

The second floor of the MUFI Community Center was designed as an extremely dynamic space capable of hosting a wide range of programmes. On any given day, the space may be used for community meetings, technical training courses, seminars, board meetings, yoga lessons, etc., thanks to flexible technologies that include movable walls and folding furniture. There are also spaces used for production, packaging and cooking.

Focussed on value-added products, this space is equipped with everything needed to prepare them (including cold bottled beverages, sauces, preserves), in keeping with the slogan 'from the farm to the store' (Table 2).

Conclusions

Investigating how and how much spatial structuring, productive organisation, implemented technologies and functions used in urban farming and its interaction within the urban environment affect the well-being of the local population requires a strong commitment in the design of participatory tools and innovative methods in the urban regeneration project. On the other hand, the verification of the effectiveness of a participatory and multidisciplinary systemic approach on these types of contexts represents an advancement of knowledge both from a theoretical and a methodological point of view, with a decisive motivational drive given by the need to create community and identity, recomposing social bonds. Research in this sense also requires investigating new relationships of solidarity, participation and reciprocity, which become fundamental precisely because they are inserted in increasingly disaggregated urban contexts. The poverty that we are experiencing in this phase of profound crisis is not only economic, but it is a poverty of values, of participation and of a common and shared conscience. The recovery of a direct relationship with the earth and the environment represents one of the few social levers to transform the misery of the present into the richness of a possible future.

References

Battisti, A. (2014). *Rinnovare le Periferie, Riqualificazione bioclimatica e ambientale delle aree urbane periferiche*. Firenze: Alinea.

Branzi, A. (2000). *1969–1972 No-stop city: Preliminary notes for a masterplan, Lotus no.107*. Milan: Edizioni Electa.

Carrière, E. A. (1890). *Montreuil-aux-Pêches: Historique et pratique et quelques coutumes de sa banlieue: Bagnolet, Rosny-sous-Bois, Fontenay-sous-Bois, Vincennes, Romainville, etc..*. Paris: Librairie agricole de la maison rustique.

Commissione Europea. (2014). *Mapping and assessment of ecosystems and their services Indicators for ecosystem assessments under Action 5 of the EU Biodiversity Strategy to 2020 second Report*—Final, February 2014.

de la Salle, J., & Holland, M. (Eds.). (2010). *Agricultural urbanism: Handbook for building sustainable food systems in 21st century cities*. Winnipeg, Manitoba: Green Frigate Books.

EEA. (2017). *Landscapes in transition. An account of 25 years of land cover change in Europe, EEA Report n. 10/2017*. Copenhagen: European Environment Agency.

Ellis, E. C., & Ramankutty, N. (2008). Putting people in the map: Anthropogenic biomes of the world. *Frontiers in Ecology and the Environment, 6*(8), 439–447.

FAO. (2014). *Building a common vision for sustainable food and agriculture. Principles and approaches*. Rome: FAO.

IPBS. (2017). *Report of the plenary of the intergovernmental science-policy platform on biodiversity and ecosystem services on the work of its seventh session*. Retrieved May 29, 2019, from https://www.ipbes.net/system/tdf/2018_ldr_full_report_book_v4_pages.pdf?file=1&type=node&id=29395.

ISPRA-SNPA. (2018). *Consumo di suolo, dinamiche territoriali e servizi ecosistemici*—Edizione 2018. Rapporti 288/2018.

Jordan, D. P. (2004). Haussmann and Haussmannisation: The legacy for Paris. *French Historical Studies, 27*(1), 87–113. https://doi.org/10.1215/00161071-27-1-87.

Kaufman, J., & Bailkey, M. (2000). *Farming inside cities: Entrepreneurial urban agriculture in the United States*. Cambridge, MA: Lincoln Institute of Land Policy.

Latouche, S. (2011). *Come si esce dalla società dei consumi. Corsi e percorsi della decrescita (Italian language edition of Sortir de la société de consummation)* (p. 69). Turin: Bollati Boringhieri.

Marsden, T., Murdoch, J., & Morgan, K. (1999). Sustainable agriculture, food supply chains and regional development: Editorial introduction. *International Planning Studies, 4*(3), 295–301. https://doi.org/10.1080/13563479908721743.

Natural Resource Conservation Service NRCS. (2013). *Soil Survey Data*. Retrieved from http://websoilsurvey.nrcs.usda.gov.

Nelson, A. C. (1995). The planning of exurban America: Lessons from Frank Lloyd Wright's Broadacre City. *Journal of Architectural and Planning Research, 12*(4), 337–356.

Ricci, L. (2018). *Building the public space between history, culture and nature. Three perspectives for an integration*. Edizioni, Roma: INU.

Savard, N., & Brunet, J. (2005). *Les Savard. Histoires de vies, 1880–1930, histoire illustrée d'une famille d'horticulteurs de Montreuil*. Montreuil: Musée de l'histoire Vivante.

Stevens, C. (2011). *Agriculture and green growth*, Report to the OECD, 2011.

UN. (2015). *Resolution adopted by the general assembly on 25 September 2015. Transforming our world: The 2030 Agenda for Sustainable Development. New York City, New York*. Retrieved from https://www.un.org/en/development/desa/population/migration/generalassembly/docs/globalcompact/A_RES_70_1_E.pdf.

UNCCD. (2017). *Good practice guidance SDG indicator 15.3.1 proportion of land that is degraded over total land area*. Retrieved from https://www.unccd.int/sites/default/files/relevant-links/2017-10/Good%20Practice%20Guidance_SDG%20Indicator%2015.3.1_Version%201.0.

van Veenhuizen, R. (Ed.). (2009). *Cities farming for the future—Urban agriculture for green and productive cities*. IDRC and IIRR: RUAF Foundation.

World Bank. (2012). *Inclusive green growth. The pathway to sustainable development*. Washington, DC, USA: The World Bank.

Greening the Urban Environment: An Integrated Approach to Planning Sustainable Cities—The Case of Greater Cairo

Khalid Zakaria El Adli and Noha A. Abd El Aziz

Introduction

Urbanization is a powerful global trend characterized by rapid and often uncontrolled urban growth through excessive emigration of people from the countryside to urban areas leading to the emergence of substantial dispersed and de-compacted cities (Brilhante and Klaas 2018). Rapid urbanization has led to the growth of numerous megacities in developing countries placing increasing pressures on resources and infrastructure while influencing the living conditions and quality of life of its inhabitants. At the outset of the twentieth century, around 12.5% of the world's population (200 million inhabitants) lived in cities (Höjer and Wangel 2014). Today, the twenty-first century has been dubbed as the "urban century" with more than half of the world's population (52% or 3.6 billion people) living in towns (Höjer and Wangel 2014). This surge in urban population is expected to prevail and it is anticipated that by 2050 urban inhabitants will account for 67% of the global population (Höjer and Wangel 2014).

Moreover, cities are facing numerous challenges including economic, social, political, and environmental problems. Floods, heat waves, hurricanes, earthquake, and tsunami hazards are all warning signs of global warming (Keivani 2010). Additionally, cities currently account for some 75% of the world's energy use and are responsible for over 70% of the world's carbon dioxide emissions. Most megacities, especially in developing countries, suffer to different degrees from deterioration of urban environmental performance. Rapid urbanization and sprawl have led to the decline in biodiversity, destruction of urban natural resources and green fields,

K. Z. El Adli (✉)
Urban Planning and Design, International Programs, Faculty of Urban and Regional Planning, Cairo University, Cairo, Egypt

N. A. Abd El Aziz
Urban Design, Faculty of Urban and Regional Planning, Cairo University, Cairo, Egypt

A. Sayigh, A. Trombadore (eds.), *The Importance of Greenery in Sustainable Buildings*, Innovative Renewable Energy,
https://doi.org/10.1007/978-3-030-68556-0_3

loss of agricultural land, environmental pollution, reduced water quality, population overload, increased pressures on infrastructure, and increased release of greenhouse gases (Höjer and Wangel 2014). This is further pronounced as unsustainable urban morphologies have led to congestion, extensive buildup of transport and other infrastructure, increased dependency on vehicles, and use of fossil fuel, thus increasing the city's carbon footprint (Brilhante and Klaas 2018). These ills represent a prime concern to developing countries primarily because of their institutional inability to implement regulations and manage natural resources (Assadian and Nejati 2013).

During the past 50 years, the urban population in India grew fivefold affecting air quality; this comes as their transport sector contributes ~90% of total emissions in the country (Shrivastava et al. 2019). Loss of agricultural land around cities is another aftereffect of rapid urbanization. It has been estimated that a total of 1.4 million acres of agricultural land around the Greater Cairo Region (GCR) has been lost between 1952 and 2002 (Keivani 2010). Furthermore, according to the FAO, Nigeria's deforestation rate of primary forests is the highest in the world endangering the forest ecosystems (Mfon et al. 2014).

On the social front, cities with high levels of uncontrolled urbanization are prone to social inequalities manifested in terms of relative poverty, crime, and social exclusion of particular sectors of society (migrants, youth, the unemployed, and the disabled). This is clearly witnessed in informal settlements, where access to utilities and infrastructure can be challenging. The illegal status of these settlements impedes governments from providing needed services, thus aggravating environmental pollution while leading to urban inequity and declined community health (Keivani 2010).

Yet, cities are dynamic and complex; they are sources of both environmental opportunities and threats. Cities have an essential economic function; they are the economic engines contributing up to 55% of gross national product in low-income countries, 73% in middle-income countries, and 85% in high-income countries (UN-Habitat 2006). Moreover, cities are places for commercial, cultural, scientific, industrial, and social activities. They are centers of political power and administration, places to enable global economic functions and to offer quality life opportunities. It is but critical that cities adopt environmentally, socially, and economically responsible green growth strategies (OECD 2013).

Sustainable Green Cities

The concept of sustainability emerged in the 1990s tackling social equity, economic growth, and environmental preservation as related to city development. The concept paved the way for interrelated notions such as green urbanism, compact planning, and livable cities (Brilhante and Klaas 2018). Although there is no unique definition for sustainable development, the most widely accepted definition is that declared by the World Commission on Environment and Development in 1987, whereby sustainable development is defined as the "development that meets the needs of the present without compromising the ability of future generations to meet their own needs" (Assadian and Nejati 2013). In 2006, the UN-HABITAT asserted that sustainable development seeks to strike a dynamic balance among the objectives of economic, environmental, and sociocultural development within the framework of a local government whereby citizens proactively participate in the development process (Assadian and Nejati 2013; Chan and Hun Lee 2019).

One term that emanated from sustainability is "green city" with "green" signifying different things to different people. The term nowadays is widely used by private and public organizations as a brand for sustainability and eco-friendliness (Brilhante and Klaas 2018). Green sustainable cities are an integrated, multi-sector process that covers a broad range of environmental issues while linking these issues to economic and social objectives (EBRD 2016). "Green cities" minimize ecological impact and maximize high environmental performance with the objective of improving and supporting the natural environment while utilizing investments, services, regulation, and other relevant policy instruments. They seek to reduce dependency on nonrenewable energy sources, preserve (air, water, land/soil, and biodiversity), include green and resilient infrastructure and low-carbon transport, mitigate risks deriving from climate change, and deliver improved quality of life (EBRD 2016; Lewis 2015). Adopting green city strategies offers multiple opportunities to create jobs, attract businesses and investment, improve local environmental quality, and address global environmental challenges, particularly climate change (OECD 2013).

Benefits of Sustainable Green Cities

Urban form and spatial development have notable consequences on sustainable development encompassing not only environmental issues but also social and economic aspects. Based on studies by diverse scholars (Keivani 2010; Lewis 2015; Keleg 2018) a "sustainable green city" approach facilitates the realization of environmental, social, and economic benefits including but not limited to:

Environmental Benefits:
- Reducing carbon dioxide emissions by utilizing renewable energy sources
- Reducing heat island effects and carbon footprint
- Optimizing natural assets such as sunlight and wind flow
- Providing places for biodiversity in cities
- Creating a quieter and healthier microclimate
- Decreasing air and water pollution
- Reducing reliance on automobiles
- Offering adaptable and resilient urban spaces and infrastructure
- Mitigating hazards
- Producing healthier buildings
- Improving visual amenity

Social Benefits:
- Increasing livability of cities
- Improving quality of life
- Enhancing opportunities for connection with nature and understanding of the natural habitat
- Enhancing physical and mental health and well-being
- Improving community health and decreasing ailment rates

Economic Benefits:
- Enhancing economic vitality
- Increasing marketing potential
- Reducing poverty
- Decreasing operating costs
- Increasing productivity through reduced commutes and travel
- Improving service delivery

Sustainable Green City Strategies

Urban policies could play a significant role in realizing environmental and green growth goals. Cities are responsible for a substantial share of infrastructure investments, which if invested wisely could contribute to national efforts to combine growth with environmental performance leading to sustainable "green cities" (OECD 2013). Establishing a "green city" requires firm policies, a robust regulatory framework, strategic planning, and links to finance (Lewis 2015). By manipulating the built environment and infrastructure, cities could be greener and more sustainable. This involves intervening at various stages of development including design, construction, operation, and maintenance (Lewis 2015). Sustainable green strategies could be summarized to include the following:

Integrated Urban Development

Integrated urban development is crucial to build "green cities" as it enables the planning and development process to consider green sustainable principles while keeping in mind the interaction between sectors, services, and aspired outcomes. It takes into consideration multiple sectors and objectives within a planning and development process while seeking to reconcile conflicting development objectives (Lewis 2015).

Sustainable Land-Use Planning

Urban form and land-use planning could play a key role in reducing environmental impact. Sustainable land-use planning targets energy optimization and integrated transport networks while focusing on enhancing mobility through public transport, cycle routes, and greater social inclusion (Keivani 2010). This is manifested in compact city planning which focuses on mixed land-use development, short commutes, and high residential density.

Sustainable Building Practices

Energy-efficient buildings consider both building and operating phases. The first phase embodies energy required to extract, process, transport, and install building materials. The second phase comprises operating energy to provide services such as heating, cooling, and powering equipment. Sustainable building practices seek to reduce reliance on nonrenewable energy sources, incorporate alternative energy sources, and improve operating efficiency (Lewis 2015). This involves adopting approaches and tactics such as solar access, water capture, treatment, and reuse.

Sustainable Mobility and Transport

A sustainable transport policy seeks to minimize unnecessary trips, encourage behavior change, and improve the operational efficiency of motorized transport (Lewis 2015). A proclaimed approach to achieving sustainable mobility and transport system is through transit-oriented development (TOD). TOD considers mass transit/multimodal transport hubs supported by high-density residential and mixed-use development along corridors and activity nodes. The system represents a low-carbon transport solution stressing nonmotorized transport, thus reducing dependency on petroleum-based modes (Lewis 2015). Increasing the efficiency of the transport system in cities not only benefits the environment and contributes to equitable access, but also enhances the city's attractiveness to investment, skilled human capital, and economic growth (Assadian and Nejati 2013; OECD 2013).

Greening the Urban Environment

Greening is one strategy cities strive to achieve in order to enhance its livability and quality of life. Urban green open spaces provide many benefits environmentally, socially, and economically. City greening considers the provision of a network of green open space, multifunctional, natural, as well as seminatural areas (Lewis 2015). Urban greening has been an aspiration since Howard's green city movement and has served as the paradigm for such notions as green infrastructure, greenways, green roofs, pocket parks, urban parks, and green corridors (Keleg 2018).

Green Infrastructure

A green infrastructure is an approach that incorporates both the natural environment and engineered systems with the objective of protecting, restoring, and/or mimicking the natural cycles. This involves implementing an urban strategy that considers the infrastructure as a green sustainable network offering to balance natural and man-made systems (Abdou et al. 2016). Green infrastructure is practical and economical, conserves ecosystem values, and enhances community safety and quality of life.

Energy Conservation and Renewable Energy

By and large, energy strategies engage in four aspects: access, security, reliability, and affordability. Green energy strategies consider the same aspects; however it adds alternative energy sources, low-carbon options, and approaches to lower carbon footprint produced by energy consumption while focusing more on renewable sources of energy as well as energy efficiency to serve the rising demand in the built environment, transport, as well as sectors of industry and trade (Assadian and Nejati 2013; Lewis 2015).

Water and Waste Management and Recycling

The water sector comprises three main elements: water supply, wastewater and sanitation, as well as stormwater management and drainage (Lewis 2015). On the one hand, managing water and wastewater networks constitutes the most important domain for building a healthy greener city. Access to safe drinking water and sanitation has an intense effect on increasing public health. Sustainable water policies in "green cities" ensure combating the contamination of surface water in rivers, springs, and underground water. A well-designed sewage system will significantly minimize the pollution of surface water and thus increase water supply. New water consumption techniques should also be considered and could include using smart systems to conserve water (Assadian and Nejati 2013). Collection and reuse of storm as well as gray water should also be considered.

On the other hand, solid waste management involves the collection, transfer, and disposal of waste. The "3R" approach (reduce, reuse, and recycle) represents the core of solid waste management for "green cities" (Lewis 2015). Green industry considers the multiple life cycle of various products and by-products; it encourages recycling where the by-product of an industrial process becomes the inputs for another, thus generating a closed loop to insure reserving the natural resources and managing it in a sustainable matter (Lewis 2015).

Resilient Strategies

Resilient strategies enhance city capacity to respond, adapt, function, and recover from shocks, stresses, disasters, and/or changing climate. It considers both the built form and human elements of a city. In "green cities" the ability of people particularly the poor and vulnerable to live, work, survive, and recover is augmented as part of building up its resilience (Lewis 2015). The concept applies to infrastructure as a part of the built environment in a way that enables such infrastructure to withstand critical events such as flooding, disasters, earthquakes, and climate change.

Greenways: An Approach to Planning Sustainable Green Cities

Greenways are more than just parks or amenities, but a response to the physical and psychological pressures of urbanization (Searns 1995, p. 65). They play a significant role in the development of urban and suburban areas providing much-needed natural corridors in urban settings while establishing the essential counter form to the built environment. Greenways constitute an adaptive response, a way to offset and mitigate the effects of rapid urbanization. They could be considered lines of opportunity, environmental corridors for the purpose of stitching together fragmented sites while providing for environment, ecology, education, and exercise (Bischoff 1995, p. 317). Greenways have been defined by Ahern (Ahern 1995, p. 134) as networks of land planned, designed, and managed for multiple purposes including ecological, recreational, cultural, aesthetic, and/or other purposes compatible with the concept of sustainable land use. Yet, greenways went beyond their initial amenity-oriented objectives of connection, movement, aesthetics, and recreation to address more environmentally sensitive issues including protecting the natural habitat, sustaining threatened ecosystems, preserving and highlighting area culture and heritage, as well as safeguarding against environmental hazards (Searns 1995, p. 72, 73). They represent the platform where much social life and learning take place and are thus educative and informative presenting the relationship of people with natural systems. Greenways could be utilized to help heal the human psyche by providing alternative corridors that offer attractive visual form, greenery, and solace (Searns 1995, p. 72).

Greenways thus constitute an important part of the urban landscape providing an attractive opportunity for stimulation and for offering a wealth of sensual variation. They have the potential to restructure city regions and improve connectivity of open public space, making cities greener and friendlier to pedestrians and cyclists (Abdou et al. 2016; Bousemma et al. 2018). Hence, greenways are a noteworthy tool for conceiving "green cities," for realizing congruence, safeguarding nature, enhancing economic development, and upholding societal values (Ahern 1995). The articulation of greenways and open-space networks in the planning and development of sustainable “green cities” is all but common. Globally, the concept has been increasingly used, especially in cities experiencing deindustrialization process (Qu et al. 2015).

Typologies of Greenways

Scale Based

The greenway network could be seen as a framework which varies greatly in scale starting from a national framework right down to the neighborhood and narrow ribbons of green space linking together regional ecological corridors, city parks, community gardens, and pocket parks (Bousemma et al. 2018; Qu et al. 2015).

Function Based

Greenways can serve various functions with multiple goals at national, regional, city, and neighborhood levels. Fabos classified greenways into three categories: recreational, ecological, and heritage greenways (Fabos 2004). The most common typology is that of the recreational greenways, which emphasizes public access and recreational provision. These greenways are usually formed along routes and corridors passing through recreational spaces linking water and landscape resources with high visual value (Bousemma et al. 2018; Fabos 2004). On the other hand, natural corridors of ecological importance are formed by spaces along rivers and valley sides. These corridors facilitate protecting and developing of natural resources and wildlife, migrating of species, sustaining of biological diversity, and natural hiking (Ahern 1995; Bousemma et al. 2018; Fabos 2004). Cultural and heritage greenways have visual and historical value. They link cultural and historic resources attracting tourists, providing economic and educational benefits while affording seasonal accommodation (Ahern 1995; Fabos 2004).

Guiding Principles

Strategically planned and designed, “greenways” should consider the following guiding principles:

- Specify the function, appropriate uses, and activities as well as proposed design of the greenway. In some cases, determining the greenway function can be challenging particularly when proposed functions contradict one another and spatial as well as functional determinants are conflicting.
- Plan the greenway to include an origin, a destination, and significant attractions.

- Balance environmental, social, and economic objectives.
- Make use of vacant lots, unused corridors, and spaces in the urban settings.
- Create a public space network connecting scattered open spaces and destinations.
- Recognize the character and distinctiveness of different sites while ensuring that it is reflected in the programs.
- Seek equity by ensuring accessibility to the entire community, to all socioeconomic classes and neighborhoods (Crawshaw 2009).
- Reflect the needs, demands, aspirations, as well as social and cultural values of regular green open-space users in the design (Qu et al. 2015).
- Avoid spreading invasive species into protected areas (Crawshaw 2009).
- Provide services and amenities including lighting, drinking water, restroom facilities, and parking to encourage use of the greenway (Chen et al. 2017).
- Provide for clarity, safety, security, and visual interest (Chen et al. 2017).
- Connect existing green routes such as waterways and green corridors and slow-traffic routes in linear paths.
- Enhance accessibility to the greenway network through public transportation, allocating transit nodes along existing public transportation routes (e.g., metro lines, bus lines).
- Connect local public facilities including schools, shops, and public transportation nodes to the network.
- Offer multifunctional resources capable of delivering a wide range of environmental and quality-of-life benefits (Abdou et al. 2016).
- Engage key partners and diverse stakeholders in the planning and design process.
- Monitor performance, with funding from government and private partnerships (Abdou et al. 2016).

Global Exemplars and Best Practices

North America is the leading region in implementing the greenway concept and has inspired Europe, Australia, and China to follow in its footsteps. In recent years, attempts have been made to transform derelict sites and open spaces into urban greenways with the objective of improving environmental quality and restructuring the city fabric. The notion has since developed to include numerous success stories globally affording a spectrum of best practices. The following case studies represent a selected sample of exemplars that were chosen based on their positive contribution to social, economic, and ecological functions. They reflect a range of planning levels, from the national to city scale, in addition to highlighting successful planning, design, and implementation processes.

East Coast Greenway

Overview

The East Coast Greenway is one of the most popular greenways in the USA. It represents an exemplar for national scale greenways as it connects 15 states and 450 cities and towns along a 3000-mile stretch from Maine to Florida (see Fig. 1). The greenway is a protected cycling and walking route serving to improve public health, environmental sustainability, economic development, and civic engagement. The trail was created on sewer and riverfront easements, abandoned railways, as well as widened sidewalks. Currently, it accommodates pedestrians, joggers, cyclists, in-line skaters, horseback riders, as well as wheelchair users (see Fig. 2) (Keleg 2018).

Best Practices

A critical review of the East Coast Greenway illustrates a number of guiding principles and best practices. These include the following directives:

- Pursue funding from numerous sources and at different levels. This may include seeking governmental funds, specific programs, private foundations, as well as local funding.

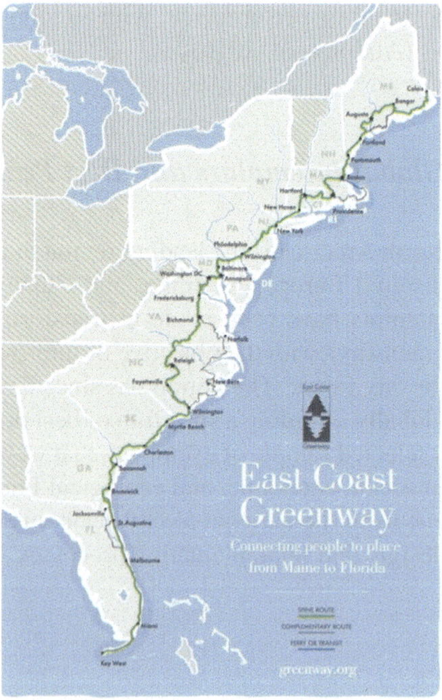

Fig. 1 The East Coast Greenway. Source: East Coast Greenway Alliance (2019)

Jogging lanes *Signs to guide users* *Bike lanes*

Fig. 2 The East Cast Greenway design features. Source East Coast Greenway Alliance (2019)

- Use durable, easy-to-maintain materials.
- Minimize and where possible avoid conflict with vehicular traffic.
- Ensure path accessibility and comfort particularly for the physically disabled.
- Ensure high connectivity with no missing links.
- Reduce distances between activities.
- Ensure continuity and cohesion by minimizing distances between parallel and intersecting routes.
- Direct users amidst visually attractive and lively zones.
- Provide unbroken flow by limiting long waits at traffic lights.
- Offer sufficient signage to help identify route directional changes.

The Sauvegarde/Puurs/Baasrode Greenway, Belgium

Overview

The municipality of Puurs in Belgium decided to develop a greenway within its municipal boundary along a derelict railway line connecting Antwerp and Termonde. The regional greenway is planned to cater to both pedestrians and cyclists stretching 6.5 km and continuing eastwards towards Baasrode. The goal is to foster coexistence between different municipalities while facilitating access to schools. However, in 1997 the National Railway Company of Belgium decided to revive the Eastern railway section of Puurs—Antwerp. Measures to ensure safe coexistence between the active railway line and the greenway were thus undertaken (see Fig. 3) (European Greenways Association 1998).

Best Practices

The Puurs Greenway offers a multitude of best practices; among them are the following:

- Lease or expropriate land as a public utility to create greenways.

Fig. 3 The Puurs Greenway represents a good example of coexistence between an active railway line and a greenway. Source: European Greenways Association (1998)

- Coexistence between railway lines and greenways is possible provided that security measures are implemented to safeguard greenway users. These may include among others dividing barriers, safe zones, as well as signal visibility at level crossings.
- Use anti-slip shock-absorbing surfaces for the comfort of pedestrians and cyclists alike.
- Include suitable water drainage to all cycle and pedestrian paths.

Radnor Trail, Pennsylvania, USA

Overview

Located in northern Radnor township, the 2.4-mile Radnor Trail signifies a local recreational trail connecting municipal parks, residential neighborhoods, small businesses, and an art center. This local trail caters to a number of recreational activities including leisure strolls, cycling, jogging, in-line skating, dog walking, and fishing (see Fig. 4) (Township of Radnor 2019).

Best Practices

The Radnor Trail offers the following best practices:

- Reuse derelict rail right-of-way and abandoned sites to create greenways.
- Integrate heritage sites and historical buildings in planning the trail with the aim of educating users about the city history.
- Plan greenways while affording users scenic views and distant vistas.
- Utilize site potentials (fishing spots, etc.) to attract users to the trail.

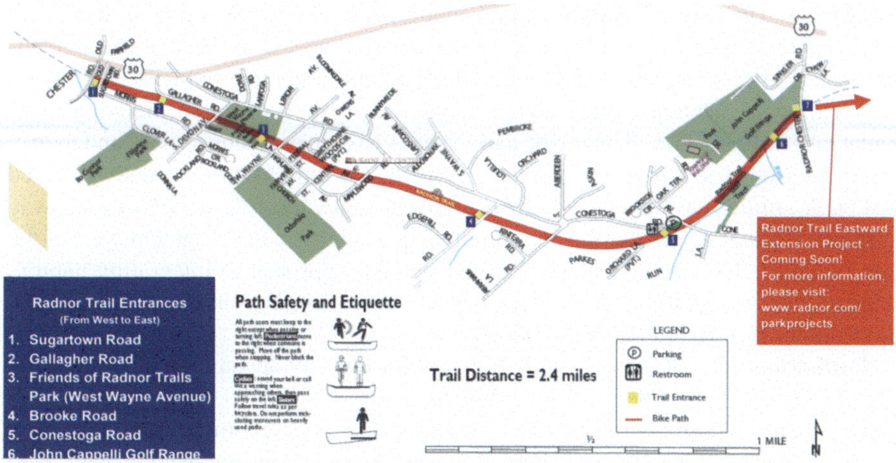

Fig. 4 The Radnor Trail map. Source: Township of Radnor (2019)

Greenway Planning in the MENA Region

Greenway planning in the MENA region is uncommon and is confronted with numerous challenges. From the limited studies that tackle greenways in the Middle East, Bousemma experiments with the concept of greenway panning in Sousse city, Tunis. His findings reveal that though there is a great potential for adopting greenway planning in Tunis, the notion of greenways and its foundations are absent from Tunisian environmental policy and are nonexistent in town planning strategies (Bousemma et al. 2018). On the other hand, Turk in 2017 developed a strategy for defining, prioritizing, and selecting greenway alternatives utilizing an analytical hierarchy process (AHP) based on an empirical study in the urban region of Trabzon, Turkey. The study concludes that the AHP has a promising capability in prioritizing and selecting the most favorable alternative (Türk 2017).

In Egypt, Fathy focuses on adopting the greenway concept in Historic Cairo, a rich heritage site celebrated with ancient buildings and monuments (Fathy 2006). Additionally, Abdel Salam in 2009 explores the possibility of transforming vacant and neglect spaces within the city of Cairo to open green spaces linked together via an interconnected green network (Abdel Salam 2009). Furthermore, in his study, Hussein investigates planning and management characteristics of green areas in Borg El Arab assessing the potentials and challenges facing the greenway concept. The object of his investigation is to develop a comprehensive vision for the development of green networks in new desert cities (Hussein 2018). In 2010, another study was undertaken by graduate students from Cairo University, the American University in Cairo, and the University of California, Berkeley, under the direction of faculty from all three universities with the objective of developing a vision for reconnecting the city to the river Nile (Kondolf et al. 2011). The study involved four regions in Greater Cairo, namely downtown Cairo (Kedivial Cairo), Old Cairo, Athar El Nabi, and Maadi. The prime objective of this enquiry was to develop a plan to reconnect local residents with the Nile river through increased access to the waterfront, environmental improvements, pedestrian pathways, and attractive public spaces. On a different scale, Mahmoud and Sayed propose a method of green planning in El Sadat city utilizing GIS. The investigation seeks to enhance connectivity and reduce fragmentation through an integrated greenway system (Mahmoud and El-Sayed 2011). A similar case study by El Adli demonstrates the role of urban greenway systems in planning residential communities in a suburban development west of the city of Cairo. The findings identify a step-by-step approach to integrating natural, recreational, and cultural greenways and corridors in planning future residential developments in Egypt (El Adli 2006).

Furthermore in 2002, the first GCR greenway concept appeared in the Structural Plan of Giza City, yet its recommendations were never realized due to feeble law enforcement and urban development pressures. The plan recommended reclaiming the Nile banks for tourist and recreational activities while connecting new open green space to existing urban parks such as the Orman and the Giza Zoo via an interconnected network of green corridors. It emphasized safeguarding visual corridors to the Nile while enforcing the environment law. In 2012, the strategic "Cairo 2050 Vision" embraced the augmentation of green spaces in the city as part of an interconnected network of open space (see Fig. 5). The vision relied on three cross-cutting pillars that translated into eight strategic actions (Ministry of Housing, Utilities and Urban Development, General Organization of Physical Planning and UN Habitat 2012):

- Developing social equity by improving living conditions for all
- Enhancing the region's transportation infrastructure
- Revitalizing the inner city

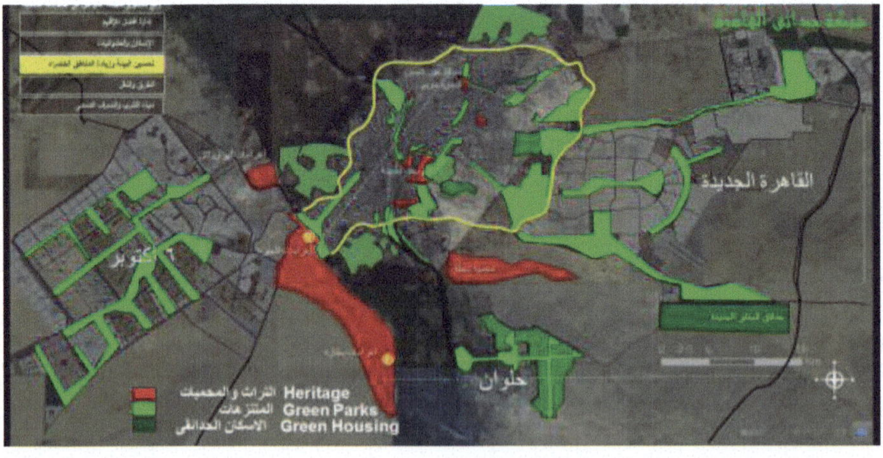

Fig. 5 The proposed green network in Cairo's 2050 Vision. Source: Ministry of Housing, Utilities and Urban Development, General Organization of Physical Planning and UN Habitat (2012)

- Strengthening new urban communities as vibrant and diverse hubs
- Affording conditions for tourism to flourish
- Providing a competitive environment for a knowledge-based economy
- Adopting more environmentally friendly policies and approaches
- Setting up efficient governance to steer the development

The 2050 Vision resulted in the identification of 22 guiding projects that were meant to facilitate vision realization (Ministry of Housing, Utilities and Urban Development, General Organization of Physical Planning and UN Habitat 2012).

The vision was criticized for being too ambitious and unrealistic, as it is built on relocating concepts unsupported by legislative frameworks. Relocation strategies included transforming cemeteries and unsafe settlements into public parks, relocating all polluting industries to the city outskirts, and converting informal settlements into public open spaces and service hubs (Ministry of Housing, Utilities and Urban Development, General Organization of Physical Planning and UN Habitat 2012). Spatial planning studies for a handful of projects from "Cairo 2050 Vision" have since been developed and include Dahab and El Warraq Islands as well as Cairo's waterfront and Bolak Abo El Ela development project. Yet, attempts to green the urban environment remain ineffective due to absence of a comprehensive vision to formulate an interconnected green network instead of isolated green islands dispersed within the city.

The Greater Cairo Region (GCR): A Case Study

As a metropolitan city, the Greater Cairo Region (GCR) faces tremendous urban pressures as a result of its phenomenal growth since the turn of the twentieth century. Encompassing three governorates, Cairo, Giza, and Qalyubia, the GCR suffers from high population (up to 24.7 million and a density of 38,247 inhabitants/km^2) (Central Agency for Public Mobilization and Statistics 2019). Hence, despite the concentration of services and job opportunities, the city ranks relatively low on the livability index rank (121 from 140 countries in 2015) (The Economist Intelligence 2015). Ills of rapid urbanization are manifested in environmental pollution, congestion, pressures on infrastructure, and loss of agricultural land. This is further pronounced due to the acute shortage, degradation, and inequitable distribution of public open space in general and that of green open space in specific. Unbalanced distribution of open space coupled by their disconnection within the larger network fails to contribute to social equity (Keleg 2018). Nowadays, Cairo's share per capita of green spaces stands at 0.5 m^2/capita remaining relatively lower than that of international standards, even when compared to other cities in the MENA and Gulf region (Keleg 2018). To further exacerbate this predicament, much of the green open space and promenades along the banks of the river Nile have been in recent years and continue in the present day to be consumed by development. Trees are being cut off and open land is appropriated to the benefit of public services and infrastructure including water treatment facilities, cafeterias, social clubs, floating restaurants, and plant nurseries (see Fig. 6). As a result, a considerable percentage of city dwellers are deprived from direct contact with nature lacking access to natural environments, recreational opportunities, and a learning experience (Keleg 2018).

Nevertheless, the GCR possesses promising potentials for developing greenway networks as riparian greenways, parkways, and farmland greenways owing to its diverse natural features. The most prominent feature is the river Nile that represents an ecological blueway and an environmental corridor affording multiple opportunities for tourism, leisure, and education (Abdou et al. 2016). This articulated ecological corridor could serve as the green spine that links urban districts via an interconnected network of smaller greenways. Additionally, the GCR is rich in heritage sites, such as the pyramids and the citadel; green open spaces as El Kanater park, one of the largest regional parks in Egypt; and natural reserves such as Wadi Degla. Moreover, desolate sites including derelict railways, abandoned storage sites, deserted areas, and open spaces could potentially be linked to one another via green corridors forming an interconnected network of greenways.

Furthermore, the Nile's riverbanks have the potential of becoming recreational green corridors, affording residents and visitors alike a wealth of experiences. As an ecological corridor, the Nile could represent the spine of an interconnected green network linking heritage sites to leisure hubs via an intricate system of greenways. In recent years, numerous proposals and project schemes have suggested the development of the river Nile and its banks to include new activities and land uses while advocating expanding the role of the river as a connecting corridor linking the city with its peripheries.

Fig. 6 The Nile riverbanks are consumed by development. Source: Barada et al. (2005)

The potential of integrating greenway planning in the Greater Cairo Region (GCR) is thus nothing less than promising. Yet, prospects of successful implementation are confronted with numerous challenges and impediments. A key challenge involves dealing with competing land uses in high-density urban environments.

Greening the Greater Cairo Region (GCR)

A recent undertaking by urban specialists from Cairo University in 2005–2006 proposes a comprehensive approach to dealing with the river Nile in the GCR while offering a strategic vision towards creating an interconnected network of greenways along the river (see Fig. 7) (Barada et al. 2005). The study's prime objective is to safeguard the river Nile as an ecological corridor while affording multiple opportunities for tourism, education, recreation, social interaction, and economic development.

Divided into three sections, the study illustrates the existing conditions and challenges facing the development of the Nile riverbanks, establishes the needs and aspirations of the local community, and concludes with a strategic vision and a greenway master plan for the GCR. The plan emphasizes the connection to nature, vegetation, and abundant wildlife as epitomized in the Nile river; underlines economic vitality as highlighted in recreational and commercial functions; promotes accessibility and mobility via an articulated transport system including BRT, cycling, and water taxi; and highlights sociocultural values via linkages connecting multifunctional green spaces to heritage sites. The planned greenway network connects the city to the Nile and its differentiated districts while advocating interventions to resolve traffic congestion and enhance the visual image of the GCR (Keleg 2018).

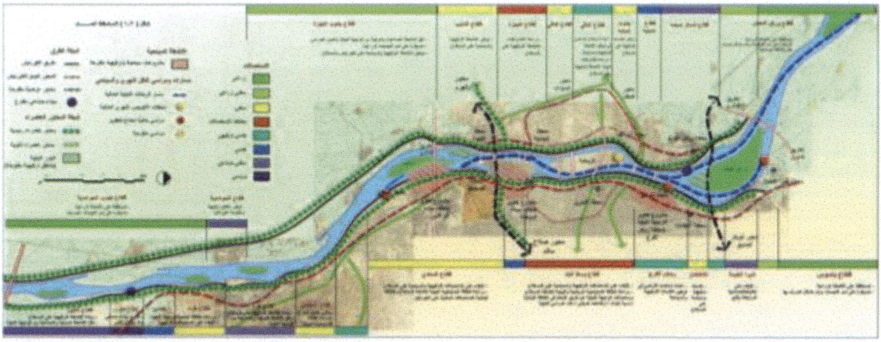

Fig. 7 Nile river master plan. Source: Barada et al. (2005)

Proposed Strategy

The strategic vision for the Nile riverbanks was formulated upon a critical review of international greenway case studies in the USA, France, China, Tokyo, Canada, and the UK, in addition to a detailed appraisal of local legislation as well as national structural and transport plans. The vision ensured integrating transport and mobility goals, environmental and recreational goals, as well as social and economic goals. The goals include the following (Barada et al. 2005):

Transport and Mobility Goals

- Transfer traffic along the Cornish Boulevard (adjacent to the river) to substitute trajectories.
- Introduce an integrated multimodal transport system.
- Ensure connectivity and continuity of the Cornish promenade (see Fig. 8).
- Safeguard pedestrians from vehicular traffic.
- Reduce the density of activities along the riverfront.

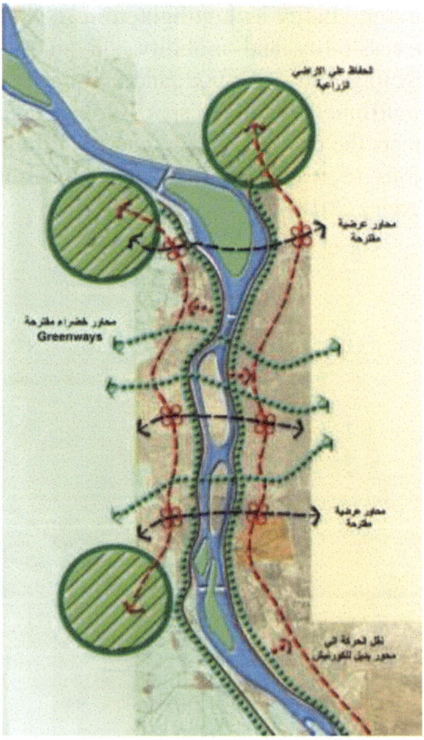

Fig. 8 Connecting the city to the river Nile. Source: Barada et al. (2005)

Environmental, Recreational, and Visual Goals

- Safeguard ecological, recreational, and aesthetic attributes of the river Nile corridor.
- Protect and preserve agricultural land in and around the city and along the Nile corridor.
- Transform the Cornish Boulevard into a recreational corridor with ample opportunities for recreation, social interaction, and learning.
- Connect the Cornish and river trail with the city, its heritage sites, and recreational, retail, and touristic nodes.
- Increase the per capita share of green open space.
- Safeguard and enhance visual corridors.

Social and Economic Goals

- Offer opportunities for social interaction, recreation, and education.
- Provide for and boost the economic vitality of the greenway network.
- Introduce new activities along the river Nile corridor and attract tourism.

The Greenway Master Plan

The green network in GCR was formulated by replanning streets and paths within the city with the objective of expanding green open spaces and introducing new functions without compromising accessibility. Figure 9 illustrates proposed linkages radiating from the heart of the region connecting urban parks, public spaces, natural and cultural resources, historical sites, and tourist destinations in an articulated green network. The proposed plan establishes four categories of hubs and two categories of links. Proposed hubs incorporate:

- Ecological hubs including natural reserves such as Wadi Degla, Mokattam Hills, as well as water canals.
- Recreational hubs including touristic, sports, and recreational uses: These embrace among others urban parks, Ain El Sira pond, garden palaces, heritage parks, as well as Cairo stadium.
- Cultural and educational hubs including historic and cultural sites such as Giza Pyramid, Sakkara Pyramid, Megra El Eion historic wall, Salah El-Din Citadel, Opera House, religious complexes, the Egyptian museum, and Cairo University.
- Central and commercial hubs including such sites as the Khedivial city center, Tahrir Square, and Opera Square.

Proposed links include:

- The Nile river including the riverbanks, as the main spine and blueway.
- Gray-ways including transportation corridors which could be utilized as parkways including such streets as Haram street, Salah Salem street, the Auto-strad, Gesr El-Suez street, El Orouba street, Ramsis street, and Port Said street.

Fig. 9 Proposed greenway network in GCR. Source: Barada et al. (2005)

The master plan suggests six specific greenways, the Nile recreational greenway, four historical greenways, and one commercial greenway connecting the downtown with the waterfront. It focuses on connecting the banks to the surrounding neighborhoods by diverting traffic to alternative routes while affording the public accessibility to linear green recreational corridor along the river Nile. The study presented some solutions to maximize the use of riverbank slopes (see Figs. 10, 11, 12, and 13).

Implementation Challenges

Planning greenways in a condensed urban center such as in the GCR often encounters numerous hurdles. The most obvious are the scarcity of vacant land and conflict with existing physical barriers such as major roads, railroads, buildings, and residential/

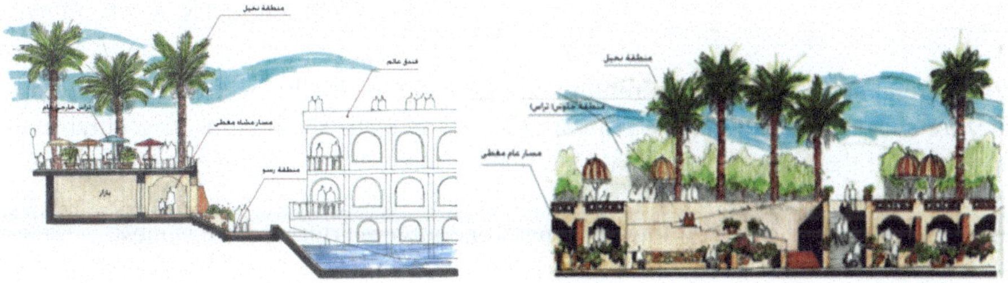

Fig. 10 Utilizing elevation differences to accommodate services and amenities while preserving scenic views of the Nile. Source: Barada et al. (2005)

Fig. 11 Creating attractive settings for recreation. Source: Barada et al. (2005)

Fig. 12 Connections between the riverbanks and the surrounding environments. Source: Barada et al. (2005)

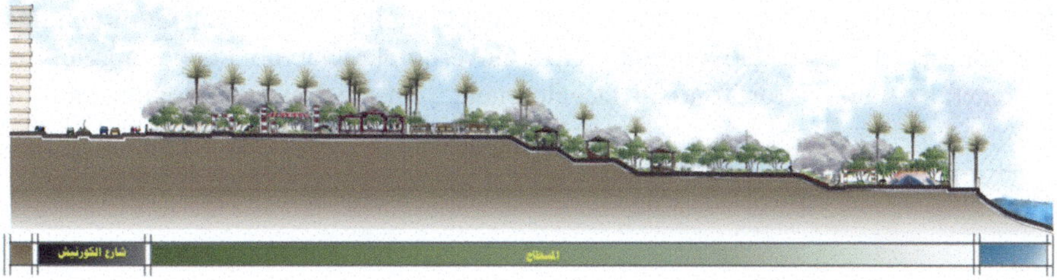

Fig. 13 Creating parks along the riverbanks and connecting them with the city. Source: Barada et al. (2005)

commercial areas. Yet, such stumbling blocks could readily be overcome by employing strategic and master greenway plans. Greenway plans should thus be accurate and applicable if they are to be translated to reality. They should identify prevailing difficulties and impediments in both the planning and implementation stages. The following illustrates the most common challenges encountered during both stages.

Challenges in the Planning Process

As practitioners, we endure numerous adversities to produce accurate, applicable, and at the same time flexible plans. Overcoming these challenges is a must to move from the planning stage to the implementation and execution stages. The most common challenges facing the planning process include[1]:

- **Competence and experience of the planning team.** The General Authority for Urban Planning commissions experts or consulting firms to provide strategic plans; yet, occasionally the quality of the product prevents the execution of the plan due to imprecise or irrelevant data (El Said 2015).
- **Insufficient/accurate database.** This is a common problem in the Egyptian planning process. Performing robust analysis and taking informed decisions warrant the availability of up-to-date data, which is not necessarily always the case (El Said 2015). Such shortcomings reflect negatively on the applicability of both the strategic and master plans limiting implementation procedures.
- **Ignoring land ownership.** Occasionally, strategic plans ignore land tenure, thus delivering unrealistic strategies, especially when monetary compensations are required for land acquisition (Ibrahim 2016). Private property may thus halt implementation of a greenway plan due to concerns relating to loss of privacy, liability, illegal parking, access, and pollution.
- **Poor representation and involvement of stakeholders.** Though expert opinion is greatly valued in developing strategic and master plans, public and stakeholder involvement is equally crucial (Ibrahim 2016). Nevertheless, seldom is the local community efficiently represented—only a limited sample of the general public is heard. Yet, local knowledge and expertise is fundamental in ensuring that proposed strategies reflect local conditions and values.
- **Lack of coordination between planning organizations.** In the GCR numerous governmental bodies are mandated with green open spaces, including but not limited to the General Organization for Physical Planning, New Urban Communities Authority, the Cairo Governorate, Giza Governorate, Qalyubia Governorate, the General Authority for Tourism Development, and the National Organization of

[1] The GCR greenway master plan encountered several of these problems.

Urban Harmony. This is further exacerbated due to the lack of coordination among these bodies, with each advocating different goals and strategies.
- **Poor connection with other plans.** A critical criterion for the lack of effectiveness of greenway master plans is its disengagement and disconnection with other plans such as economic, infrastructure, and utility plans.

Challenges in the Implementation and Execution Process

While implementing an approved plan issues arise that could hamper implementation. These include:

- **Multiple jurisdictions.** The implementing of greenways, especially across multijurisdictional boundaries, can be challenging if not impossible. Implementing an urban greenway plan in the GCR across three governorates involves numerous stakeholders and implementing authorities. This could include several ministries, governorates, local authorities, city councils, as well as local community groups, residents, and landlords. Coordinating among the different units and stakeholders is a bewildering task especially if they hold conflicting agendas or inconsistencies in their planning policies and information systems.
- **Absence of an action plan.** Many strategic plans are not translated into detailed plans that are applicable, and thus they remain in the conceptual phase.
- **Poor law enforcement.** Weak enforcement of bylaws and regulations on the local level is critical in crippling the implementation of urban plans particularly when it involves relocating or changing land uses as a part of the plan (Ibrahim 2016).
- **Lagging behind schedule.** Due to rapid change in land uses, densities, utilities, and other related urban environments, diverting away from the set schedule could hinder the implementation process.
- **Limited awareness of green cities.** Governmental institutions lack the resources to plan, implement, and manage greenways solely. Instigating legitimate channels to embrace the involvement and the support of both the public and NGOs should thus be undertaken for conceptual planning, detailed design, maintenance, and monitoring of greenway systems. Yet, ignorance of the holistic importance of greenways to sustainable green cities contributes to unproductive public participation, unsatisfied decision makers, and limited political support.
- **Insufficient funding.** Project funding is possibly the most crucial challenge for greenway projects. Funds should be secured beforehand for construction and maintenance. Currently, funding for greenways predominantly comes from the local government (Ibrahim 2016); however it could be thought from both private businesses and nonprofit organizations.

Conclusion and Recommendation

Greenways are linear corridors found in rural and urban contexts serving both pedestrians and cyclists. They function at a range of scales from the regional to the neighborhood, facilitating access to nature. Greenways are connectors linking parks, natural reserves, and historical and recreational sites and are able to tackle economic, social, and environmental concerns simultaneously. They seek to boost the quality of life and mitigate exacerbated conditions in cities where rapid urbanization and large concentrations of people and activities have created a myriad of complex socioeconomic challenges with often severe environmental consequences. Due to their varied benefits, greenways are a successful tool for sustaining green cities. Yet, and though the greenway concept is well acknowledged, its implementation is limited and inadequate.

It is thus highly recommended that innovative practical approaches be considered to foster the likelihoods of greenway implementation. On the one hand, targeting a greener national framework would facilitate addressing city-specific challenges and ensure coherence and consistency between national and local policies while integrating greenway opportunities into planning policies. Providing effective planning legislation system is necessary to coordinate all levels of the planning process, through a combination of enforcement and incentives to ensure compliance and inter-municipal cooperation to manage urban development. On the other hand, measuring and monitoring tools that cross administrative boundaries should also be made available, as well as a body to collect and disseminate cross-sector data at the microscale. A detailed local plan is crucial to translate greenway strategies while allowing different stakeholders besides professional planners, to partake in the planning process from start to implementation and management.

Furthermore, it is fundamental to establish sustainable financing options and to diversify domestic funding sources. A well-designed property tax and development fee could tackle urban sprawl and raise money for funding greenways. Real estate developers could be charged to link their new projects to existing greenway networks. Local governments could also seize part of the value increases of real estate dues benefiting from greenway networks to maintain existing trails. Another alternative is to adopt public-private partnerships (PPPs), whose success in similar projects has been substantiated.

Moreover, green strategies should capitalize on the great potential for NGOs, private sector, and investors to change the current status of green spaces in the GCR. Trained professionals and leadership with clear mechanisms for gathering and updating physical, social, economic, and environmental data could cause a colossal uplift in the quality of urban plans. To conclude, further endeavors are essential to create a comprehensive and pertinent vision for greenways in the GCR with huge prospects for enhancing the quality of life for millions of residents and visitors alike.

References

Abdel Salam, A. (2009). *Maximizing greenway network in Egypt, Master thesis. Faculty of Regional and Urban Planning*. Cairo: Cairo University.

Abdou, A., Abd Elgawad, I., & Fouad, T. (2016). Green infrastructure to achieve sustainability in urban design Nile corridor in Greater Cairo as a case study. *Journal of Al Azhar University Engineering Sector, 11*(40), 1055–1066.

Ahern, J. (1995). Greenways as a planning strategy. In J. G. Fabos & A. Jack (Eds.), *Greenways: The Beginning of an International Movement* (p. 134). New York: Elsevier.

Assadian, A., & Nejati, M. (2013). Chapter 10: Challenges faced by megacities in the future. In *Managing information resources and technology: Emerging applications and theories* (pp. 138–151). Hershey: IGI Global.

Barada, A., Steno, M., Tulan, N., & Yousry, M. (2005). *The planning and urban studies of River Nile: Master plan of River Nile banks of Greater Cairo. Faculty of Regional and Urban Planning (FURP)*. Cairo: Cairo University.

Bischoff, A. (1995). Greenways as a vehicle for expression. In J. G. Fabos & A. Jack (Eds.), *Greenways: The Beginning of an International Movement* (pp. 317–318). New York: Elsevier.

Bousemma, S., Allouche, F., Saidane, I., Mechmech, Y., & Bettaieb, T. (2018). State of the art of greenway concept application in Tunisian green policy: A case study of an urban landscape in Sousse city. *International Journal of Environment and Geoinformatics, 5*(1), 1–18.

Brilhante, O., & Klaas, J. (2018). Green city concept, a method to measure green city performance over time applied to fifty cities globally: Influence of GDP, population, size and energy efficiency. *Sustainability Journal, 10*, 1–23.

Central Agency for Public Mobilization and Statistics. (2019).*Greater Cairo Information.*

Chan, P., & Hun Lee, M. (2019). Developing sustainable city indicators for Cambodia through Delphi Processes of panel surveys. *Sustainability, 11*, 3166.

Chen, Y., Gu, W., Liu, T., Yuan, L., & Zeng, M. (2017). Increasing the use of urban greenways in developing countries: A case study on Wutong Greenway in Shenzhen, China. *International Journal of Environmental Research and Public Health, 14*(6), 554.

Crawshaw, P. (2009). *The future of greenways in Sydney, Faculty of the Built Environment University of New South Wales.*

East Coast Greenway Alliance. (2019). *Greenway Criteria & Design Guide Webinar.*

EBRD. (2016). *Green cities program methodology*. London: European Bank for Reconstruction and Development.

El Adli, K. (2006). Role of urban greenway systems in planning residential communities: A case study from Egypt. *Landscape and Urban Planning, 76*(1), 192–209.

El Said, M. (2015). *Problem of strategic plans of cities and villages in Egypt. Master thesis, Faculty of Engineering*. Azhar: Azhar University.

European Greenways Associating. (1998). *The European Greenways Good Practice Guide: Examples of actions undertaken in cities and the periphery.*

Fabos, G. J. (2004). Greenway planning in the United States: Its origins and recent case studies. *Landscape and Urban Planning, 68*, 321–342.

Fathy, H. (2006). *Greenways in Egypt as a conservation approach for natural and cultural, historical heritages landscape and resources*, PhD. Dissertation, Faculty of Regional and Urban Planning, Cairo University.

Höjer, M., & Wangel, J. (2014). Smart sustainable cities: Definition and challenges. In L. M. Hilty & B. Aebischer (Eds.), *ICT innovations for sustainability. Advances in intelligent systems and computing* (pp. 333–349). New York: Springer International Publishing.

Hussein, M. (2018). *Sustainable regeneration of urban green areas in Egypt's desert cities: Adopting green infrastructure strategies in New Borg El-Arab City. Master thesis of Science in Resource Efficiency in Architecture and Planning*. Hamburg, Germany: HafenCity University.

Ibrahim, I. (2016). *Obstacles to the implementation of urban plans through methods of preparation and implementation of detailed plans. Master thesis, Faculty of Regional and Urban Planning.* Cairo: Cairo University.

Keivani, R. (2010). A review of the main challenges to urban sustainability. *International Journal of Urban Sustainable Development, 1*, 5–16.

Keleg, M. (2018). Review of green spaces provision approaches. In *Cairo, Egypt Research Student Conference, Faculty of Technology, Design and Environment*. Oxford: Oxford Brookes University.

Kondolf, G. M., et al. (2011). *Connecting Cairo to the Nile: Renewing life and heritage on the river. 2011. IURD Working Paper No. WP-2011-06. Department of Landscape Architecture & Environmental Planning*. Berkeley: University of California.

Lewis, E. (2015). *Green city development tool kit*. Mandaluyong: Asian Development Bank.

Mahmoud, A. H. A., & El-Sayed, M. A. (2011). Development of sustainable urban green areas in Egyptian new cities: The case of El-Sadat City. *Landscape and Urban Planning, 101*(2), 157–170.

Mfon, P., Akintoye, O., Mfon, G., Olorundami, T., Ukata, S., & Akintoye, T. (2014). Challenges of deforestation in Nigeria and the millennium development goals. *International Journal of Environment and Bioenergy, 9*(2), 76–94.

Ministry of Housing, Utilities and Urban Development, General Organization of Physical Planning and UN Habitat. (2012). *Cairo 2050, Greater Cairo Region Urban Development Strategy, Volumes I and 2.*

OECD. (2013). *Green growth in cities, OECD green growth studies*. Paris: OECD Publishing.

Qu, L., Fu, B., & Calabrese, L. (2015). Beyond the greenways: A people-centered urban planning and design approach for Shenzhen, the 'World Factory' in transition. *Proceedings of the 8th Conf. Int. Forum of Urbanization.*

Searns, R. (1995). The Evolution of greenways as an adaptive urban landscape form. In J. G. Fabos & A. Jack (Eds.), *Greenways: The Beginning of an International Movement* (pp. 65–79). New York: Elsevier.

Shrivastava, M., Ghosh, A., Bhattacharyya, R., & Singh, S. (2019). Urban pollution in India. In *Urban pollution: Science and management* (1st ed., pp. 341–356). Hoboken: John Wiley & Sons Ltd.

The Economist Intelligence. (2015). *Livability ranking and overview*. https://media.heraldsun.com.au/files/liveability.pdf.

Township of Radnor. (2019). *Radnor trail map.*

Türk, E. (2017). Multi-criteria decision-making for greenways: The case of Trabzon. *Turkey Planning Practice & Research, 33*(3), 326–343.

UN-Habitat. (2006). *State of the world's cities 2006/7*. London: Earthscan.

Urban Vegetation and Microclimatic Comfort in Warm Climates

Gustavo Cantuaria and Manuel Correia Guedes

Introduction

Since the beginning of civilisation, good and healthy places have been inevitably linked to spaces containing different forms and variations of greenery. This led to the fictional concept and the incessant search of "paradise". Depictions of man-made paradise have been recorded throughout history by architectural visionaries. In the lack of such a perfect place in reality, paradise was sought in paintings, literature, and music. In all the artistic forms it is always linked to nature. It is part of our well-being to be connected to vegetation, with the environment.

An environmental friendly architecture relieves us from stress, arouses our senses, attenuates the thermal environment, and gives a notion of peace, of paradise. The partnership with trees creates spaces we long to inhabit and share. We perceive not only visually, but also acoustically, aromatically, emotionally, and especially thermally. By controlling the environment around us we can tempt our senses. Architecture, as other arts, can enliven our experiences with the surrounding environment and provide pleasant moments and emotions.

Trees have always been a part of civilisation as a source of shelter, food, medicine, and clothing. Its contribution to architecture by defining, modifying, and improving the environment is not well recognised. This work attempts to diminish this dent by portraying the benefits of tree microclimates in popular urban settlements by improving

G. Cantuaria
The Graduate and Undergraduate School of the Architecture Department, University Centre of Brasilia, UniCeub, Brazil

M. C. Guedes (✉)
Department of Civil Engineering and Architecture, Instituto Superior Técnico, Lisbon, Portugal
e-mail: manuel.guedes@tecnico.ulisboa.pt

A. Sayigh, A. Trombadore (eds.), *The Importance of Greenery in Sustainable Buildings*, Innovative Renewable Energy,
https://doi.org/10.1007/978-3-030-68556-0_4

the outdoor thermal conditions around the house, and consequently improving life quality. Trees also modify behaviour and expectations, and create a range of experiences to our advantage.

Contemporary Overview

Importing the architectural ideas and concepts of foreign countries whose geography, environment, and climate are totally different from its own has led to improper architectural solutions and typologies. Furthermore, any interesting lessons of what were once the most proper way of building are being ignored and forgotten. When money and resources are abundant, how, where, and when one builds are frequently overlooked. On the other hand, if one considers a low-income population, where basic needs are just about sufficient for the day, it is important, not to say essential, that one makes the most benefit of its living environment, in an intelligent and sustainable manner.

Fortunately, vegetation cannot be globalised. Some species may adapt to a different environment, and others just die. Plants and trees say much about a place. Along with telling generic things such as season changes, they can also tell more specific things. The crooked trees of the Brazilian savannah, for example, indicate that there is too much iron in the soil. Plants and trees to survive in certain conditions regulate and adapt themselves. These adaptations, as the research suggests, may come to the advantage of our own living conditions. In a world that is fast becoming stereotyped, it should be the objective of the urban context to engender diversification, starting with vegetation.

Environmental design contributes to diversity and addresses formal and spatial issues that are not universal, but on the contrary are related to a determined context and microclimate. This incentive to diversity enriches spaces and lives. It admits architecture to be rich and interesting instead of vague, unsubstantiated, and indistinct. Contextual architecture permits uniqueness rather than superficiality. Vegetation reflects this richness through its innumerable distinct variations found in distinct places.

The complexity of a living element and its response to different situations and climates raise questions of considerable interest for our understanding of any type of vegetation. As no human being can be totally predictable in any kind of environment, the same can be said about a tree. “Despite all the variations, there remains substantial constancies. The strongest of these is the importance of nature itself” (Kaplan and Kaplan 1989).

Trees, and vegetation overall, have long been known for its cooling effects and improving unfavourable microclimatic conditions around buildings. Nevertheless the potential of vegetation has been and continues to be ignored in recent attempts of building energy-efficient dwellings.

This chapter promotes the idea of microclimatic design as an essential issue of environmental architecture. In this sense, trees are of great significance, not only for its visual effect alone, but also by promoting an unparalleled experience and awareness of life and space. The sensory responses to a determined place, in general, cannot be drawn; they are just felt. Trees contribute to all our senses, making them totally perceivable in a positive manner. They inform, improve, and upgrade spaces and consequently lives, and therefore are an essential element in designing architecture. Taking into consideration site conditions including its vegetation typology addresses the idea of "continuity" and also "contextualism" which are words that define sustainability. It shows that architecture is not constrained by environmental planning, nor economically dependent. In fact it makes the architect more aware of the surrounding qualities such as visual and environmental comfort, contributing to the essential beauty of space, building, and environment.

Theme and Problems

Cities and urban centres overall have come to the end of the twentieth century with a paradox to discuss. As the cities look to reach the skies with their growing skyscrapers, they also reach out to attract and embrace as many people as possible. Aiming to become an important financial, commercial, cultural, technological, or touristic place, and therefore attract investments, it needs to grow. The more it grows, the more people it will need to make it grow. The more people it has, the more infrastructure is necessary. With this ongoing trend, problems such as transportation and health become an increasing snowball. In the process of looking for a sustainable city, it is equally important to find sustainable solutions.

Every human being, with its activities and implications, affects its surroundings. The urban centres are one of the most important works produced by humanity, and undoubtedly the one that has caused most environmental changes. Deforestation, heat island effect, and climatic changes are examples that blame the rapid growth of the industrialised city.

For example, in the Federal District, where Brasilia, the capital city of Brazil, is located, the process of urbanisation has been one of "land cleaning", which means all the vegetation and its natural covering are removed in an irresponsible attempt to simplify the urban implementation. This process of clearing the land has enormous impact on the environment leaving the land vulnerable to erosions, lack of shading, and a lot of dust. A major problem is the castigating excessive solar radiation. More importantly, the combination of devastated lands with large asphalt areas produces dry environments with very low relative humidity, reaching values of 10% (INMET 1996). This is considered very hazardous for the health by the World Health Organization. These conditions make daily urban tasks simply unbearable and improper for certain times of the year, taking a toll both physically and psychologically.

The use of vegetation other than water for evaporative cooling is chosen because it is a more sustainable solution not only for Brasilia, which is arid and lacks water, but also on a bigger scale, in many different countries worldwide, characterised by warm climates. According to the UN, 70% of the planet is covered by water. Of this, 98% is salty water, which is improper for human consumption. Of the remaining

2%, only 0.014% is available for consumption in rivers, lakes, and streams. The rest is found in subterranean reserves. From this, 73% goes for agriculture. Therefore only 27% of 0.014%, or 0.00378% of the total water, is for human beings. This means that 3.4 of 6.0 billion inhabitants have 50 L of water available per day. Curiously, this is 1/7 of the North American average. Consequently 40,000 children die each day because of diseases due to the lack of water. Needless to say that a more sensible solution is needed as there is no room for wasting water, especially in low-income social housing popular settlements where there is no abundance (Novaes 1992). Another solution, such as tree planting, is necessary to ameliorate and humidify the dry and hot air of the region.

Passive Cooling and Microclimatic Effects of Trees

The presence of trees in the urban environment is of invaluable help in regions with hot-dry seasons, in providing more suitable conditions for human comfort in outdoor microclimates. The wind comes from the weather systems, and makes convection the link of microclimate with the macroclimate. The increasing distance away from a surface will decrease the effect that surface can have on the ambient temperature. Trees act as barriers, limiting the influence of the macroclimate on the microclimate. During the hot-dry season, especially in low-income urban settlements, human discomfort is aggravated. A thoughtful plan of strategic planting of trees may help overcome the undesired discomfort. A single tree per site is enough to create a more comfortable microclimate. Trees can provide relief in two ways for the local microclimate. First through a direct shading effect, where incoming solar radiation is intercepted by the canopy and may be either reflected or absorbed, hence diminishing the temperature. Second through evaporative cooling, where energy is used for transpiration rather than heating the surface and the air. Outdoor spaces sheltered by trees create ideal conditions for enjoyment of the ambience by people attuned to such relaxation.

The proper choice and placement of trees can be an effective control of solar radiation impinging on their surrounding spaces and consequently on horizontal and vertical surfaces.

Furthermore, "buildings located in sheltered spaces are more comfortable to live in or work in, with less need to rely on artificial climate and greater economy" (Dodd 1988). Although the cooling effects of a tree microclimate are not directly related to the indoors, it is possible to make a few recommendations, based on both the case studies and literary research, to benefit the most from the outdoor microclimate in indoor comfort, in warm climates:

- A single tree should be located to provide maximum shade to west or northwest facades, due to more solar radiation exposure, alternatively east and southeast.
- The facades with largest window area, preferably with the above orientations, should be privileged in shading.

- The cooling potential of the shade tends to diminish with distance to its trunk. Trees should be planted considering that when mature, the outer part of the canopy is close to the facade. This should obviously be in accordance with restraints, such as root system, and safety considerations as branch resistance.
- The shading of roofs by tall and large canopies: Damage to the building or to the walls can be avoided by selecting the correct species for the available space.
- The planting of evergreen species, with rapid growth: The need for cooling and humidifying the environment is present most of the year.

There is no excuse to fail to maximise the site potential for the best tree shelter with limited insolation. Landscape planting is no substitute for good architecture and site planning, but should be part of an integrated design. "The common sense is the faculty to judge, to distinguish the true from the false. Common sense, however, is not sufficient to judge well. It is necessary a method which makes use of analysis and of deduction and which includes the truth by the criteria of evidence" (Descartes, *Principles*).

It is impossible to precisely simulate trees. Leaves are like fingerprints; they are all unique. In each leaf there is a whole new microcosmos with different action going on. The number of stomata is unequal. Membrane thickness, stomatal behaviour, protoplasm cells, and chlorophylls may all differ (Gomes 1982). Therefore, predicting the actual exact value of a certain leaf or canopy evapotranspiration, for example, becomes secondary. On the other hand, it is important to predict the pattern and range in which microclimatic transformations occur, and be able to measure its reliability, to see if it occurs by chance or nature.

In places with a hot-dry climate as in the region of Brasilia, solar radiation must be avoided. In popular urban settlements, sites are usually long and narrow. Luckily this form helps to minimise ventilation which is carrying heated air. The front yard should be shaded by a tree. Buildings should be as close as possible to the shade of the tree, to gain the most from its cooling potential. They should not be too close as to compromise foundations, but not too far as not to benefit from the tree shade. In case studies investigated, all tree microclimates had a consistent beneficial effect by increasing humidity and lowering temperature. The greatest cooling capacity occurs in the early afternoon, while in the late afternoon and early evening, vapour pressure continues to decrease as evaporative demand induces the trees' stomata to close. Humidity is higher under the trees than in the open. However, there is a smaller effect at the edge of the tree canopy and a greater effect in the middle of the canopy itself.

Urban forest is the group of either naturally occurring or cultivated or arboreal vegetation in a city's landscape. It looks at environmental architecture as not only methods of energy conservation, but also a way to use the potential of the urban surroundings to ameliorate the quality of life of its inhabitants. The potential of the interaction between shading and evapotranspiration is the aim of strategic urban planning and design. A tree is a powerful element of design, and can act like an air conditioner in open space. It also looks at the social and cultural impact that greenery has on the inhabitants.

Energy Savings by Tree Planting

Environmental concern about global warming, heat island effects, and air pollution has brought attention to the potential of trees in ameliorating climate and conserving energy. As stated earlier, climate and human comfort benefit from trees basically through shading, which reduces the amount of radiant energy absorbed and stored as well as that radiated by the surrounding surfaces, and evapotranspiration, which converts radiant energy into latent energy, consequently reducing the heat that warms the air. Airflow modification may also affect the diffusion of energy.

About one-sixth of all electricity generated in the United States is used for air conditioning in buildings. That is about 40 billion dollars. Strategies for heat island mitigation and energy savings were studied by Rosenfeld et al. (1995). With strategies of reroofing and repaving in lighter colours and planting shade trees in Los Angeles, annual residential air-conditioning savings could be reduced by 100 million dollars. By shading buildings, trees are most effective, but there are significant savings even from simply cooling the air by evapotranspiration. In the target city, it is stated that 1.5 GW of peak power for air conditioning can be avoided, representing more than 15% of the city's air conditioning. By generalising the estimate to the whole country it was estimated that 25 GW can be avoided, and a potential five billion dollar annual savings. Savings in air conditioning were estimated by assuming that five million houses in Los Angeles were coastal and not air conditioned, leaving only 1.8 million as inland air-conditioned houses; an estimated roof area per house of 200 m^2, and that all five million houses would change to lighter colour roofs; and that 11 million trees are "planted" resulting in three shade trees per air-conditioned house and one shade tree for each non-residential building and leaving 4.6 million trees to shade non-air-conditioned houses, streets, and parks. Calculations and simulations were done using DOE-2 programme combined with meteorological models. When temperature itself is at its highest at 14:00, an estimated cooling average for the heat island effect of 3 °C was reached, clearly demonstrating the potential of the cooling strategies.

Charles McGinn (1983) investigated the potential of suburban microclimate in Davis, a small city in California, to diminish summer cooling needs. The research monitored the effects of the microclimate on energy use of two small buildings. One building was surrounded by trees while the other was not. From the data collected it was observed that greater canopy density leads to an increase in wind speed, standard deviation of wind speed, and direction. Average daily air temperatures in new and intermediate canopy densities were higher than in mature densities. An inverse behaviour was reported for relative humidity, therefore reducing below younger canopies.

A mathematical model simulating the thermal behaviour of the building indicated that the energy use in new suburbs without trees could increase by 10–15%, while the microclimate with mature canopies could decrease around the same amount. If the site had been planned in advance, the energy reduction could reach around 40–50% with the proper geometry of canopy shading and building orientation. Results clearly show that if priority is given in the process of urban planning and landscape design, there can be a significant payoff in energy savings.

Reducing urban summertime air temperatures and saving cooling energy use in buildings by increasing urban vegetation hold great potential. Akbari et al. (1997) monitored peak power and cooling energy savings from shade trees in two houses in Sacramento, California. Seasonal cooling energy savings of 30% were reported, corresponding to an average daily savings of 3.6 and 4.8 kWh/day. Peak demand savings were about 27% for one house and 42% for the other.

Studies of urban forest impacts on regional cooling in Sacramento were also done by Simpson (1998). Annual cooling savings of approximately 157 GWh (US$ 18.5 million) per year, or 12% of total air conditioning in the county, was stated. These savings are credited to shading, air temperature, and wind speed reduction.

Findings from simulations by McPherson et al. (1993) suggest that a single tree around 7.5 m tall can reduce annual cooling and heating costs of a typical residence by 8–12%. In warm climate cities, reduced solar gains from tree shade were responsible for most of the cooling savings. Evapotranspiration also had an important role in cooling with an increased potential in regions with more cloud cover.

Strategic landscaping for air-conditioning savings was also documented by Alan K. Meier (1990). By using different types of vegetation as grass, shrubs, and trees, in different locations, air-conditioning energy savings of up to 80% were measured. Trees provided the best cooling effects. Savings of 25–50% were more common, especially in hot dry and humid climates, through processes of shading from the sun and shielding from infiltration. Evapotranspirative potential of greenery was clearly noticed in studies in Arizona where turf alone provided cooling savings of 15–25% (Fig. 1). Different studies by Simpson and McPherson (1996) show that shade from a single well-placed, mature tree, around 25 ft. crown diameter, would reduce annual air-conditioning use by 2–8%. Furthermore, studies by Huang et al. (1987) state that a 25% increase in urban tree cover could save up to 40% summer cooling energy use in Sacramento, and 25% in Lake Charles and Phoenix, all in the USA.

Studies by McPherson et al. (1989) reviewed a range of experimental designs, building types, and landscaping, and found that measured air-conditioning savings ranged from 25% to 80%. Larger savings were associated with more dense and

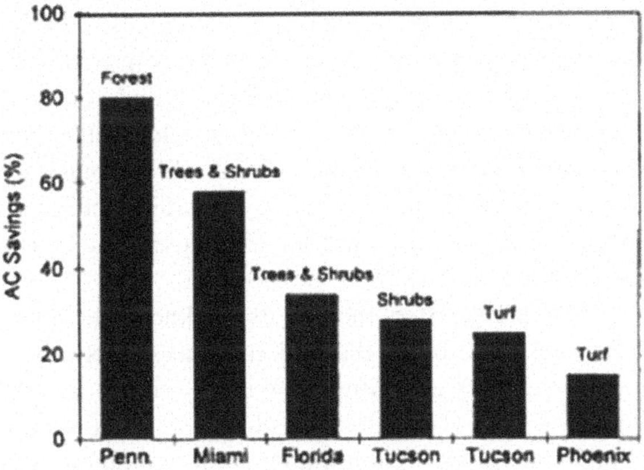

Fig. 1 Air-conditioning electricity savings due to vegetation (Meier 1990)

extensive shading, by large compact canopies. Other studies by Simpson and McPherson (1996) using simulation evaluated the potential effects of tree shade on residential air conditioning and heating energy for different tree orientations, building insulation levels, and climate zones for California. Cooling load reductions associated with tree shade were always greater than any increase in heating. Larger air-conditioning savings were in warmer climates and uninsulated buildings. Shading on the west wall and next on the southwest also produced the largest savings. Reductions of 10–50%, 200–600 kWh, on the annual energy use for cooling were registered.

Studies by Holm (1989) investigated the use of creepers instead of trees. Best results were obtained in low-mass building in hot-arid climates with equator-facing walls of high radiation absorbance in winter, but covered by vegetation during summer, where all other outside walls were covered with evergreens. A constant of 2.5 °C cooling effect in the room temperature facing the equator was found. Indoor temperature was reduced from a 17–33 °C range to an 18–28 °C range in an ambient temperature range of 21–31 °C dry bulb. The thickness of the foliage acts as a canopy where the outer layer filters and reflects the solar radiation while deeper layers act like insulation material.

Akira Hoyano (1988) also investigated climatological uses of different vegetation forms to control solar radiation, improving indoor and outdoor thermal environment. Experiments and impressive results proving the amelioration of microclimates by the use of vegetation were documented by case studies which included a pergola composed of horizontal wisteria sunscreen, a vine sunscreen designed for a south-west veranda, a row of evergreens planted next to a west wall, an ivy sunscreen covering a west wall, and rooftop turf-planting layers. Temperature differences by means of solar radiation control only were the focus. The studies show that an exterior wall without an ivy sunscreen allows a maximum heat of 838 kJ/m^2 to flow in. On the other hand, by providing an ivy sunscreen, this was reduced to about one-quarter. Therefore the outdoor surface temperature of the external wall with an ivy screen was lower than that without an ivy sunscreen. In a different study, the external wall shaded by a row of trees 60 cm apart was also lower by 3 °C than the surface temperature of the exposed wall.

Del Barrio (1998) developed a mathematical model to analyse the thermal behaviour and cooling potential of green roofs in buildings. Parametric elements such as leaf area index, foliage characteristics, and soil characteristics were introduced in the green roof design. The leaf area index determines the effectiveness of the canopy as a shading device along with the leaf angle distributions, which determine both the long- and short-wave transmittance of the canopy. Vegetation with a horizontal distribution and large foliage is more likely to mitigate solar radiation transmission.

Soil characteristics include its thickness, its apparent density, and moisture content. They determine the soil thermal diffusivity, which will augment with the apparent density and diminish with the soil moisture content. Thermal conductivity and weight are also reduced with lighter soils. Studies showed that the garden roof acted better as insulation than a cooling element, mainly by reducing the heat flux through the roof. Evapotranspiration had a secondary effect as a green cooling device. Nevertheless the air exchange rate between tree canopies and the surrounding free air is more important and effective to the microclimate.

Takakura et al. (2000) also studied the cooling effect of greenery cover over a building. Four concrete roof models were constructed and each arranged with different coverings: bare concrete, soil layer, soil layer with turf, and soil layer with ivy. It was reported that the cooling effect increased with the increase in leaf area per unit roof, or leaf area index (LAI), and the same consideration can be applied to trees. Canopies which have greater foliage area have larger effective areas for evapotranspiration than areas with smaller leaf area.

Sociological Benefits

Planting design, considering climatic factors, can help to moderate and improve outdoor microclimates. Benefits are not only physiological but also psychological. Daily activities, especially in uncomfortable conditions, have a mental and physical toll. Stress and weariness lead to loss in productivity, efficiency, and healthiness. To counterbalance, it is vital that mind and body recuperate through rest and recreation.

In warm climates the outdoors is an important aspect to the local style of life. People socialise, work, entertain, and play outdoors. By creating more pleasant and attractive outdoor spaces with trees, people will be less exposed to solar radiation and physical problems that may recur from it such as dehydration, stress, and strain. More comfortable spaces are an incentive for more neighbour interaction. Trees provide the perceptual stimulation needed, be it visual, tactile, or aural. It ensures diversification and avoids monotony. This will result in a bigger community spirit and a safer place to live as well as a more attractive and healthier environment. Furthermore, because tree resources are renewable, they can become a permanent part of any environment program designed to increase well-being, environmental consciousness, and education. By planting trees in urban spaces, it is the primary intention to create more comfortable microclimates, and secondarily offer the benefits of an urban forest, designed to intensify ecologically, economically, and aesthetically the life quality of its users.

Physiologic Characteristics of Trees

In arid and semi-arid regions, and also in other areas where drought can bring about restrictions on watering, the use by trees and vegetation of water is an important issue. It is therefore indispensable to choose a tree species having a cooling potential independent of high water use. The internal water balance, as indicated by the turgidity of cells and tissues, is the most important aspect of plant-water relations. All other processes connected with water relations are important mainly because they affect the internal water relations of trees and consequently modify conditions that affect growth and physiological processes. The lack of available soil moisture, or drought, is the most common cause of internal water deficit. Atmospheric factors such as high temperature, low humidity, and wind which lead to high rates of transpiration increase the damaging effects of drought which can even reduce tree growth. Trees with protoplasm having some capacity to endure dehydration, together with anatomical and morphological characteristics which postpone the development of critical internal water deficit, are drought resistant. Mango trees, as fig and almond trees, have high rates of transpiration but a low water deficit due to their deep-root systems (Koslowski 1987). A large leaf area and structure also help transpiration. The mango tree falls in this category. Other characteristics which influence drought resistance include stomatal behaviour, osmotic pressure, and efficiency of the water-conducting system. Its cell size and shape also enable a tree to survive dehydration. The mango tree maintains its green canopy in the arid and moistureless landscape as in Brasilia, Brazil, during the hot-dry season, which is very beneficial and effective in keeping its cooling potential year-round. Case studies shown next all involve mango trees, and help illustrate the promising capability that it has in ameliorating microclimates below its tree canopy.

Case Studies with Special Reference to the Mango Tree: Mango Tree Description

The mango tree, above anything else, was chosen for its vast and dense shade, translated into a space of comfort, relaxation, and well-being, a true blessing of nature. People even forget that it produces one of the most tasty and rich fruits, which is another strong point in its favour.

Natural from India, it was brought by the Portuguese when Brazil was still a colony. Today it is widely planted and spread especially in hot places with a humid season with a rainfall of 75–250 cm, followed by a dry season, which is totally indispensable for the growing of its fruits. The mango tree is naturally adapted to tropical lowlands between 25°N and 25°S of the Equator, and elevations of around 1000 m (Morton 1987). This is compatible to the climate and region of Brasilia, where mango trees have been planted, but only for ornamental purposes. Brazil is one of the biggest mango producers in the world, losing out only to India, and the bulk of the crop is for domestic consumption.

The mango tree is a beautiful medium- to big-size tree, reaching from 10 m up to 25 m when mature in height. Its crown is rounded, uniform, with a nice compact foliage. The leaves are alternate, ranging from 10 to 30 cm in length (Fig. 2). When pressed between the fingers they have a pleasant characteristic smell. Overall, it is a very nice and respectful tree, easy to identify in a landscape (Fig. 3). The mango tree is long-lived, some specimens being known to be 300 years old and still fruiting. It is fast growing; trees less than 8 years old may flower and fruit regularly every year.

It is naturally adapted to Brazil. It grows readily from seed, not needing any special care, fertilising, or watering (Morton 1987). The mango is a delicious fruit; very nutritious; and rich in vitamins A, B, B2, PP, and C, and makes for a good part of a healthy diet (Table 1). It is also used as a natural medicament against scurvy, dysentery, and haemorrhage.

Vegetation Typology Experiment

Experiments were performed in the ecological park of Embrapa on the outskirts of Brasilia. Sites of the local vegetation typology in two different density situations (Cerrado and Cerradinho) were compared to a mango tree site situated in the same park. The intention of this experiment was to compare the microclimatic performance of indigenous vegetation in different densities with that of a non-indigenous species. The mango tree, although not native to the region, is very well adapted to the area. These microclimates were also compared to a common regional tree species, which is the sibipiruna (Fig. 4). This tree is tall and has a wide and sparse canopy with beautiful yellow flowers. The native vegetation of the area is something similar to savannah. This type of vegetation is characteristic for its crooked and sinuous tree trunks and very light canopies. What is called "Cerradinho" (Fig. 5) is the indigenous vegetation in a natural environment, where the trees are scarce. The "Cerrado" consists of the same trees but agglomerated together to create a more compact and dense forest-like environment.

Since the mango tree canopy is much denser and therefore provides a bigger and more effective shade, it was thought that the mango microclimate would provide a more comfortable microclimate than the "Cerradinho". On the other hand it was thought that it would be less effective than the "Cerrado" due to its denser and higher configuration which barely allowed light through.

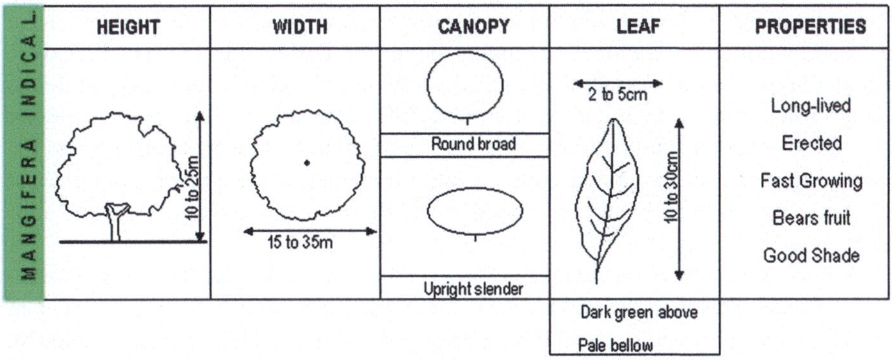

Fig. 2 Mango tree properties

Fig. 3 Mango tree in Brasilia

Table 1 Mango fruit levels of food constituents (Manica 1997)

Calories	62.1–63.7 g	Iron	0.20–0.63 mg
Moisture	78.9–82.8 g	Vitamin A (carotene)	0.135–1.872 mg
Protein	0.36–0.40 g	Thiamine	0.020–0.073 mg
Fat	0.30–0.53 g	Riboflavin	0.025–0.068 mg
Carbohydrates	16.20–17.18 g	Niacin	0.025–0.707 mg
Fibre	0.85–1.06 g	Ascorbic acid	7.8–172.0 mg
Ash	0.34–0.52 g	Tryptophan	3-6 mg
Calcium	6.1–12.8 g	Methionine	4 mg
Phosphorus	5.5–17.9 mg	Lysine	32–37 mg

Both temperature and humidity data loggers were placed at a height of around 1.5 m. The objective was to compare measurements of only the four subcanopy microclimates, to observe which tree typology was more effective. The days were clear and warm, having rained one night (the ninth). Differences of up to 30% were found. On the eighth, Cerradinho reached a low of 40% relative humidity, while the mango microclimate kept cooler at around 70%. The shade of the mango canopy also showed cooler temperatures. Differences of up to 9 °C were registered on the eighth too. Overall, the temperatures of the two better shades, mango canopy and the dense jungle-like Cerrado, recorded lower values than the Cerradinho and sibipiruna which had similar patterns (Tables 2 and 3).

Wilting affects the water balance in leaves and consequently affects evapotranspiration. A frequent result of the midday decrease in water content is wilting. Different leaf and stem compositions cause different levels of wilting, which could be incipient, temporary, or permanent. The low evapotranspiration of the sibipiruna is characteristic

Fig. 4 Sibipiruna tree

Fig. 5 Cerradinho typology

of incipient wilting where the water loss barely exceeds the water intake and the decrease of water content is too small to produce visible loss of turgor (Kramer 1960). On the other hand, the mango tree is related to temporary wilting, due to water deficit from rapid transpiration. It is likely to happen daily in warm climatic conditions. Regardless of the size of the water deficit developed, visible wilting does not occur because most of the structure of trees contains a large amount of lignified tissue. Overnight, trees recover from temporary wilting or whenever absorption is higher than transpiration. The water deficit is eliminated more easily with trees with a profound and radial root system as the mango tree. Furthermore, temporary wilting progresses

Table 2 Relative humidity (%) in four distinct subcanopies—Cerradinho, Cerrado, mango tree, and sibipiruna (Embrapa, Federal District, Brazil, 6th–10th January (rainy season))

	Cerradinho	Mango tree	Cerrado	Sibipiruna
N	1486	1486	1486	1486
Median	89.4	87.3	90.85	81.7
25–75 percentiles	68.1–97.6	77.1–93.2	81.1–96.7	70.6–89.0
Minimum	34.3	44.7	47.7	41.7
Maximum	100.0	100.0	100.0	100.0

Kruskal-Wallis one-way ANOVA on ranks: $p < 0.001$
All pairwise multiple comparisons (Dunn's method): $p < 0.05$ in all pairs except Cerradinho vs. mango tree (non-significant—*NS*)

Table 3 Temperature (°C) in four distinct subcanopies—Cerradinho, Cerrado, mango tree, and sibipiruna (Embrapa, Federal District, Brazil, 6th–10th January (rainy season))

	Cerradinho	Mango tree	Cerrado	Sibipiruna
N	1486	1486	1486	1486
Median	22.3	21.85	20.6	23.6
25–75 percentiles	19.8–26.1	20.6–23.7	18.5–22.7	20.6–27.5
Minimum	17.0	18.4	16.0	18.4
Maximum	37.7	31.8	30.7	39.8

Kruskal-Wallis one-way ANOVA on ranks: $p < 0.001$
All pairwise multiple comparisons (Dunn's method): $p < 0.05$ in all pairs except Cerradinho vs. mango tree (non-significant—*NS*)

gradually into permanent wilting when absorption becomes slower each day, wilting longer and failing to recuperate turgor overnight and recover the water deficit. Consequently not much evapotranspiration occurs. Leaves and stems remain wilted throughout droughts, and closure of stomata also happens. Drastic wilting seriously interferes with many physiological processes, and even growth. This is common to shallow-rooted trees as can be seen with the Cerradinho. The different tree species, with contrasting root systems and distinct leaf characteristics, lead to opposed wilting degrees which help explain the differences in humidity of the subcanopy microclimates, as well as the cooling potential.

At night the mango canopy had a warmer microclimate which was more similar to that of the sibipiruna and Cerradinho than that of the Cerrado. The radiation received during the day by the leaves of the canopy reradiates the space and makes it warmer than the overshaded Cerrado, where not much sun reaches the leaves due to a high number of trees clustered together and thin canopies.

Different One Tree Site Experiment

An experiment was conducted in Brazlandia, on the outskirts of Brasilia, during the rainy summer period, which involved comparing the single mango tree site with a similar neighbouring site with a single different tree typology. Plot sizes and orientations were the same. The sites consisted of 8 m-by-12 m plots with the houses located at the end of it. Both trees were west orientated which allowed a direct comparison. Measurements were taken under the canopy and at the external veranda to see if the tree was also affecting that space. The mango site was more effective. Although the verandas differed slightly, statistical tests showed that these differences are significant and should be considered. The extension to which benefits of the tree canopy are effective diminishes with distance but should not be neglected. Cooling of bigger canopy densities and size is more perceptible and will stretch further.

The mango tree site showed better results, offering a cooler microclimate under its canopy than the pata-de-vaca tree at the other tested house. The bigger, denser, and rounder canopy was usually more than 10% higher than in the other planted site. Temperatures were also lower at the mango site. They were up to 4 °C lower than the site with the pata-de-vaca tree. During all day the mango tree had the best microclimate, while at night all temperatures were fairly similar.

In all the experiments described above, one tree can have a very important and noticeable effect on its immediate surrounding, working as an artificial air conditioner.

Peak Time Experiments

Growing in moist or dry soil, either way the leaves of most trees present a daily cycle in water content. Usually in early afternoon minimum water content occurs. It is about the time of maximum evapotranspiration and the warmest hour of the day. On the other hand maximum water content which would be expected to occur towards morning occurs around the middle of the night. It is thought that this behaviour could be related to the translocation of water from leaves to other tree parts, probably due to redistribution of carbohydrates within during the night.

In general, evapotranspiration rapidly increases during the afternoon when the stomata are open and the rising temperatures accentuate the steepness of the vapour pressure gradient from the intercellular spaces to the surrounding external air. Evapotranspiration then decreases rapidly late in the afternoon as the stomata close and temperature falls, as can be seen in all the next study cases. Water absorption follows, until the diffusion pressure of the tree is eliminated.

Four more experiments were carried out to reinforce the idea that the tree canopy has an enormous cooling capacity, even a single one, and to illustrate its range of effect. These experiments have the intention to analyse the differences during peak time and maximise their benefit. All experiments were performed in typical sunny dry season days with high temperatures in the range of 35 °C and low humidity values reaching below 20%. Measurements were taken only in the afternoon because it is the most uncomfortable time of the day as well as the most active in water variations for the leaves, as explained before, and therefore the most important to be dealt with.

The objective of these studies was to quantify the effects of trees in the urban area on air temperature and humidity as well as to determine if the effects are significantly different for different species of trees in the same environment; trees of the same species in different canopy sizes, or maturity; and to what extent do the effects range on a vertical and horizontal plane in relation to the mango tree canopy.

Horizontal Measurements of Tree Canopy Shade

The first case consisted of measuring the range and effect of a single mango tree shadow on a horizontal plane. Four pairs of data loggers (humidity and temperature) were placed in a straight line, all at the same height of approximately 1.50 m. The first pair was placed on the tree trunk, the second pair at a distance of 3 m, and a third pair at a distance of 5.0 m. A last pair was placed outside the shaded canopy to be used as a control of the external temperature as well as control of external humidity (Fig. 6).

It was thought initially that not much difference would be perceived in this experiment due to the fact that the distances from one measurement to the other were not much and therefore any breeze would be able to easily mix the air. Curiously the Tinytalks registered differences in both humidity and temperature from one measuring spot to the other.

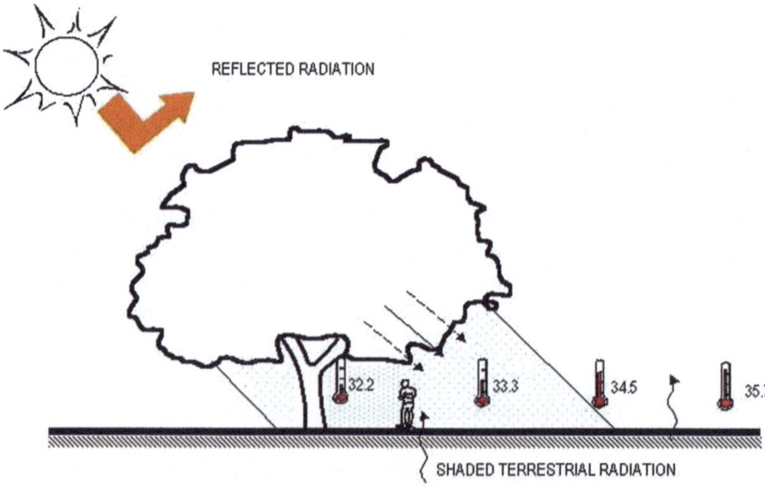

Fig. 6 Maximum temperature variations of a shaded mango microclimate

In the temperature graph it can be perceived that the closer towards the trunk, the cooler the measurements. The medium temperature of the data loggers set on the tree trunk was 31 °C, differing from that at 3 m in the canopy's shade, which was around 33 °C. Measurements at 5 m also registered a medium temperature of 33 °C, except when it reached peaks of 34 °C at 3 o'clock in the afternoon. The exposed spot was the highest of them all with peaks of 35 °C (Tables 4).

As can be seen, there is a gradient away from the tree, and the edge is less affected by shade and more by small-scale advection of heat. The tree canopy seems to lose the cooling potential as the distance increases from the trunk that is the canopy's centre. These temperature differences are perceptible due to almost non-existent breezes, which permits the leaves to evapotranspirate and block solar radiation providing consequently a cooler shade (Table 5).

A similar cooling pattern could be observed with the humidity measurements. Once again the closer to the canopy centre the higher the humidity. Measurements located at the trunk reached humidity levels around 17%. At 3 m they were around 16% while at 5 m they registered 14%. As expected, the exposed was marked lowest with measurements around 12%. This also shows that the cooling potential of the canopy, which includes evapotranspirative effect, decreases the further it is from the tree trunk.

Vertical Measurements of Tree Canopy

An experiment was performed measuring temperature differences in the vertical direction. Tinytalks were placed along the trunk axis. The first pair of temperature and humidity data loggers was allocated at a height of 1.5 m. The second pair at approximately 3.0 m which was where the mango canopy initiated and the third pair was placed in the middle of the tree canopy at around 5.0 m. The last pair was used as external control and also fixed at 1.5 m (Fig. 7).

Confirming expectations, the best measurements were of those placed in the middle of the tree canopy at 5.0 m, with peaks around 30 °C. Then temperature rose gradually. At 3.0 m, peaks were around 31.5 °C, and at 1.5 m they reached approximately

Table 4 Relative humidity (%) in different horizontal distances from the tree (trunk, 3.0 and 5.0 m) and exposed (Lago Sul, Federal District, Brazil (hot-dry season))

	Trunk	3 m	5 m	Exposed
N	340	340	340	340
Median	16.9	16.3	14.1	13.5
25–75 percentiles	16.1–17.8	15.4–16.3	13.3–15.6	11.8–16.1
Minimum	14.3	13.6	10.1	9.2
Maximum	24.6	23.4	21.5	22.1

Kruskal-Wallis one-way ANOVA on ranks: $p < 0.001$

All pairwise multiple comparisons (Dunn's method): $p < 0.05$ for all the pairs except 5 m vs. exposed (NS)

Table 5 Temperature (°C) (trunk, 3.0 and 5.0 m) in different horizontal distances from the tree and exposed (Brasilia, Brazil (hot-dry season))

	Trunk	3 m	5 m	Exposed
N	406	406	406	406
Median	31.4	32.6	32.6	33.3
25–75 percentiles	30.7–31.8	31.4–32.6	30.7–33.0	31.1–34.1
Minimum	28.1	27.7	27.7	27.7
Maximum	32.2	33.3	34.5	35.7

Kruskal-Wallis one-way ANOVA on ranks: $p < 0.001$
All pairwise multiple comparisons (Dunn's method): $p < 0.05$ for all the pairs except 3 m vs. 5 m (NS)

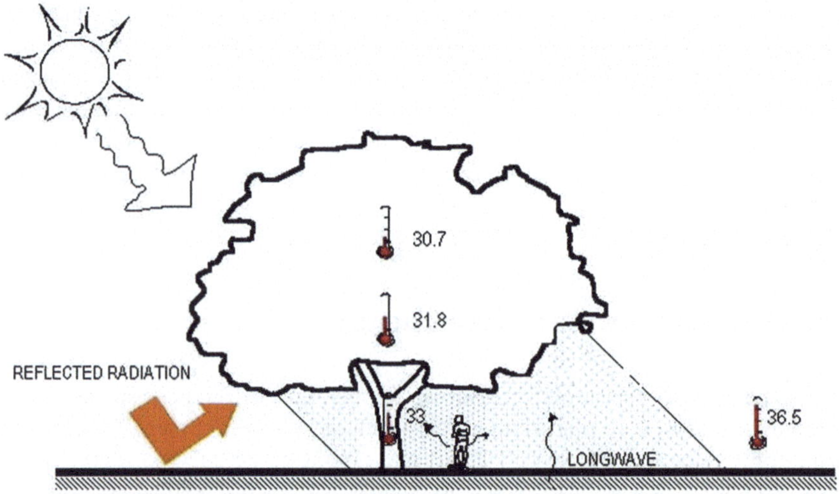

Fig. 7 Maximum vertical temperature variations of a mango canopy

32.5 °C. This shows that the more an object is surrounded by green leaves, the more it is protected from solar radiation, and the more it is influenced by cooling effects of evapotranspiration. The vertical humidity graph also verifies that there is more humidity where there is more leaf area, due to its own influence. Therefore it is to no surprise that the exposed pair of data loggers had the lowest relative humidity values followed by those at 1.5 m, and then 3.0 and finally 5.0 m (Tables 6 and 7).

It is interesting to observe that when crossing the information of both the horizontal and vertical measurements a hyperbola can be drawn (Fig. 8). This is by connecting the coolest vertical spot with the coolest horizontal spot. Therefore the hyperbola represents the cooling capacity of the canopy's shade. When connected with another hyperbola representing another tree shade on a plane, it is possible to visualise where the effect will be less effective, and what distance between trees would be the best to maximise the cooling potential (Fig. 9).

Table 6 Relative humidity (%) in different heights in relation to a tree (1.5, 3.0, and 5.0 m from the ground) and exposed (Lago Sul, Distrito Federal, Brazil (hot-dry season))

	1.5	3.0	5.0	Exposed
N	345	345	345	345
Median	25.8	27.6	30.7	20.6
25–75 percentiles	23.9–29.9	24.4–30.9	28.5–34.9	18.7–23.4
Minimum	21.1	22.4	26.5	16
Maximum	33	35.6	40.4	33.3

Kruskal-Wallis one-way ANOVA on ranks: $p < 0.001$
All pairwise multiple comparisons (Dunn's method): $p < 0.05$ for all the pairs

Table 7 Temperature (°C) in different heights in relation to a tree (1.5, 3.0, and 5.0 m from the ground) and exposed (Lago Sul, Federal District, Brazil (hot-dry season))

	1.5	3.0	5.0	Exposed
n	330	330	330	330
Median	31.4	30.7	29.9	34.1
25–75 percentiles	30.7–32.2	30.3–31.4	28.8–30.3	32.1–34.9
Minimum	28.4	28.4	28.4	29.8
Maximum	33	31.8	30.7	36.5

Kruskal-Wallis one-way ANOVA on ranks: $p < 0.001$
All pairwise multiple comparisons (Dunn's method): $p < 0.05$ for all the pairs

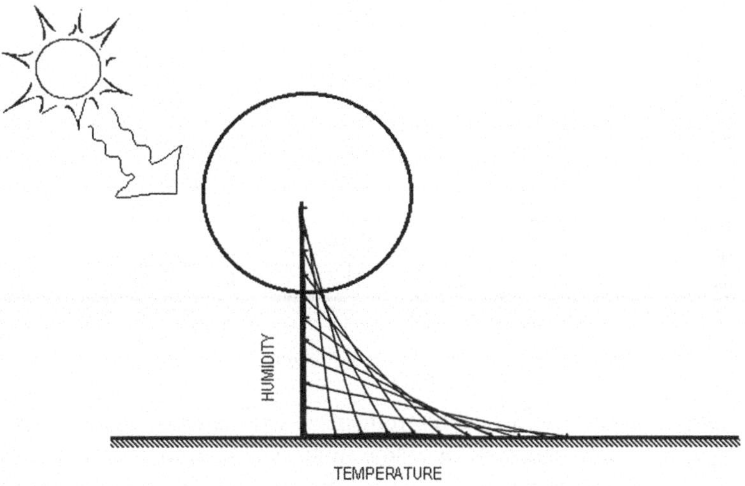

Fig. 8 Suggested "tempidity" hyperbole

A cluster of trees close together also indicate the overlapping of cooling capacity and therefore higher cooling potential of a group of trees in relation to a single one. Figure 10 illustrates on a graphic plan this possibility with the tree trunks represented by dots and their cooling potential maximised with the proximity to the canopy and therefore shown from the centre outwards, resulting in a hyperbolic form.

Fig. 9 Graphic plan of the hyperbolic cooling

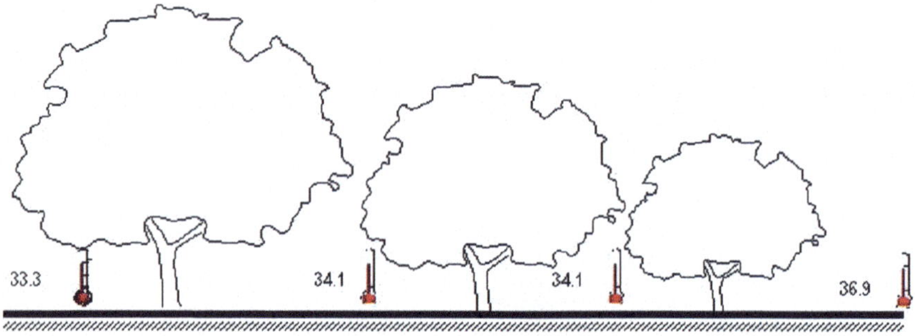

Fig. 10 Maximum temperature variations of different mango canopy sizes

These suggestions would benefit from more experiments, recording both the horizontal and vertical values at the same time. Although these results cannot be generalised, there is a strong indication of the cooling effect of the tree canopy, being at its highest potential the closer and inner to it possible. It acts like an air conditioner in an open space, where the effects will be more perceptible to the body, the closer one is to it. These values of "tempidity" (temperature and humidity) on the hyperbola graph indicate the possibility of a green unit scale (GUS) for means of comparison and to observe microclimatic tendencies. The direct relation between temperature and humidity could also lead to a hyperbolic equation. Additional studies are necessary and would be very beneficial.

Tree Canopy Maturity Studies

The third study case of the peak time experiments had the intention to verify the differences in temperatures beneath the tree canopy with different mango tree sizes, considering three different levels of maturity. Three heights were chosen. All data loggers were allocated at the same height of around 1.5 m and attached to the tree trunk. The first mango tree had approximately 6 m, the second 9.0 m, and the third 12.0 m (Fig. 11). A pair of data loggers was used once again as the external control, being located at the same site but not close to any influence of tree shade. Results suggest that physical characteristics, such as leaf area and canopy size, are important in determining the microclimate under a tree canopy (Tables 8 and 9).

The more mature canopy presented the coolest microclimate. It was 1–2 °C lower than both the small and medium canopies which had similar values. The large mango tree also had a higher relative humidity of 3–4% than the less mature canopies. Nevertheless, all canopies presented cooler microclimates than the exposed site. The large mango tree was up to 9% higher while the small and medium canopies were up to 7% higher in relative humidity than the exposed site. Temperature values were also lower in relation to the exposed temperature by up to 3 °C. This shows that the mango tree has an effective cooling potential even when not fully mature, improving with time.

Different Species

The last case of the peak time studies consisted of comparing the temperature and humidity measurements of three different tree species but with similar sizes and shades. Again the data loggers were allocated at a height of 1.5 m and under the shaded tree canopies. The dense, round, and large mango canopy was compared to the also dense but conical canopy of the jaca tree and to the very light and spread-out

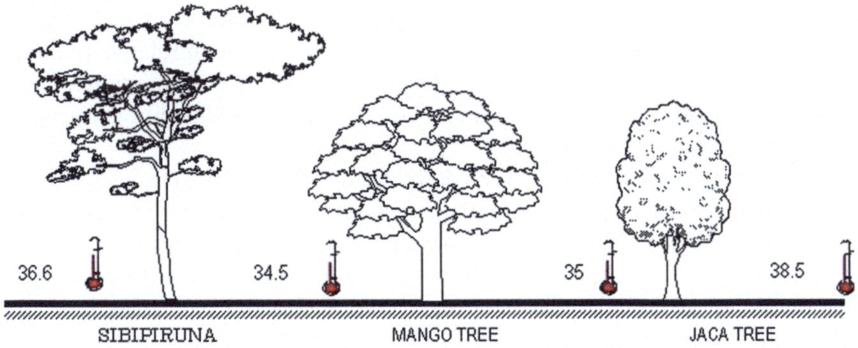

Fig. 11 Maximum temperature measurements of different species

Table 8 Relative humidity (%) measured at 1.5 m under different canopy sizes of the mango tree, small, medium, and large, and exposed area (Lago Sul, Distrito Federal, Brazil (hot-dry season))

	Exposed	Small	Medium	Large
N	326	326	326	326
Median	12.3	16.5	16.3	19.8
25–75 percentiles	10.2–15.2	14.8–17.5	15.4–18.3	18.8–21.7
Minimum	5.6	11.7	11.8	14.4
Maximum	23.4	23.0	25.4	25.5

Kruskal-Wallis one-way ANOVA on ranks: $p < 0.001$
All pairwise multiple comparisons (Dunn's method): $p < 0.05$ for all the pairs except small vs. medium (NS)

Table 9 Temperature (°C) measured at 1.5 m under different canopy sizes of the mango tree, small, medium, and large, and exposed area (Lago Sul, Distrito Federal, Brazil (hot-dry season))

	Exposed	Small	Medium	Large
N	345	345	345	345
Median	33.0	32.2	32.2	31.4
25–75 percentiles	32.2–34.1	31.4–33.3	31.7–33.0	30.7–32.2
Minimum	28.8	28.8	28.8	28.4
Maximum	36.9	34.1	34.1	33.3

Kruskal-Wallis one-way ANOVA on ranks: $p < 0.001$
All pairwise multiple comparisons (Dunn's method): $p < 0.05$ on all the pairs except small vs. medium (NS)

canopy of the sibipiruna. The jaca tree is also a tropical fruit tree while the sibipiruna is only ornamental with its yellow flowers. The goal of the study is to qualify the shades of tree canopies of different species with different solar permeabilities (Tables 10 and 11).

"Over 50% of the fresh weight of a tree consists of water, but the water concentration varies widely in different parts of a tree and also with species, age, site, season and even the time of day" (Kramer 1960). The mango tree with its bigger and denser canopy proved to be the most efficient in cooling the subcanopy microclimate. The mango tree was around 1 °C cooler than the jaca tree, which in turn was around 2 °C cooler than the sibipiruna. Furthermore, all three trees had temperatures below the exposed microclimate. The same occurred with relative humidity. All tree microclimates had higher values than the exposed relative humidity, but the mango tree performed the best, recording twice as much humidity during the warmest hours than the bare microclimate.

The above peak time studies were done during typical dry-season afternoons, which is the most crucial situation, with temperature values in the mid-30s and very low humidity values. Microclimates beneath the canopies showed differences among each other, but all showed a better performance compared to the exposed space. Once again the potential of a single tree as a natural air conditioner was verified.

Table 10 Relative humidity (%) of the exposed area and on the trunk of different tree species, at 1.5 m from the ground (Brasilia, Brazil (hot-dry season))

	Exposed	Mango	Jaca	Sibipiruna
n	300	300	300	300
Median	6.3	11.8	10.1	9.0
25–75 percentiles	5.6–13.5	10.9–16.5	10.1–14.1	9.0–14.5
Minimum	3.4	10.1	8.5	7.3
Maximum	20.6	21.7	19.3	20.3

Kruskal-Wallis one-way ANOVA on ranks: $p < 0.001$
All pairwise multiple comparisons (Dunn's method): $p < 0.05$ on all the pairs

Table 11 Temperature (°C) of the exposed area and on the trunk of different tree species, at 1.5 m from the ground (Lago Sul, Distrito Federal, Brazil (hot-dry season))

	Exposed	Mango	Jaca	Sibipiruna
n	308	308	308	308
Median	35.7	33.3	34.1	35.3
25–75 percentile	33.6–36.5	32.2–33.9	32.2–34.5	32.6–35.7
Minimum	29.1	29.6	29.6	29.6
Maximum	38.5	34.5	35	36.6

Kruskal-Wallis one-way ANOVA on ranks: $p < 0.001$
All pairwise multiple comparisons (Dunn's method): $p < 0.05$ for all pairs

Conclusions

Large amounts of solar radiation can be converted into latent heat by evapotranspiration, which does not cause the temperature to rise. By planting trees in urban areas, the green shade and cover over a space has a cooling effect on the microclimate and surroundings, and may also eventually reduce the heating load on houses.

A desirable set of microclimatic conditions may be created when the knowledge of appropriate trees which can be grown in a particular area is used. The landscape quality of trees depends upon their shape, form, texture, and foliage characteristics, as well as physiology. They alone may transform an external space. Although the results are constrained to the context in which they were studied, they are very encouraging in bringing more liveable and dignifying conditions to these places.

Various case studies were conducted with four species, as well as the native savannah-like vegetation. The objective of these studies was to quantify the effects of trees in urban areas on temperature and humidity to determine if the effects are different for different species of trees, trees of the same species with different maturity, and whether the effects could be explained by the physical characteristics of the trees. Due to the fact that they are representative in size, shape, and density of sound typical specimens, and that the experiments were conducted in representative (typical) days of summer and winter, the data measured and the values calculated are reliable and important reference.

Case studies in Brasilia have proved that microclimates may ameliorate with the use of a single-tree canopy. Benefits are subject to shade quality which depends on tree species, canopy size and permeability, and tree maturity. The case studies researched may not be sufficient to make any general rules, but it allows nevertheless to point out some considerations. The scientific data provided information, which brings perspectives and substantiates certain trends. Through fieldwork measurements of humidity and temperature it has been computed that:

- A site with a single mango tree planted has a cooler and more comfortable microclimate than a neighbouring and similar bare site.
- Different trees provided different microclimates under canopies. The mango tree microclimate provided the best results in relation to other species and exposed microclimate.
- Temperature is lowest in the middle of the tree canopy and gradually increases further down. Relative humidity behaves in an inverse manner.
- Temperature is lowest closer to the trunk and gradually increases further from it to the outside of the canopy shade. Relative humidity behaves in an inverse manner.
- Denser and more mature mango trees decrease average daily temperatures more than intermediate sizes and even more than new canopies. Nevertheless, all have a cooling potential.
- The increase of the percentage of green in a site does not proportionally decrease the thermal differences.
- The shading and cooling potential of tree plantings are determined by factors which include species composition, size, growth, crown density, spatial arrangement, and water use.
- Temperature variations are affected by the geometric configuration of trees.

In arid and semi-arid regions, overall places where drought can bring about restrictions on watering, the urban landscape trees is an important issue. It is therefore necessary for the chosen tree species to have a cooling capacity independent of regular irrigation or high water use, as seen with the mango tree in Brasilia. Although there is no way to accurately simulate the self-regulating temperature of vegetation, we know that it exists and has a very important role in the microclimate and its energy mechanisms. By analysing and comparing different tree microclimates, it is possible to predict certain tendencies. The mango tree has proved to be the equivalent of a natural air conditioner and humidifier. It is the best alternative in creating thermally more comfortable microclimates, and more bearable outside spaces. It can be said about the mango tree that:

- The range of the cooling effect of a single mango tree was measured and perceivable up to 5 m from the trunk. The more mature the canopy, the longer the extension of its effect. Therefore its cooling potential will increase with time.
- The cooling effect of a single tree is very significant, especially during the hot-dry season. The maximum temperature difference reached 7 °C and humidity 13%.

- The cost of maintenance for the mango tree is minimal, if required, and does not need regular watering.
- The studies suggest one shade tree, as the mango tree, per house to offset the dryness and hot weather.

Furthermore, along with being vested with an inherent cooling potential, trees, particularly the mango tree, are an instrument of social, economical, and ecological dimensions, especially in popular, low-income urban settlements. Some of the additional benefits for the inhabitants are:

- They can be designed to accommodate the informal jobs and tasks of the inhabitants, and the recreational needs of young children and elderly. The space for the proposed tree is found in all sites of the settlement.
- It provides nutritious fruits. They can be an important and essential complement to their diet. Fruitful trees also attract birds and enhance biodiversity.
- The provision of agreeable and comfortable spaces for socialising stimulates more contact and neighbourhood interaction, consequently reflecting on a safer place.
- The greenery of leaves and the colours of the flowering fruits provide beauty and paint a nicer and more exciting picture of the neighbourhood instead of the monochromatic dullness of the desert and lifeless scenario.
- In short, it provides psychological and physiological welfare.

Final Comments: Environmental Discourse

It is necessary to plant a tree before building a house. Frequently, in warm climates, low-income housing is always built without any type of planning or order. The only thing in mind when creating social urban settlements is politics. There is no planning whatsoever, or any interest in public welfare. All that is thought is settling down as many people as possible, in the least time and at a minimum cost. People are ignored as living beings (never mind human).

In warm climates, especially those with a hot-dry season, it is unconceivable that a house is built in an urban settlement deprived of trees and all its natural green surroundings. It is essential to reach a minimum of outdoor comfort, and a minimum of decency. Used as shelter, a protector against solar radiation, a natural thermal regulator, as well as a psychological attenuator, a tree is the perfect architectural form from nature.

It is necessary to stop thinking about architecture as a business or investment, but consider it thoughtfully as a space which involves and shelters lives. It is essential to give it attention, affection, life, and health. The good architecture has always existed, that is, the tree. The challenge is to incorporate it to our needs and necessities without destroying or harming it. In humble scenarios like social housing, due to the bare land, people and children are exposed to harsh conditions. The excess solar radiation, scarce humidity, and lack of shade dry up people's bodies and minds, wearing out people's health, and in many cases unconsciously their hopes.

The actual irresponsible and incompetent urbanisation process needs to change. Hypocrisy has to give way to simple and effective solutions. That is defining a tree. It is a simple yet very effective natural air conditioner. Studies with vegetation in Brazil and other warm climates globally showed that even a single tree makes a big difference. It provides more dignifying, human, and comfortable microclimates. It is necessary to plant a tree before building a house. To give land but take away health is not a fair deal.

One of the obligations of architects is to create spaces either interior or exterior, which are adjustable to man in both physical and psychological ways, being determined by the needs of the occupants. Once man has decided to build a shelter, it must satisfy all purposes, which include functional, aesthetical, social, economical, cultural, technological, and last but not least environmental purposes. Although there is no problem more important than the other, many times it seems that the architect focuses mainly on the aesthetics of the envelope than its climatic response to the environment. Despite the people who can afford the "look", one must not forget the millions of other humble people who are willing to do some kind of sacrifice in exchange of a habitat which will provide shelter, security, and comfort. It is important to understand that one does not do architecture by focusing uniquely at environmental, functional, cultural, or any other aspect. Architecture is done when all of the purposes are linked and accomplished in a united solution according to its use and occupants' expectations. For this architecture is so complex, and for this architecture is gratifying.

Impressive technological progress during the last decades has given architects options of new materials and new construction methods. The formal aspects tend to monopolise and overcome basic principles and concepts. Insistence upon this point has led to buildings constructed in ignorance and to the occupant's disadvantage.

Why consider new glazing and not a tree canopy? Why think of insulation and not leaves? Why choose highly reflected materials and not the tree shade or foliage? Why waste energy and elevate costs instead of passive cooling with vegetation? To what extent is modern sustainable? These are questions not as much to be answered as to be pondered. Certain materials and certain construction methods can only be used in certain climates and proper places, whereas a tree and vegetation can be beneficial in any place, in any climate.

Natural air conditioning is the method in which the living space presents comfort conditions required by man without the use of any extra energy consumption. This is of significant importance in times like now where the energetic crisis has touched world economy and there is an increasing preoccupation towards the need of a sustainable future. In developing countries the elevated costs of equipment and technology are also strong statements to reinforce the need of natural conditioning, and their intelligent interrelationships will make it possible to achieve.

Vegetation should not be seen as mere ornamental elements, but understood as a natural conditioner with extreme potential. With no better shade than its own, and having insulation characteristics, vegetation is capable of creating very pleasant microclimates in not so pleasant climates. The careful siting of it, on and around buildings, can make energy savings possible. Vegetation by providing shade from the sun and shielding from air infiltration may modify microclimates. Leaves, along with acting as insulation for external building surfaces, transpire and humidify the surrounding air in the process of evapotranspiration, another microclimate modifier. By benefitting from the changes in microclimate, the reliance on artificial mechanical conditioning systems may be reduced as well as energy consumption. Landscape and the built form designed together can make use of the vegetation potential and improve also the microclimate beyond the building skin. Although less explored, vegetation is probably the most efficient natural element for improving the thermal performance of a space. The use and manipulation of this element are essential to achieve the balance equilibrium between nature and architecture, and consequently a healthier environment.

To benefit the most from vegetation is nothing more than reasonable, and confirms one's dignity and responsibility towards man and the future. There is no doubt that life is moulded by its surroundings. If it is healthy, stepping stones are laid for decency and kindness.

> Trees are the friends of man, symbols of every organic creation; a tree is an image of a complete construction. A delightful spectacle which appears to us in the most fantastic, yet perfectly ordered arabesques; a mathematical measured play of branches multiplied each spring by a new life-giving hand. Leaves with finely placed nerves. A cover over us between earth and sky. A friendly screen close to our eyes. A pleasant measure interposed between our hearts and eyes and the eventual geometries of our hard constructions. A precious instrument in the hands of the city planner. The most concentrated expression of the forces of nature. The presence of nature in the city, surrounding our labours or our pleasures. Trees are the millenary companion of man.
>
> (Le Corbusier 1947)

References

Akbari, H., Kurn, D. M., Bretz, S. E., & Hanford, J. W. (1997). Peak power and cooling energy savings of shade trees. *Energy and Buildings, 25*, 139–148. Netherlands.

Cantuaria, G. A. C. (1995). *Microclimatic impact of vegetation on building surfaces. MA Thesis*. London: Architectural Association.

Cantuaria, G. A. C. (2000a). *Analysis of the cooling potential of a tree canopy: With special reference to the mango tree. Proc. of PLEA 2000* (pp. 537–538). Cambridge, UK: Architecture City Environment.

Cantuaria, G. A. C. (2000b). *Microclimatic impact of trees in suburban Brasilia. Proc. of PLEA 2000* (pp. 539–540). Cambridge, UK: Architecture City Environment.

Cantuaria, G. A. C. (2001). *Trees and microclimatic comfort. PhD Thesis*. London: Architectural Association.

Del Barrio, E. P. (1998). Analysis of the green roofs cooling potential in buildings. *Energy and Buildings, 27*, 179–193. Netherlands.

Descartes, R. (2004). *Principles of philosophy*. Whitefish, Montana: Kessinger Publishing.

Gomes, P. (1982). *Fruticultura Brasileira*. Sao Paulo: Ed. Nobel.

Holm, D. (1989). Thermal improvement by means of leaf cover on external walls—a simulation model. *Energy and Buildings, 14*, 19–30.
Hoyano, A. (1988). Climatological uses of plants for solar control and the effects on the thermal environment of a building. *Energy and Buildings, 11*, 181–199. Netherlands.
Huang, Y. J., Akbary, H., Taha, H., & Rosenfeld, A. H. (1987). The potential of vegetation in reducing summer cooling loads in residential buildings. *Journal of climate and Applied Meteorology, 26*, 1103–1116. Netherlands.
INMET. (1996). *Dados climáticos de Brasília*. Brasflia: Inmet.
Kaplan, R., & Kaplan, S. (1989). *The experience of nature: A psychological perspective*. Cambridge, England: Cambridge University Press.
Koslowski, T. T. (1987). Soil moisture and absorption of water by tree roots. *Journal of Arboriculture, 13*, 39–44.
Le Corbusier. (1947). *Quan les cathédrales etaient blanches. (When the cathedrals were white). Trans. Francis E. Hylop Jr.* London: Routledge.
Manica, I. (1997). *Fruticultura em áreas urbanas: Arborização com plantas frutíferas, o pomar doméstico, fruticultura comercial*. Porto Alegre: Editora Cinco Continentes.
McGinn, C. N. (1983). *The microclimate and energy use in suburban tree canopies. PhD Thesis*. Davis: University of California.
McPherson, E. G., Simpson, J. R., & Livingston, M. (1989). Effects of three landscape treatments on residential energy and water use in Tucson, Arizona. *Energy and Buildings, 13*, 127–138. Netherlands.
McPherson, E. G., Simpson, J. R., & Rowntree, R. A. (1993). Energy conservation potential of urban tree planting. *Journal of Arboriculture, 19*(6), 321–331. Urbana, Illinois.
Meier, A. K. (1990). Strategic landscaping and air-conditioning savings: A literature review. *Energy and Buildings, 15–16*, 479–486. Netherlands.
Morton, J. F. (1987). *Fruits of Warm climates*. Miami, Florida: Creative Resources System.
Novaes, W. (1992). *A Terra pede agua* (pp. 14–23). São Paulo: Imprensa.
Rosenfeld, A. H., Akbari, H., Bretz, S., Fishman, B. L., Kurn, D. M., Sailor, D., & Taha, H. (1995). Mitigation of urban heat islands: Material, utility programs, updates. *Energy and Buildings, 22*, 255–265. Netherlands.
Simpson, J. R. (1998). Urban forest impacts on regional cooling and heating energy use: Sacramento county study. *Journal of Arboriculture, 24*(4), 201–209. Urbana, Illinois.
Simpson, J. R., & McPherson, E. G. (1996). Potential of tree shade for reducing residential energy use in California. *Energy and Buildings, 13*, 127–138. Netherlands.
Takakura, T., Kitade, S., & Goto, E. (2000). Cooling effect of greenery cover over a building. *Energy and Buildings, 31*, 1–6. Netherlands.

From Vernacular to Sustainable Contemporary Architecture: Urban Green and Patios

Carolina Ganem Karlen

High Efficiency and Complex Features of Greenery

In human-inhabited spaces, the use of vegetation has always had multiple functions, from symbolic, aesthetic or ornamental to production and regulation of the microclimate. The integration of greenery with the constructed spaces is of particular importance today, especially as a valid solution to environmental discomfort and pollution load of urban spaces.

Internationally, urban greening constitutes an important trend in architecture and urban planning. Integrating a greener environment with human-made structures constitutes a key strategy for improving the quality of life for urban dwellers while achieving sustainable development (Santamouris 2001; Smardon 1988; Oke 1990; Nikolopoulou 2003; Georgi and Zafiriadis 2006; Guo-yu et al. 2013; Correia Guedes and Cantuária 2017). The efficiency of the greenery systems depends on the energy balance, plant characteristics and latent heat from the moisture content.

First, the energy balance of any building envelope consists of three modes of heat transfer: radiation, convection, and conduction. When the vegetated layer is added, it receives short-wave radiation from the sun, and it also exchanges long-wave thermal radiation between the ground, sky and surrounding surfaces. The vegetation absorbs some of the incident radiation. Direct radiation impacting on walls and windows is the primary source of heat gain, but two other factors are also important: heat from ambient air and indirect long-wave radiation from the immediate surroundings. All three of these factors can be moderated by vegetation growing close to a building's surface. The greenery works like a second skin to roofs and walls, acting as insulation.

C. Ganem Karlen (✉)
CCT-CONICET-Mendoza, Mendoza, Argentina
e-mail: cganem@mendoza-conicet.gov.ar

A. Sayigh, A. Trombadore (eds.), *The Importance of Greenery in Sustainable Buildings*, Innovative Renewable Energy,
https://doi.org/10.1007/978-3-030-68556-0_5

Second, the heat flow through the vegetation is transferred by convection, if there is a layer of airspace, or transferred by conduction to the substrate layer. The energy balance also includes an additional term for the radiation exchange between leaves of the plant layer. These parameters include leaf absorptivity, which is the fraction of incident solar radiation absorbed by a leaf (Campbell and Norman 1998), leaf dimensions which affect vapour conductance and convective heat between plant leaves, and leaf area index which is defined as the total projected area of leaves per unit surface area. The parameters also include the radiation attenuation coefficient, which indicates the decrease in absorbed radiation in the plant canopy. Besides, they include leaf conductance, which is the rate of water vapor leaving the plant surfaces through the pores on the leaf surface during transpiration.

Third, the latent heat is transferred through the moisture content, which comes from the amount of precipitation, according to the climate variation, evapotranspiration process of the plants and water content in the substrate layer (Seyam 2019).

Measurements of transpiration of foliage to estimate the thermal effect of plant physiological activities showed that solar radiation was the main source of heat gain in the vegetation canopy and dominated its energy balance process. The vegetation canopy could consume more than half of the absorbed solar energy during the transpiration process, supporting transpiration as one of the most important sources of cooling. Approximately 30% and 20% of the absorbed solar energy was, respectively, dissipated by long-wave radiation exchange between the canopy and surrounding environment and convective heat transfer between the foliage and the air (Zhang et al. 2019).

The complexity of the described phenomena guarantees internal space protection from direct radiation, a high level of ventilation between the leaves and building envelope and a percentage of transmitted radiation (mainly diffuse) that secures a good lighting performance with a quality of dynamic light that would be difficult and expensive to get with artificial devices. In fact, this type of phototropic regulation implemented by the foliate mantle of the vegetation is compared to sophisticated artificial systems of dynamic regulation implemented with different modulators. They can be mechanical like the diaphragms of Jean Nouvel in the Institut du Monde Arabe [Arab World Institute] (1987), or physical-chemical like transparent photosensitive materials. From the complexity level required, one could understand how important it is to develop models and technical systems that hybridize the logic of natural systems with that of technical systems like green and glass integration. Examples of these are the building for the Fondation Cartier pour l'art contemporain [Cartier Foundation for Contemporary Art] (1994) also by Jean Nouvel, and the building for the Consorcio Nacional de Seguros de Vida [National Consortium of Life Insurance] by Enrique Browne y Borja Huidobro (1993). See Figs. 1 and 2.

The building for the National Consortium of Life Insurance by Enrique Browne y Borja Huidobro (1993) was one of the first architectural examples that integrated greenery in modern architecture in South America. This building, contemporary to Nouvel's Cartier Foundation for Contemporary Art, was designed with a double

Fig. 1 From left to right: (**a**) Arab World Institute, Paris (J. Nouvel 1987); (**b**) Cartier Foundation for Contemporary Art, Paris (J. Nouvel 1994)

Fig. 2 From left to right: (**a**) Exterior view, (**b**) inner space in between the two façades and (**c**) interior view of the National Consortium of Life Insurance building, Santiago de Chile (E. Browne and B. Huidobro 1993)

façade, one technical and one natural. The interior façade is composed of thermo-panels, and in the exterior one creepers conform to a vegetation mesh.

The majority of humans are exposed to environmental conditions that often challenge human health and well-being and also threaten natural resources. The sustainable approach recalls the importance of this hybridization process, that is, to reactively use all the renewable resources of nature, also revaluating the environmental characteristics of past solutions and recognizing that they have contributed to creating many sophisticated and environmentally friendly local cultures.

From the Hanging Gardens of Babylon to the Oasis City of Mendoza

The thermal behaviour of cities is largely a by-product of urban morphology or, more specifically, the composition and three-dimensional structure of materials that constitute the urban frame (McPherson 1994). Urban greening is the most widely applied strategy to mitigate this phenomenon which can achieve huge temperature reductions (Solecki et al. 2005; Rizwan et al. 2008).

The thermoregulating function of the vegetation, in the summer period, has been recognized since the most remote times in temperate and hot climates. There are descriptions written in cuneiform on clay tablets as early as 2000 B.C. that tell us about one of the earliest kinds of garden, 4000 years ago. The Babylonians and Assyrians planted gardens in cities, palace courtyards and temples in which trees with fragrance and edible fruits were prominent for recreating their concept of paradise.

The famous Hanging Gardens of Babylon constructed around 700 B.C. was described as a remarkable feat of engineering with an ascending series of tiered gardens. The invention of the Archimedean screw is remarkable, using cast copper or bronze, for watering the gardens (Dalley 1993). The Babylonians established pipelines and a chain pump or Ferris wheel to raise the liquid element to the terraces and water all the lush vegetation.

Its name is derived from the Greek word kremastós (κρεμαστός, lit. 'overhanging'), which has a broader meaning than the modern English word 'hanging' and refers to trees being planted on a raised structure such as a terrace. According to the words of the writer Philo of Byzantium, the verge of Mesopotamia located next to the royal palace looked like this: The so-called Hanging Gardens, with its plants above the earth, grows in the air (Philo of Byzantium 2019). The roots of the trees form a roof above the ground. Figure 3a presents an Assyrian relief which shows trees hanging in the air on terraces and plants suspended on stone arches.

On the other side of the world, next to the Andes Mountains, 20 centuries apart, the urban structure of Mendoza, located in central western Argentina (32°40′

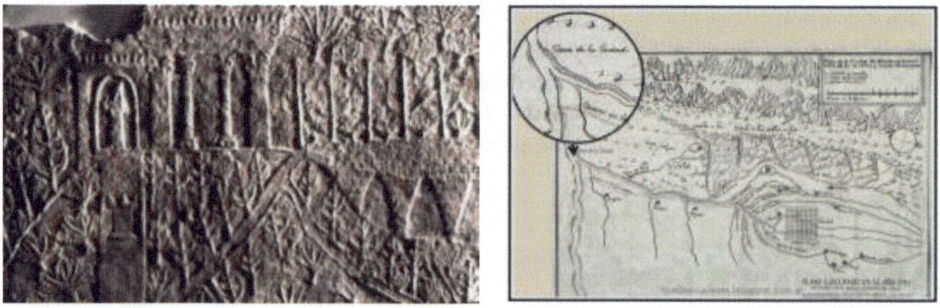

Fig. 3 From left to right: (**a**) Assyrian relief (V B.C. circa). British Museum London (2020). (**b**) Spanish map (1781) showing Mendoza's irrigation system (Ponte 1998)

southern latitude, 68°51′ western longitude, 750 m above sea level), is one of the cities with the largest agglomeration west of the Argentine Republic and one of the Spanish foundations in South America that probably contain one of the most complex urban plots.

Although Mendoza is located in a semi-arid continental climate with low percentages of atmospheric relative humidity and high heliophany (see Table 1) it does not follow a compact urban model. Its urban model is defined by its wide and tree--lined streets that form green tunnels. The chequered frame contains the buildings while the main strategy for minimizing the sun exposure is the vegetal frame. Mendoza's urban structure intermingles three types of meshes that overlap in space: a water network, the characteristic Spanish regular orthogonal grid and, finally, a green mesh that arises due to the interaction of the first two.

While the evolution of urban green in the city of Mendoza has long-standing records, no official date is known regarding when the work of forestation began, which arose practically from the possibility of the colonizers to use the irrigation system created by the Huarpes and perfected by the Incas for the initial irrigation of orchards and supply of drinking water in the plots that formed the stables of the first population settlement, around 1562 (Ponte 1998).

Foreign chroniclers and travellers from the fifteenth to eighteenth centuries tend to describe the eminently rural character of Mendoza. The incipient agricultural activity, mainly fodder, cereal and winemaking, was consolidated taking advantage of the aboriginal tradition. That agricultural city, which was already located at the

Table 1 Climatic data for the city of Mendoza, Argentina (Servicio Meteorológico Nacional 2020)

Annual values	Average maximum temperature	22.6 °C
	Average minimum temperature	11.0 °C
	Mean temperature	15.9 °C
	Global horizontal irradiance	18 MJ/m^2
	Relative humidity	54.70%
	Mean rainfall	218 mm
July (winter)	Average minimum temperature	3.4 °C
	Mean temperature	7.8 °C
	Average maximum temperature	14.7 °C
	Thermal amplitude	11.3 °C
	Mean wind velocity	7.6 km/h
	Global horizontal irradiance	9.9 MJ/m^2
January (summer)	Average maximum temperature	30.1 °C
	Mean temperature	25.3 °C
	Average minimum temperature	18.4 °C
	Thermal amplitude	11.7 °C
	Mean wind velocity	10.8 km/h
	Global horizontal irradiance	25.7 MJ/m^2
Annual heating degree-days (T_b = 18 °C)		1384
Annual cooling degree-days (T_b = 23 °C)		163

site where the city was later founded, had established an artificial irrigation system through canals and ditches.

This Huarpe culture has left a valuable and enduring legacy: the spatial unity given by the ditch-channel-path. This opens up a fact that allows Mendoza to think within the framework of an ancient history, that of irrigated civilizations. Figure 3b shows a map of the region showing the thawing water conduction system from the mountains to the cultivation areas and the city, through canals and ditches.

In 1861 the colonial city was destroyed by a massive earthquake and the reconstructed city of Mendoza ended with wider streets of 20 m instead of the former of 16.71 m. The reason was that the French surveyor Julio Balloffet prepared the layouts using meters instead of the original yards of the Spanish legislation. Another major change was in 1872 when a system of ditches was implemented with a new design to provide with fresh water and to help with the drainage. Ditches were now built in parallel to the roads and they made possible the planting of public trees.

From that time on, urban forestation is the most important bioclimatic conditioning tool for open spaces during sun hours. This forest frame creates shading and these 'green tunnels' form an authentic forest within the city of Mendoza. In this sense, the city follows the model of any other oasis city. Dimensional relationships between the tree meshes must be proportionally coordinated. The dimensional balance between the three overlapping structures results in a microclimate known as oasis city. Figure 4 presents a scheme of the overlapped meshes in the urban plot.

In the oasis city, if the architecture develops within the conditioned stratum, it benefits from its environmental qualities. There are three fundamental guiding principles essential to support the system: (1) conditioned stratum, (2) volume–space coordination and (3) dimensional coordination (Bormida 1984).

1. Conditioned stratum: The city has a horizontal dimension, corresponding to the extension of the network of trees and ditches. But it also has a vertical dimension, defined by the height of the tree mass. The stratum is the atmosphere sector environmentally conditioned due to the combined effect of the green and the water.

   The climatic variables that affect the vegetation-built interaction are the solar radiation (direct, diffuse, reflected), the 'terrestrial' infrared radiation and the intensity and direction of the wind. These climatic variables are modified by the arrangement of the vegetation and by the characteristics of the leaf mass and of the air temperature, which is modified by the sensible heat emitted by the vegetal mass, and by the latent heat of the evapotranspiration which also modifies the humidity of the air (see Fig. 5).
2. Space articulation: It consists of accompanying in a balanced way the amount of architectural volumes with forested open spaces. It provides the necessary dose of green in the construction of the plot so that it is effective; there must be green squares and patios (see Fig. 6).

   Taking into account the particularities of the city of Mendoza, it is necessary to settle the comfort conditions with the possibility of nocturnal cooling. During the daytime the solar radiation control is a key to getting comfort conditions.

Fig. 4 Scheme of the urban structure of the city of Mendoza (Bormida 1984)

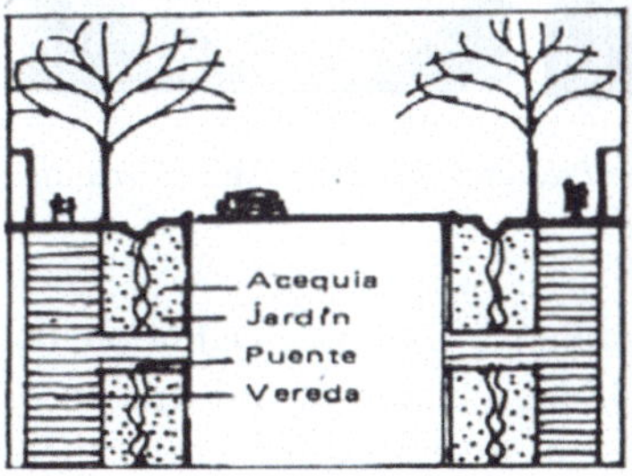

Fig. 5 The stratum is the atmosphere sector environmentally conditioned due to the combined effect of the green and the water

Fig. 6 Balance of architectural volumes and forested open spaces

Fig. 7 Forested urban canyons of the city of Mendoza. (**a**) *Platanus hispanica*, (**b**) *Morus alba* and (**c**) *Fraxinus excelsior* (Correa et al. 2010)

During the night-time, sky vision is needed for radiative cooling. In addition, the forest structure combined with the urban morphology increases soil roughness and reduces convective cooling. Therefore it is necessary to encourage these combinations of forest structure and urban morphologies that benefit both processes. Conforming to our assessment, the best combinations are (1) first magnitude species such as the 'London plane' (*Platanus hispanica*) for 26–30 m road channels (accounts for 22%); (2) the 'white mulberry' (*Morus alba*) for 20 m road channels; and (3) 'European ash' (*Fraxinus excelsior*) for 16 m road channels that are second magnitude species and represent 38% and 19% of Mendocinean trees, respectively (Correa et al. 2010, 2012) (see Fig. 7).

3. Dimensional coordination: It consists of the harmonic relationship between the height and diameter of the forest canopy, the sidewalk and street width and the building height.

In terms of building density, the city has the highest densities in the central area, which decreases progressively toward the periphery, where lower densities are found in residential areas. Low-density configurations are the most prevalent for areas already developed and for new housing (Mesa and Giusso 2014). Nevertheless, the central area high density jeopardizes the whole oasis city concept.

Oasis cities enjoy unique environmental benefits from the urban forest. These may be defined in two ways: on the one side, the situation below the tree canopy benefits low-rise buildings (3–4 stories) in the summer since the incident radiation is moderated and can even be blocked, depending on the density of the foliage. Reductions in the exterior temperature during the summer may fluctuate between 0.3 °C and 3 °C depending on the climate and context (Akbari 2001).

On the other hand, the housing units above the tree canopy are directly exposed to the climate of the region and are open to absorbing full solar radiation in the winter (desired incident energy) and in the summer (unwanted incident energy). They are also exposed to convective and radiative energy exchanges in both seasons.

Subsequently, this city model presents a microclimate that benefits low-rise buildings, which is in accord with the arid climate located in the region. Despite the differentiation of the microclimates, tall buildings tend to have a single building envelope regardless of these environmental factors of the surroundings (Balter et al. 2016) (see Fig. 8).

In this context, Ruiz, Sosa, Correa and Cantón (2015) aimed to identify the best urban configurations in terms of thermal behaviour. The following parameters were

Fig. 8 Harmonic relationship between the forest canopy and the building height

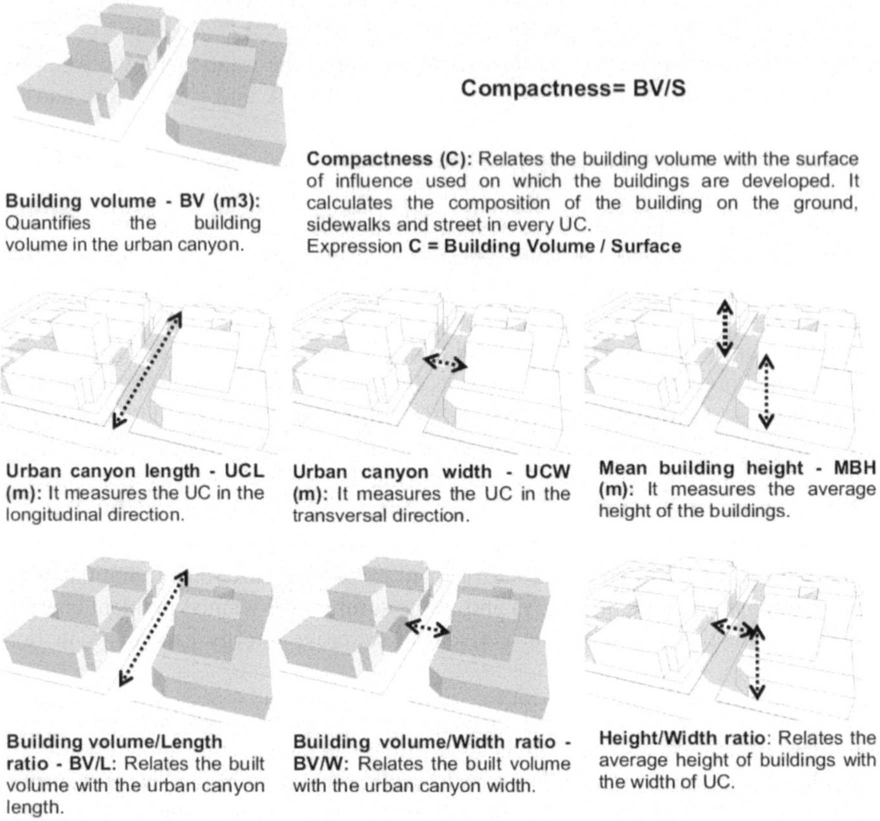

Fig. 9 Morphological structure variables (Ruiz et al. 2015)

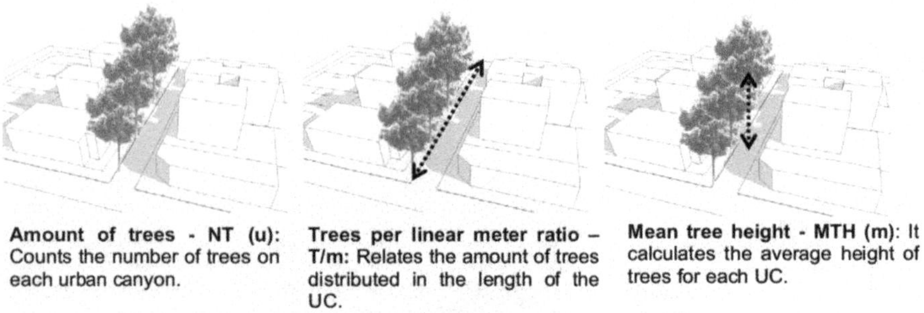

Fig. 10 Urban forest variables (Ruiz et al. 2015)

collected because they may be incorporated into building codes and used by developers and urban planners. The variables selected by category are as follows:

- Morphological structure: building volume (BV), compactness (C), urban canyon length (UCL), urban canyon width (UCW), mean building height (MBH), volume/length (BV/L), volume/width (BV/W) and height/width ratio (H/W). The explanation of each variable is shown in Fig. 9.
- Urban forest structure: solar radiation permeability of the tree species (SP), amount of trees (NT), trees per linear metre ratio (T/m) and mean tree height (MTH). The explanation of each variable is shown in Fig. 10. Solar radiation permeability of tree species is defined as a characteristic of the tree canopy, which transmits the incoming global radiation. It is a value between 0 and 1.
- Optical properties of materials: horizontal surface albedo (HA) and vertical surface albedo (VA). The explanation of each variable is shown in Fig. 11.
- Microclimatic variables: daytime air temperature (DTair), night-time air temperature (NTair), daytime pavement surface temperature (DTpav), daytime sidewalk surface temperature (DTsw) and daytime wall surface temperature (DTwall). Previous studies in the city of Mendoza show that the values of albedo of the commonly used materials vary widely between 0.42 and 0.91 while emissivity values only vary between 0.80 and 0.98 (Alchapar et al. 2014). Although the albedo is not involved in the night radiative balance, since the emissivity does not offer many possibilities for variation, it has privileged the daytime surface temperature measurements.

As a result, the best response for Mendoza is composed of a large density of trees (0.23 trees/m = 8.5 m of tree spacing), a high horizontal albedo (equal to 0.54) and wide urban canyons (30 m wide). This configuration offers cooling of 6.8 °C less at daytime and 3.5 °C less at night-time than the worst proved configuration. The worst thermal response is given by using fewer trees (0.13 trees/m = 15 m of tree spacing), a low horizontal albedo equal to 0.22 and narrow urban canyons (16 m wide) (Ruiz et al. 2015).

Horizontal surface albedo - HA : It is the ratio of reflected radiation from the horizontal surface to incident radiation upon it.

Vertical surface albedo - VA : It is the ratio of reflected radiation from the vertical surface to incident radiation upon it.

Fig. 11 Optical properties of material variables (Ruiz et al. 2015)

Urban Green: Above and Below the Tree Canopy

The height of the building is one the most jeopardizing variables to take into account in oasis cities in order to assess how much greenery will improve comfort and appeal in sustainable buildings.

An analysis of height (or relative position of the apartment within the building) is therefore mandatory. It was defined by whether the housing unit was located above treetops, which is considered as from the fourth storey (12 m) of the building. Height limits are also based on the types of trees (Cantón et al. 1994): 12 m from street level for mulberry (*Morus alba*) and ash trees (*Fraxinus excelsior*), and a limit of 15 m of height from street level for the plane tree (*Platanus hispanica*). The following indices were determined:

- Housing below the tree canopy: up to and including the fourth storey (ground floor +3), corresponds to a height of up to 12 m
- Housing above the tree canopy: starting from the fifth storey (ground floor +4), corresponds to a height greater than 12 m

Balter, Ganem and Discoli (2016) reported exterior and interior temperatures and energy consumption in four case studies. Housing units were compared according to their variations in height and materiality. In order to have clear references for the case studies, the following abbreviations were defined:

- MB: Apartment in massive building below the tree canopy
- MA: Apartment in massive building above the tree canopy
- LB: Apartment in light building below the tree canopy
- LA: Apartment in light building above the tree canopy

1. Height comparison: massive building [MA–MB]

   In summer, the MA housing unit has higher temperatures that reach 2 °C beyond the comfort zone while the MB unit is within the comfort range. The mean difference between the two cases is 2.30 °C. Energy consumption for cooling is higher in the MA with differences of 42% during the scenarios with the most demand while in the scenario with the least demand MB does not require HVAC. In winter, the MB is warmer and both housing units are below the comfort zone. The thermal average differences between the two cases are 2.40 °C. Energy consumption for heating is greater for the MB with differences reaching 23% during the scenarios with the most demand and 94% in the scenario with the least energy demand.
2. Height comparison: light building [LA–LB]

   In summer, the LA is warmer, but both case studies exceed the limit of 28 °C by 1.30 °C for the LB and 3.70 °C for the LA, which is above the comfort zone (Givoni 1991). The average difference between the housing units is 1.90 °C. Energy consumption for cooling is greater in the LA with differences of 33% in the scenario with the greatest demand and 78% in the scenario with the least demand. In winter, the LB is warmer and both apartments are within the comfort ranges. The average difference between the two cases is 2.30 °C. Energy consumption for heating is greater in the LA with differences reaching 87% in the scenario with the most demand. In the scenario with the least demand, neither housing unit required HVAC.

3. Material comparison: below the tree canopy [LB–MB]

   In summer, temperatures are higher in the light building: in the LB, temperatures rise above the comfort zone and in the MB, they fall within a comfortable range. The average thermal difference is 1.40 °C. Energy consumption for cooling is greater in the LB apartment, with differences reaching 51% in the scenario with the greatest demand, while in the scenario of the least demand the MB did not require HVAC. In winter, temperatures were higher in the light building: LB was found to be within the comfort range, while in the case of the MB, the findings were below. The differences in this case are in the order of 5.80 °C. Energy consumption for heating is greater in the MB with differences reaching 91% for the scenarios with the greatest demand, while in the scenario with the least demand, the LB did not have any consumption.

4. Material comparison: above the tree canopy [LA–MA]

In summer, as was with the case with the housing units below the tree canopy, the temperatures were higher in the light building. However, when comparing these apartments, both exceed comfort ranges. The average thermal difference between the two is 1 °C. Energy consumption for cooling is greater in the LA with differences of 44% for the scenario with the greatest requirements and 49% with the least. In winter, similarly with the previous situation, the temperatures are warmer in the massive building. The LA showed temperatures that were within the comfort zone, while for the MA, temperature fell below this range. For this comparison, the average thermal differences reach 5.90 °C. Energy consumption for heating is greater in the MA reaching differences of 47% for the scenario with the greatest demand, while for the scenario with the least demand, the LA did not require HVAC. Figures 12 and 13 show interior and exterior temperatures and energy consumption in summer and winter, respectively.

Taleghani (2018) evaluated several examples in different countries and climates, and concluded that urban surfaces play an important role in the thermal comfort of

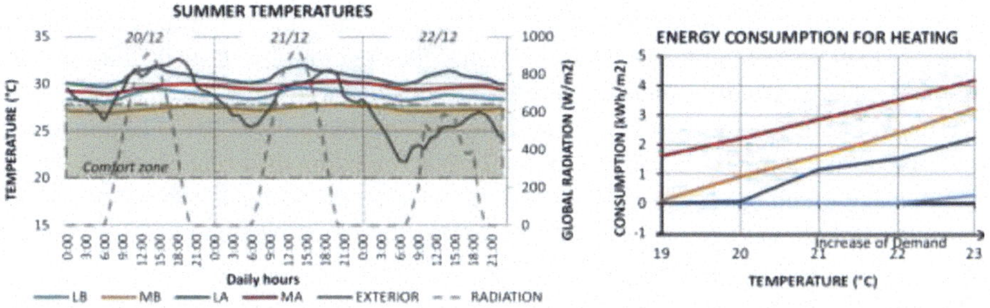

Fig. 12 Temperatures (interior and exterior), solar radiation and energy consumption for four cases in summer (December in the South Hemisphere) (Balter et al. 2016)

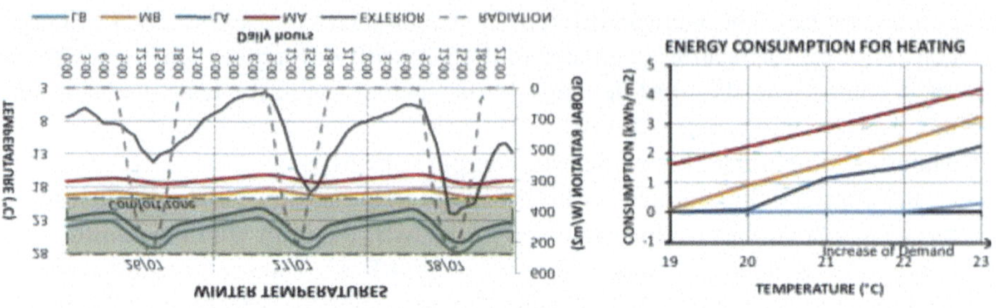

Fig. 13 Temperatures (interior and exterior), solar radiation and energy consumption for four cases in winter (July in the South Hemisphere) (Balter et al. 2016)

pedestrians. Vegetation and high-albedo surfaces showed appreciable reduction of air temperatures within urban open spaces. However, mean radiant temperature affects human thermal comfort more than the other meteorological variables. Therefore, using high-albedo materials on the ground surface causes reradiation of solar radiation to pedestrians and leads to thermal discomfort (in spite of reducing air temperature). Therefore, using vegetation in urban open spaces is a better choice for improving thermal comfort in the pedestrian level.

Coutts, White, Tapper, Beringer and Livesley (2016) measured three east-west streets during heat waves in the temperature climate of Melbourne, Australia. They considered the Universal Thermal Climate Index (UTCI) for the calculation of outdoor thermal comfort within the tree canopies. The maximum UTCI reduction by the trees was 6 °C. Moreover, they reported that the maximum air temperature reduction was 1.5 °C. A similar research done in Melbourne showed that Platanus trees led to a PET reduction of 6.6 °C during a heat wave. Regarding the cooling effect of different tree species, Doick and Hutchings (2013) discuss that the lower the foliage temperature, the greater the cooling effect.

Many other authors reported microclimate benefits bellow the tree canopy and with different types of greenery in buildings. In winter external temperature reduction varied from 5 to 10 °C and there was an increase in the vegetated wall surface temperature by 0.3–2.5 °C (Vox et al. 2018; Cameron et al. 2015; Coma et al. 2017); the negative values of external temperature reduction means that the green wall is warmer than the bare wall, which means the wall surface behind the vegetation is warmer, for instance, the minimum of −0.3 °C and −2.5 °C during the night of the winter season for green façade. That is because the vegetation acts as insulation during the winter season to protect the wall from the cold weather.

Foustalieraki, Assimakopoulos, Santamouris and Pangalou (2017) obtained the highest external temperature reduction of 21.9 °C by using different plant species in the intensive green roof system. This reduction led to 11.4% saving in heating load. However, Dahanayake and Chow (2017) obtained high external temperature reduction reaching 16.9 °C in China (Hong Kong and Wuhan), but the saving in heating load was 0.3–2% less than the previous study by constructing green wall system.

The overall thermal conductivity of the vegetated wall is less than the bare wall resulting in saving heating load during cold weather.

During spring and summer seasons, the solar radiation and the ambient air temperature significantly rise resulting in increasing of the external surface temperature more than the outdoor air temperature by about 5–10 °C due to the high absorptivity of the construction material. It was obvious that the temperature reduction for the external surfaces was higher than that of other seasons with a range from 2 °C to 20 °C. The high values were related to the exposure to peak solar radiation, which was absorbed by plants, and occurred between 14 and 16 solar hours, while the low values were related to the reduction during the night or cloudy days.

Patios: Private and Public Greenery

Traditionally, construction typologies associated to human needs—shelter, work and rest—were structured around an open space as an organization centre, usually called patio. Courtyard houses have an ancient history: examples have been

Fig. 14 From left to right: (**a**) Troglodyte Cave Dwellings in Tunisia. (**b**) Chinese underground courtyards in Honan

excavated at Kahun, in Egypt, which are believed to be 5000 years old, while the Chaldean City of Ur, dating from before 2000 B.C., also comprised houses of this form.

The most primitive and homogeneous society to build courtyard houses was probably the one that built the Troglodyte villages in the Matmata of Southern Tunisia (see Fig. 14a). Around 2000–1500 B.C., similar houses were built in the Indus Valley using the same philosophy. The houses were designed as a series of rooms opening on to a central courtyard (Nangia 2000).

In another part of the world, the early Chinese houses were highly influenced by the principles of Yin and Yang (see Fig. 14b). As we can see in Fig. 14 there is a striking similarity between the simplest form of Chinese underground settlements in Honan and the Troglodyte dwellings in Tunisia.

The Classical Age of architecture, marked by sophisticated Greek and Roman design and planning, bears evidence to the universal appeal of courtyard houses. The Greeks discovered the thermal advantage of courtyard buildings and then they designed their homes in a manner to allow low winter sun in the courtyard, while blocking the high summer sun by the overhanging eaves on the portico. The Romans were later inspired by the light and airiness of Greek peristyle houses and the atrium houses of the Etruscans. And, from the classic civilizations they arrived to almost every occidental modern culture, especially in hot-arid climates and Mediterranean temperate climates. It is still the traditional house type of many Asian, North African, South American and European countries (Nangia 2000; Hinrichs 1987).

The patio (courtyard) regulates the architectural idea from which interior spaces are then distributed. It has been developed during the last periods through the process of trial and error to a general climatic satisfying state within the available resources (Al-Azzawi 1984). One of the main reasons that courtyards have survived for more than 5000 years is their potential to provide a thermally comfortable area for living. Courtyards can be a source of fresh air, light and heat or coolness. They have been generally referred to as a microclimate modifier in the house due to their ability to reduce peak temperatures, to channel breezes and to adjust the degree of humidity (Taleghani 2014). Courtyard has contributed to improving local climate conditions and creating comfortable interior spaces.

The courtyard building is suitable for use and enjoyment if special arrangements are made at the design stage. This includes the internal envelope's finishing and materials, as well as the proportions of the physical parameters of the courtyard form by proportioning the courtyard internal envelope ensuring adequate solar radiation accessibility in winter for warming up the building and providing sufficient shadows in summer avoiding or reducing the need for cooling. Several authors have demonstrated that temperatures in those interior spaces located next to courtyards can be lower than in exterior spaces depending on the climate, microclimate and building constructive characteristics such as form, dimensions, orientation of courtyards and amount of vegetation to shadow them (Cantón et al. 2001, 2014; McPherson 1992; Aldawoud 2008).

Thermal performance of courtyard building has been investigated by many researchers such as Mohsen (1979); Meir, Pearlmutter and Etzion (1995); Taleghani

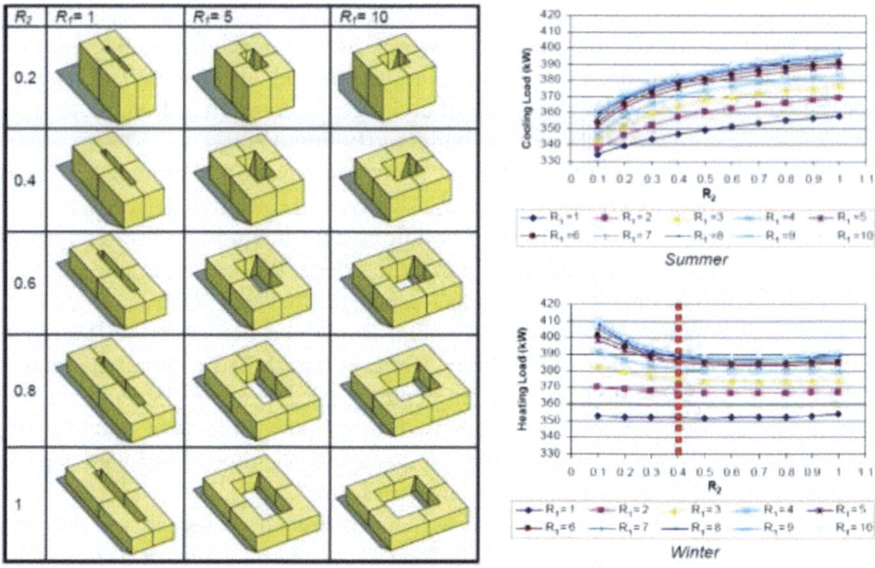

Fig. 15 Geometrical characterization of courtyards (Muhaisen and Gadi 2006)

(2014); and Zamani, Heidari and Hanachi (2018), with special evaluation concerning the influence of the geometrical and physical parameters of the courtyard on the received solar radiation. The conclusions of all these studies recommended protection of the form's surfaces and its surroundings from intense solar radiation.

Muhaisen and Gadi (2006) proposed a methodology to geometrically characterize courtyards. Courtyards present simple cubic shapes with different values of R1 and R2. The former ratio is taken as the ratio between the courtyard's floor perimeters P and the form's height H (P/H). This ratio indicates the depth of the form and it ranges between 1 and 10 in one-degree steps. The latter ratio, which indicates the elongation of the form, is the ratio of the rectangular courtyard width W to its length L (W/L). It varies between 0.1 and 1 in tenth of degree intervals. See Fig. 15.

The deeper and more elongated the courtyard building, the most preferable for reducing the required cooling load, whereas deep forms with R2 value greater than 0.4 would be desirable for a winter minimum heating load. It is observed that, in winter, changing the courtyard's plan elongation has minor effect on the building's heating requirement. The maximum percentage of increase in the minimum heating load, which may occur as a result of changing R2, at any value of R1, was found to be about 0.05% (at R1 of 10). This indicates that, for winter-efficient courtyard design, the main concern should be given to adapt the depth of the courtyard building, as the plan elongation would have very slight effect on the heating consumption (Muhaisen and Gadi 2006).

In the case of the city of Mendoza, Argentina, located in a semi-arid context, courtyards were shadowed for centuries with green 'roofs', particularly with bowers of grapevines or creepers. This type of plant climbing was used to avoid or reduce heat gain on massive surfaces that limited the patio space. A green coat fulfils, without any doubt, sun protection function; however, the difference from any other protection is that it overthrows short-wave radiation and absorbs radiant energy, especially in the red wavelength, that will no longer be transferred to the surrounding open space or to the interior of the building (Bellomo 2003).

Galleries were also employed for the same purpose, as intermediate spaces between interior and exterior environment. This type of transitional spaces, considered essential parts of the envelope of buildings with intermittent use, can act as connectors on one phenomenon and a barrier to others, thus acting as climatic 'filters', under feasible control by occupants (Ganem 2006; Coch 2003).

A research conducted by Cantón, Ganem, Barea and Fernandez (2014) compared two courtyards in the same building. Both courtyards had a similar elongation of the form as their R2 ratios are only one-tenth of a degree apart (0.4 and 0.5). The main difference between the courtyards is in the depth of the form rated by R1: courtyard 1 has a value of 4 and courtyard 2 rates 9. Therefore, courtyard 1 has a compact form and courtyard 2 presents an open form.

Courtyards present shade conditions associated to their shape and solar protection. Shadows projected by form and shadows projected by greenery were graphically obtained for each courtyard and calculated by area and by percentage of vertical and horizontal envelopes, on an hourly basis, for a summer representative day between 8:00 a.m. and 4:00 p.m. Solar position was defined using a programme called Geosol 2.0 (Hernández 2003) that allows obtaining solar coordinates, altitude and azimuth, in a given location throughout a day. See Fig. 16.

As the construction technology is permeable to exterior microclimate conditions, courtyards have a strong influence on the resulting interior temperature.

Graphics of temperature (Fig. 17a) show the following: There is an hourly displacement of maximum temperatures: Courtyard 2 starts its heating period first and reaches maximum temperatures at 1 p.m., because of the higher solar exposure of its surfaces. Such exposure is the result of the shape of the courtyard and its orientation, which generates heating conditions from the beginning of the day, with relative efficiency of the shading element. However, in courtyard 1, with less solar exposure,

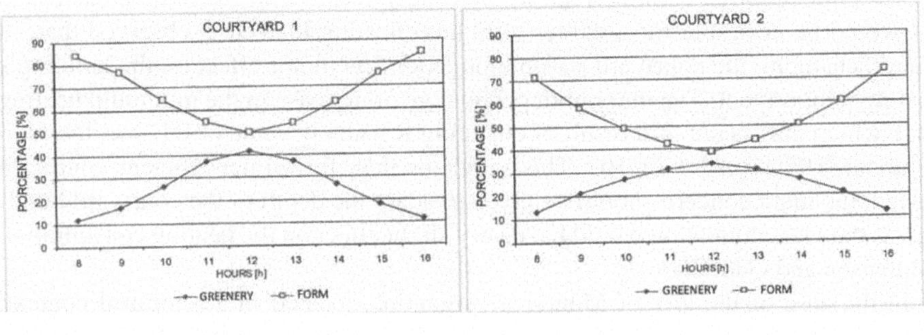

Fig. 16 From left to right: Courtyards 1 and 2 hourly distribution of shaded areas by greenery and by form

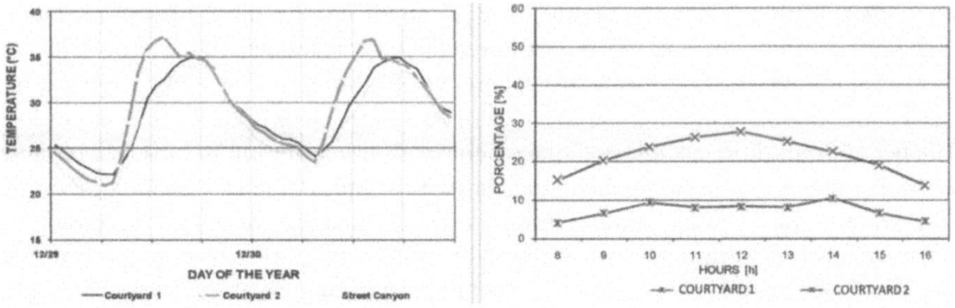

Fig. 17 From left to right: (**a**) Temperatures of courtyards and of street canyon. (**b**) Areas with solar radiation (vertical and horizontal surfaces)

the start of the heating period is delayed and thus maximum temperatures are reached at 5 p.m., 4 h later than in courtyard 2.

The comparison of the highest temperatures shows that courtyard 2 is 2 °C warmer than courtyard 1. This is explained by the combined effect of the open morphology and the relative efficiency of the green shading. In courtyard 1 the relative efficiency of the green shading is compensated by the shape that is more effective for solar control. Therefore, shape plays a decisive role in the thermal behaviour of courtyards. Also, in Fig. 17b it can be observed that courtyard 1 has a lower percentage of areas with solar radiation throughout the day. At minimum temperatures, differences of about 0.5 °C are observed. This variation results from the better cooling capacity of courtyard 2. Potential cooling of courtyard 1 is conditioned by the larger amount of accumulated mass and the smaller sky view. These on-site measurement results are consistent with the ones found by Muhaisen and Gadi (Zamani et al. 2018) in analytical analyses. In their research, which addresses solar heat gain and energy requirements in summer, they concluded that when the courtyard buildings are deeper and more elongated (when R1 is closer to 1 and R2 is nearer 0.1), the more adequate strategy is to reduce the required cooling load.

Results indicate that the courtyard geometry has its highest influence on shade conditions during the first morning hours and late afternoon hours, reaching values between 70% and 85%. At solar noon the courtyard geometry has a reduced impact. A reverse phenomenon is observed in the shade projected from the green galleries, where a larger area is shaded at solar noon, between 32% and 40%. This behaviour is consistent with the solar angle with respect to the vertical parapets during the first morning hours and last afternoon hours and to horizontal protections at solar noon.

Regarding shape, the geometry and materials used in open spaces play a primary role in their thermal behaviour. In the case of North-South-oriented courtyards, and for equal materiality of the courtyard, low geometric ratios of R1 and R2 limit solar access to a greater extent in the morning and in the afternoon. For solar protection, the green gallery shows relative efficiency because it controls direct radiation but fails to prevent the rise in exterior air temperature, since it also diminishes air movement because the heated air remains under the canopy. However, simulated data shows that, from the energetic point of view, small interior air temperature reductions imply significant energy savings to reach comfort temperatures.

At night, courtyard performance is perceptibly different due to the higher convective and radiative cooling capacity of open spaces with high R1 and R2 ratios. This benefit only impacts interior space behaviour if ventilation is used as a passive cooling strategy. All in all, results suggest that in semi-dry areas, characterized by cold winters, hot summers and a large number of clear sky days, it is necessary to adopt compromise solutions that combine blockage of the solar resource in summer with guarantee of full access in winter. To suit both strategies, exterior spaces should be designed as an open typology with efficient shading devices for summer. In this way an adequate solar access in the cold period is achieved as well as a minimization of the radiation heating effect in the hot season. When appropriate solar protection is not guaranteed, the use of a self-shading form to ensure less radiation in summer is recommended, even if this reduces access to the resource in winter.

Conclusions: Greenery in Sustainable Contemporary Architecture

The main results obtained from this review are that the greenery systems can be used as insulation, block solar radiation and reduce the wall and roof surface temperature resulting in a reduction in heat transfer through building envelopes, which can be as high as 80–90%. The greenery systems have a significant impact especially in hot and dry climates since they were able to block high solar radiation and produce the cooling effect from the evapotranspiration process.

Plants also have a significant effect during the daytime more than in night-time and summer more than in winter seasons. More foliage coverage on the wall or roof structure can have a more significant effect of external temperature reduction than no or limited coverage—this yields to save building energy consumption.

To enrich all the described benefits regarding the use of greenery to improve thermal behaviour and energy efficiency of buildings, it seems significant to conclude this chapter by highlighting how much greenery improves comfort and appeal in sustainable buildings.

In the case of cities in temperate climates such as Mendoza, Argentina, the climate is characterized by its changing conditions along the day and through the year. Cold winters and hot summers are separated by intermediate seasons in which in most part of the day exterior conditions are comfortable. With temperature differences up to 20 °C within a day, building's envelopes must integrate an interesting number of architectural filters, barriers and connecting elements to allow adaptation to these varying conditions. The architectonic result of these climatic variations is evident in the great amount and variety of buffer spaces (patios, balconies, galleries, pergolas, bowers) where life can be developed in contact with nature. Those elements confer our architecture and cities a repertoire of shapes and green solutions that are repeated and adapted over time and confer our living surroundings identity.

It is widely admitted that plants around buildings alter the adverse microclimates and make the thermal environment more pleasant and liveable. (Hoyano 1988). Architecture is the form of art with most impact on people and their perceptions. Architecture reflects, materializes and eternalizes ideas and images of ideal life. Buildings and towns enable us to structure, understand and remember the shapeless flow of reality and, ultimately, to recognize and remember who we are (…). Our living environment strengthens the existential experience, one's sense of being in the world, essentially giving rise to a strengthened experience of self (Pallasmaa 1996).

Particularly in the case of Mendoza, the city itself is structured by tree canyons conforming an urban forest. This constant relationship with greenery is built within its citizen's idiosyncrasy. After all, plants that integrate green systems are alive and their harmonic growth depends on the constant care and maintenance provided by the user. It is a significant cultural election to include greenery in our city lives. The closer we bring nature to our habitats, the better use we can make of it, gaining in physiological and psychological aspects.

References

Akbari, P. T. (2001). Cool surfaces and shade trees to reduce energy use and improve air quality in urban areas. *Solar Energy, 70*, 295–310.

Al-Azzawi, S. H. A. (1984). *A descriptive, analytical and comparative study of traditional courtyard houses and modern non-courtyard houses in Baghdad, in the context of urban design in the hot-dry climates of the sub-tropics*. London: University College.

Alchapar, N. L., Correa, E. N., & Cantón, M. A. (2014). Classification of building materials used in the urban envelopes according to their capacity for mitigation of the urban heat island in semiarid zones. *Energy and Buildings, 69*, 22–32.

Aldawoud, A. (2008). Thermal performance of courtyard buildings. *Energy and Buildings, 40*, 906–910.

Balter, J., Ganem, C., & Discoli, C. (2016). On high-rise residential buildings in an oasis-city: Thermal and energy assessment of different envelope materiality above and below tree canopy. *Energy and Buildings, 113*, 61–73.

Bellomo, A. (2003). *Pareti verdi*. Roma: Sistemi Editoriali.

Bormida, E. (1984). *Mendoza, una ciudad oasis. Revista de la Universidad de Mendoza*.

British Museum. (2020). London.

Cameron, R. W., Taylor, J., & Emmett, M. (2015). A Hedera green façade—energy performance and saving under different maritime-temperate, winter weather conditions. *Building and Environment, 92*, 111–121.

Campbell, G. S., & Norman, J. M. (1998). *An introduction to environmental biophysics* (2nd ed.). Berlin: Springer.

Cantón, M. A., Cortegoso, J. L., & De Rosa, C. (1994). Solar permeability of urban trees in cities of western Argentina. *Energy and Buildings, 20*, 219–230.

Cantón, M. A., Cortegoso, J. L., Fernández Llano, J., & de Rosa, C. (2001). Environmental and energy impact of the urban forest in arid zone cities. *Architectural Science Review, 44*, 3–16.

Cantón, M. A., Ganem, C., Barea, G., & Fernández Llano, J. (2014). Courtyards as a passive strategy in semi dry areas. *Renewable Energy, 69*, 437–446.

Coch, H. (2003). *La utilidad de los espacios inútiles. PhD Thesis*. Barcelona: UPC.

Coma, J., Pérez, G., de Gracia, A., Burés, S., Urrestarazu, M., & Cabeza, F. L. (2017). Vertical greenery systems for energy savings in buildings: A comparative study between green walls and green facades. *Building and Environment, 111*, 228–237.

Correa, E. N., Ruiz, M. A., & Cantón, M. A. (2010). Morfología forestal y confort térmico en "ciudades oasis" de zonas áridas. *Ambiente Construído, 10*(4), 119–137.

Correa, E. N., Ruiz, M. A., Cantón, M. A., & Lesino, G. (2012). Thermal comfort in forested urban canyons of low building density. An assessment for the city of Mendoza, Argentina. *Building and Environment, 58*, 219–230.

Correia Guedes, M., & Cantuária, G. (2017). The increasing demand on high rise buildings and their history. In A. Sayigh (Ed.), *Sustainable high rise buildings in urban zones*. Berlin: Springer.

Coutts, A. M., White, E. C., Tapper, N. J., Beringer, J., & Livesley, S. J. (2016). Temperature and human thermal comfort effects of street trees across three contrasting street canyon environments. *Theoretical and Applied Climatology, 124*(1), 55–68.

Dahanayake, K. W. D. K. C., & Chow, C. L. (2017). Studying the potential of energy saving through vertical greenery systems: Using EnergyPlus simulation program. *Energy and Buildings, 138*, 47–59.

Dalley, S. (1993). Ancient Mesopotamian gardens and the identification of the Hanging Gardens of Babylon resolved. *Garden History Journal, 21*(1), 1–13.

Doick, K. J., & Hutchings, T. (2013). *Air temperature regulation by urban trees and green infrastructure*. London: Forestry Commission Ed.

Foustalieraki, M., Assimakopoulos, M. N., Santamouris, M., & Pangalou, H. (2017). Energy performance of a medium scale green roof system installed on a commercial building using numerical and experimental data recorded during the cold period of the year. *Energy and Buildings, 135*, 33–38.

Ganem, C. (2006). *Rehabilitación ambiental de la envolvente de viviendas. PhD Thesis*. Barcelona: UPC.

Georgi, N. J., & Zafiriadis, K. (2006). The impact of park trees on microclimate in urban areas. *Urban Ecosystems, 9*, 195–209.

Givoni, B. (1991). Comfort, climate analysis and building design guidelines. *Energy and Buildings, 18*, 11–23.

Guo-yu, Q., Hong-yong, L., Qing-tao, Z., Wan, C., Xiao-jian, L., & Xiang-ze, L. (2013). Review effects of evapotranspiration on mitigation of urban temperature by vegetation and urban agriculture. *Integrated Agriculture, 12*(8), 1307A–1315A.

Hernández, A. (2003). GEOSOL: A computational tool for calculating solar coordinates and estimating hourly solar irradiation. *Avances Energ Renov Medio Ambiente, 7*(11), 19–24.

Hinrichs, C. L. (1987). *The architecture of the courtyard house as generated by passive solar energy and other factors*. Manchester: Manchester University.

Hoyano, A. (1988). Climatological uses of plants for solar control and the effects on the thermal environment of a building. *Energy and Buildings, 11*, 181–199.

McPherson, E. G. (1992). Accounting for benefits and costs of urban greenspace. *Landscape and Urban Planning, 22*, 41–51.

McPherson, E. G. (1994). Cooling urban heat islands with sustainable landscapes. In R. H. Platt et al. (Eds.), *The ecological city: Preserving and restoring urban biodiversity*. Amherst: University of Massachusetts Press.

Meir, I. A., Pearlmutter, D., & Etzion, Y. (1995). On the microclimatic behaviour of two semi-enclosed attached courtyards in a hot dry region. *Building and Environment, 30*(4), 563–572.

Mesa, A., & Giusso, C. (2014). La urbanización del piedemonte andino del área metropolitana de Mendoza, argentina. Vulnerabilidad y segmentación social como ejes del conflicto. *Revista Iberoamericana de Urbanismo, 11*, 63–77.

Mohsen, M. A. (1979). Solar radiation and courtyard house forms—I. A mathematical model. *Building and Environment, 14*(2), 89–106.

Muhaisen, A. S., & Gadi, M. B. (2006). Effect of courtyard proportions on solar heat gain and energy requirement in the temperate climate of Rome. *Building and Environment, 41*, 245–253.

Nangia, A. (2000). *Architecture of India: Indus Valley Civilization*. http://www.boloji.com/architecture/00002a.htm. India Nest.

Nikolopoulou, A. D. M. (2003). Vegetation in the urban environment: microclimatic analysis and benefits. *Energy and Buildings, 35*, 69–76.

Oke, T. R. (1990). *Boundary layer climates* (3rd ed.). London: Methuen.

Pallasmaa, J. (1996). *The eyes of the skin—architecture and the senses*. London: Academy Ed.

Philo of Byzantium. (2019). (V B.C.) De septem mundi miraculis. Retrieved August 29, 2019, from https://books.google.com.ar/books?id=2ZYAAAAcAAJ&printsec=frontcover&redir_esc=y#v=onepage&q&f=false.

Ponte, R. (1998). *Mendoza, Aquella ciudad de barro. Historia de una ciudad andina desde el siglo XVI hasta nuestros días* (1st ed.). Buenos Aires: CONICET.

Rizwan, A. M., Dennis, Y. C. Y., & Liu, C. (2008). A review on the generation, determination and mitigation of Urban Heat Island. *Journal of Environmental Sciences, 20*, 120–128.

Ruiz, M. A., Sosa, M. B., Correa, E. N., & Cantón, M. A. (2015). Suitable configurations for forested urban canyons to mitigate the UHI in the city of Mendoza, Argentina. *Urban Climate, 14*, 197–121.

Santamouris, M. (2001). *Energy and climate in the urban built environment*. London: James & James.

Servicio Meteorológico Nacional. (2020). Fuerza Aérea Argentina.

Seyam, S. (2019). The impact of greenery systems on building energy: Systematic review. *Building Engineering, 26*, 100887.

Smardon, R. C. (1988). Perception and aesthetics of the urban environment: Review of the role of vegetation. *Landscape and Urban Planning, 15*(1–2), 85–106.

Solecki, W. D., Rosenzweig, C., Parshall, L., Pope, G., Clark, M., Cox, J., & Wiencke, M. (2005). Mitigation of the heat island effect in urban New Jersey. *Environmental Hazards, 6*, 39–49.

Taleghani, N. (2014). *Dwelling on Courtyards. Exploring the energy efficiency and comfort potential of courtyards for dwellings in the Netherlands. Architecture and the Built Environment no. 18*. Delft: Delft University of Technology.

Taleghani, N. (2018). Outdoor thermal comfort by different heat mitigation strategies—A review. *Renewable and Sustainable Energy Reviews, 81*, 2011–2018.

Vox, G., Blanco, I., & Schettini, E. (2018). Green façades to control wall surface temperature in buildings. *Building and Environment, 129*, 154–166.

Zamani, Z., Heidari, S., & Hanachi, P. (2018). Reviewing the thermal and microclimatic function of courtyards. *Renewable and Sustainable Energy Reviews, 93*, 580–595.

Zhang, L., Deng, Z., Liang, L., Zhang, Y., Meng, Q., Wang, J., & Santamouris, M. (2019). Thermal behaviour of a vertical green facade and its impact on the indoor and outdoor thermal environment. *Energy and Buildings, 204*, 109502.

Rooftop Greenhouses: Smart and Inclusive Design

Lucia Ceccherini Nelli

Introduction

Energy is one of our greatest challenges, partly due to the increasingly urgent realities of climatic change and partly due to a failure of strategic planning. Energy needs include and exceed the scale of buildings, so this chapter explores how architecture fits into the idea of energy planning. Applying greenery to rooftop buildings (as for energy planning) inevitably leads to an energy saving as a sustainable solution.

Ideas about waste and ecosystem thinking are relevant in the design of green rooftop and it suggests six principles for energy planning rooftop solutions:

- Demand reduction through radical increases in efficiency as the first priority
- A source of energy that will last indefinitely
- Resilience through diversity and distributed networks
- Resource flows that are non-toxic and compatible with a wide range of other systems
- Larger scale schemes in master planning and urban design
- Green suitable technologies and their building integration

Contemporary urban and green designers work in urban environments with a better understanding of ecology and sustainable strategies, integrating a good quality of life for its occupants with clean air, healthy and km 0 food, access to nature, and so on.

It becomes a great challenge to design a city like an ecosystem when a lot of infrastructure already exists; the most important points to follow are:

L. C. Nelli (✉)
Department of Architecture DIDA, Nelli University of Florence, Florence, Italy
e-mail: Lucia.ceccherininelli@unifi.it

A. Sayigh, A. Trombadore (eds.), *The Importance of Greenery in Sustainable Buildings*, Innovative Renewable Energy,
https://doi.org/10.1007/978-3-030-68556-0_6

- Use waste as a resource.
- Diversify and cooperate to fully use the habitat.
- Collect and use energy efficiency.
- Use materials sparingly.
- Use greenhouses for growing plants.
- Social inclusion.
- Resilience regarding local interactions.
- Use solar energy resources for greenery production.

Designing cities for the ecological age is, in many ways, the defining challenge of our time. Greenery can help us to transform existing cities, and in some case create new sustainable environments that could be inclusive and productive.

Agriculture in the City

Roofs are one of the most unused spaces in cities, and urban rooftop farming is diffusing around cities driven by the growing interest in urban agriculture. In recent years interest in rooftops as a resource for food production is increasing, especially in the big cities, such as in the United States, Canada, and Europe. Great interest has arisen about the potential for creating new productive roofs, for transforming unused rooftops, and for adapting existing green roofs into productive green spaces.

In the last few years, productive green roofs have become a diffuse practice and so rooftop gardening and farming are evolving into practice. Open-air rooftop farming experiences are found worldwide and a range of educational and commercial projects are grown.

With the fast diffusion of many different design rooftop experiences, used for food growing, designers have created many solutions for food growing on rooftops such as:

- Rooftop farming
- Vertical farms
- Greenhouse rooftop farms

This chapter offers some reflections based on a few experiences of green planning and design. Exemplary cases of designing for urban agriculture, as well as their knowledge of productive roofs in many countries, offer a base of discussion for strategic design and green projects, interesting implications for the built environment, social inclusion, and sustainability in buildings.

Research on these forms of urban agriculture has mainly focused on theoretical and agronomic aspects. Saporito et al. (2015) reviewed current urban rooftop farming projects and discussed their contribution to sustainable and inclusive urban agriculture.

Building-integrated agriculture includes a new form of rooftop agriculture which includes rooftop greenhouses. Integrating greenhouses with buildings means to improve agriculture production with a protected space, where air, energy, and water exchanges are regularly controlled among them, creating an indoor climate, suitable for the fast growth of plants.

There are numerous advantages given by the complementarity of the technological system: integrated energy, water, and gas management; improvement of the building metabolism; renewable energy contribution if integrated into the greenhouse; and management of water resources. Rooftop productive greenhouses can be very suitable from industrial parks to commercial and office buildings (Figs. 1 and 2).

A new solar panel for greenhouse roofs that not only generates renewable electricity, but, thanks to a light-altering dye too, also helps to optimize photosynthesis in plants.

The greenhouse is integrated with LUMO panels, which are wavelength-selective photovoltaic systems (WSPVs) that feature narrow photovoltaic strips embedded in a bright magenta luminescent dye that can absorb some of the sunlight's blue and green wavelengths while converting some green light into red light, which has the highest efficiency for photosynthesis in plants. Another advantage of WSPVs is their lower cost, which is said to be about 40% less than conventional solar panels.

Location, orientation to the sun, and climate are all important factors to consider—before designing—a productive and sustainable rooftop with nature-inclusive design, and the rooftop greenhouses can easily solve climatic problems for a faster growing agriculture model. Rooftops can be turned into lush gardens with attractive functions and connection to nature, with a good insulation, stormwater recovery system, climate control, energy savings, better views, and added real estate value, and they are all factors involved in the design of the rooftop urban farm. From a social perspective, gardening on a common area such as residential building rooftop, if shared by tenants, provides an opportunity for meeting people and sharing knowledge on gardening. From an ecological perspective, rooftops can attract pollinators and birds and create a more biodiverse ecosystem in the city skyline (Orsini et al. 2014). Gardening in buildings with the creation of green roofs can do more

Fig. 1 Photovoltaic integration in a greenhouse for urban farming

Fig. 2 Greenhouse integrated with solar energy panels; red modules help plants to optimize photosynthesis

than just providing food, with vegetable and fruits, especially for socialization and social inclusion. Roof gardens can also contribute to sustainability in building with climate control, water recycling, and reducing heat island in cities. Shading a roof surface can reduce the roof temperature; the green roof soil can absorb rainwater reducing surface runoff and improve the thermal insulation of the building rooftop, leading to lower energy consumption and thus lower carbon emissions (De Filippi and Saporito 2018).

Edible Roof

Designing green roofs is an opportunity for urban farming to grow local vegetables and fruits for the surrounding area. Depending on the size, structure, and amount of production, these productive roofs can supply local restaurants and markets, or be used directly by the building occupants. Many of them born for the production of the restaurant on-site can open the garden to the clients and let them know the km 0 vegetable production.Intensive roof gardens grow as rooftop farms with many environmental benefits:

- Energy saving
- Stormwater management
- Creation of habitats for biodiversity

As high-tech greenhouses, these year-round production facilities save water and support reliable and predictable food security. With greater climate control, inorganic fertilization and pesticides can be avoided. The local vegetable and fruit production produces environmental benefits without transportation, packaging, and refrigeration.As for the edible rooftop agriculture, we can summarize five principal models:

- Community rooftop farm
- Rooftop farming and restaurant
- Rooftop farming and market
- Rooftop hôtellerie-restaurant-café
- Integrated rooftop

The community rooftop farm is a sort of economic model of growing for agriculture and food distribution, which is directly connected to the local farmers and consumers. The members pay a membership fee at the start of the growing season or yearly for a share of the expected harvest. Once the season begins, members can pick up their share of vegetables, eggs, honey, etc. each week. They also have chances to sow seeds and take care of the plants through the season and take part in the harvest. Denmark's first-ever *rooftop farming* brings local food production to the city with support from the municipality and the local *community*. The rooftop farm was established with the support of the Copenhagen Municipality, but nowadays the project is self-sustainable and financed through revenue educational activities and membership fees from the families involved among other revenue streams.

From a social perspective, the community garden acts as a shared meeting place for getting together and making friends. A sort of horticulture therapy and built socially inclusive and resilient communities through effective collaborative efforts and partnerships. Volunteers participate in the community garden programme for urban food production, showing their commitment to social responsibility and creating new collaboration forms to grow the community. People naturally interact with a natural setting and feel connected, revitalized, relaxed, and inspired (Wilkinson and Dixon 2016) (Fig. 3).

ØsterGRO was started in 2014 and is the first rooftop farm in Denmark, established on top of the old car-auction house, Nellemannhuset, in the heart of Copenhagen's Climate Neighbourhood. ØsterGRO covers 600 m^2 with fields of organic vegetables, herbs, and eatable flowers; a greenhouse; henhouse; and three beehives. ØsterGRO is organized as community-supported agriculture (CSA) and sells its produce, in cooperation with the urban agriculture Seerupgaard in Dragør, to 40 members who pay for one year at a time for a share of the harvest. But ØsterGRO is not only for the members and volunteers who participate in the project. It is a green breathing space for the whole city. Situated on the roof is the restaurant Gro Spiseri; as part of the experience, guests receive a tour of the farm. This is just one of the many ways in which the green urban space educates its 13,000 annual visitors about food waste and sustainable food production. At the same time, **the farm contributes to climate proofing the municipality.** Below the raised flower and vegetable beds, there is a 350 m2 water reservoir where rainfall is collected for the irrigation of plants during the growing season, https://www.oestergro.dk/ (Fig. 4).

Fig. 3 Aerial view of the rooftop of the community farm ØsterGRO in Copenhagen

Fig. 4 The greenhouse in the middle of the roof with the restaurant

Rooftop farming and restaurant, a diffuse model in these years, is a local food production to provide fresh and good choice quality and variety to eat together as a community, a km 0 production without packaging. In the case of MOSS, clients and associate members can pair up with chefs from the early phase so they can plan out the garden based on the menu's seasonal taste (Figs. 5, 6, and 7).

Another example of a roof with edible plants protected by a glass greenhouse is the terrace of the Triennale building in Milan, Italy—the Palazzo dell'Arte is a big terrace. The rooftop garden emphasizes the importance of food, all kinds of edible plants are used, and their fragrances and colours enrich the elegant rooms of the glass pavilion by marking the passage of the seasons. DESIGNERS Studio Antonio Perazzi (Figs. 8, 9, 10, and 11).

Fig. 5 Restaurant Vermeer. A Michelin star restaurant in Amsterdam designed with rooftop production of herbs with a circular urban farming system

Fig. 6 Detail of seasonal plants to use in the Vermeer Restaurant

Fig. 7 Palazzo dell'Arte for the Triennale in Milan, internal view of the terrace

Fig. 8 View of the terrace with the edible plants, vegetation is protected by the greenhouse

Los Angeles called SynthE AeroGardens, on a building known as The Flat, which, by the way, is the home of the famous Blue Velvet restaurant. The roof was designed by Southern California Institute of Architecture (SCI-Arc) professor Alexis Rochas. As one can see from the pictures, there are furrows which are filled with edible plants such as herbs and lettuces.

The garden is maintained and enjoyed by both the residential tenets and the chefs and patrons of Blue Velvet restaurant. Additionally, the waste products from the garden will be composted and returned to the garden.

Fig. 9 Edible plants for the Los Angeles Blue Velvet restaurant

Fig. 10 Architectural integration of the green roof with edible plants, example of the complex roofs

Rochas's plan was to use the space to reduce the heating and cooling costs of structure, as well as for better management of the stormwater runoff this building produces. Instead of using regular soil, the garden uses a synthetic growing medium.

Rooftop farming and market produce agriculture in cities; products are sold directly to supermarkets, farm stands, and farmers' markets. Now there is a rising interest for supermarkets to offer locally grown, organic food. The market model of

Fig. 11 The chef of the Blue Velvet restaurant in Los Angeles, organic Garden Edible Green Roof SynthE in Los Angeles on the Blue Velvet restaurant

urban farming business is successful in numerous cases, particularly in North America and Canada.

Gotham Greens Whole Foods in Brooklyn, NY: This greenhouse is technologically advanced; data-driven, climate-controlled facilities use the most efficient production systems. This greenhouse is one of the highest-yielding farms around, and uses less energy, land, and water than other farming techniques. Gotham Greens' produce is grown using hydroponic systems in 100% renewable electricity-powered greenhouses that use 95% less water and 97% less land than conventional farming. Further, controlled-environment agriculture (CEA) producers have formed a coalition to work with the government, academics, and industry leaders to strengthen food safety standards for leafy greens. The greenhouse is 20,000 square feet atop the new supermarket building. The greenhouse and the market are full of energy-saving, sustainable improvements. The market operates 60% more efficiently than the average US grocery store, with yearly cost savings of $369,300, according to the New York State Energy Research and Development Authority (https://www.archdaily.com/636587/research-center-icta-icp-uab-h-arquitectes-dataae/5567b7d2e58ececc6c0000d5-research-center-icta-icp-uab-h-arquitectes-dataae-photo). The greenhouse acts similarly to vegetated green roofs, by helping to mitigate the urban heat island effect while insulating the building below and is LEED accredited. By growing and marketing its crops locally, it eliminates the need for long-distance food transportation and sharply reduces transportation fuel consumption and carbon emissions associated with food miles (Figs. 12, 13, 14, 15, 16, 17, 18, 19, and 20).

Fig. 12 Gotham Greens Whole Foods. Main entrance in Brooklyn, NY

Fig. 13 Detail of the greenhouse

Rooftops and Horeca: This model is extended to all buildings that host hotels, restaurants, and cafés. This model is created to host guests in a roof garden to have a good view over the city, combining comfort, accessibility, professional maintenance, and good, fresh food and drink, a mixed concept design within a rooftop bar or restaurant

Fig. 14 Internal view of the greenhouse of Gotham Greens Whole Foods. Emissions are eliminated to ensure product freshness, quality, and nutrition for thousands of customers in the area

Fig. 15 Internal view of the hydroponic system of growing

Integrated Rooftop Greenhouse

This model integrates building rooftop with greenhouses for a harvest urban farming agriculture, using extreme agronomic techniques of growth and technological system integrated with renewable energies and through the use of saving energy strategies to reduce energy consumption in the production of green products.

The greenhouse (IRTG) is a research centre (Marius et al. 2019), connected with the interior of the building by controlled flows of energy, oxygen, carbon dioxide,

Fig. 16 View of the rooftop greenhouse with the photovoltaic plant. The electrical needs of Gotham's current Greenpoint greenhouse facility are met, in part, by a 56 kW on-site solar PV system that produces 70,000 kWh of electricity annually

Fig. 17 General view of the building with greenhouses and the PV plant

and water. The main IRTG feature is the tight human-plant symbiosis ensured by a two-flow ventilation system, conveying O2-enriched air from RTG to building and CO2-enriched air from building to RTG. To be applicable at large scale with no particular infrastructure demands, IRTG systems are provided with renewable energy devices suited to the local climate, heat pumps (water to water for building's basement and air to air for greenhouse), solar panels, etc., and also with the network

Fig. 18 HOPP at CASA400 Amsterdam. Thanks to an excellent marketing and trendy food and beverage concept, this project earned its ROI within two 7-month seasons, showing how an attractive roof garden magnetizes business

of things equipment. In such way IRTGs can harvest local renewable energy resources (geothermal, solar, wind, etc.), store them, and manage them together with the water resources (Figs. 21 and 22).

Due to the smart use of renewable energies, iRTG energetic footprint is very low so the existing buildings could be upgraded to iRTGs without perturbing the existing urban infrastructure, according to their condition. This is leading us towards the possibility to extend the use of iRTG at the scale of a whole city.

Edible Roofs and Nature-Inclusive Design Study Cases

LAVA Designs Zero-Carbon 'LIFE Hamburg' with Waterholes and an Edible Roof

LAVA (laboratory for visionary architecture) has collaborated with urban agricultural collective city plot to design LIFE Hamburg, a new self-sufficient 'education campus' with a new type of landscape; the structure will open in 2023; the infinity loop-shaped building comprises three levels that will accommodate 1600 students, bringing them together upon the backdrop of a continuous landscape that merges both inside and out.Status: invited competition 2019; feasibility 2019; construction finished 2023

Size: 12,000 m^2

Fig. 19 Aerial view of the productive rooftop

Partners: SBP (Schlaich Bergermann Partner); Transsolar image credits: LAVA (Figs. 23, 24, and 25).

The architects have designed the three-storey campus with a load-bearing wood structure and a highly insulated glazed shell that permits a lot of natural light and the envelope of the building uses a highly insulated glazed shell.

A continuous garden landscape flows from outside to inside—first to the central agora and then on the open terraces and up to the rooftop garden; there are many edible gardens of herbs, views, and space for outdoor learning. The building includes a carbon dioxide-absorbing facade made up by seasonal greenery balconies.

Zoku (Metro Pool Building) Roof Garden Zoku (Metropoolgebouw—Metro Pool Building) Amsterdam, The Netherlands 5,382 sf Greenroof

Located on the Weesperstraat, one of Amsterdam's main arteries, the Metro Pool Building stretches over seven levels and it is a designated national Dutch monument which recently underwent renovations, including a spacious roof garden. In 2016 a new long-stay hotel type emerged that embraces the new international life and work concept.

Zoku offers smart lofts with ingenious use of space and common areas of the hotel to work, connect, and facilitate 24/7. Zoku, Japanese for family, tribe, or clan,

Fig. 20 Atrium view with the internal garden

Fig. 21 Research Center ICTA-ICP

Fig. 22 The integrated rooftop greenhouse (IRTG), resulting after connecting a rooftop greenhouse to the underneath building, by controlled flows of energy, water, and gases: O_2-enriched air from rooftop greenhouse to building and CO_2-enriched air from building to the rooftop greenhouse. IRTG supports an efficient integrated management of energy, water, CO_2, and O_2 that collects the disposable renewable energy resources: sun (by greenhouse effect), wind (by air-to-air heat pumps), geothermal (by water-to-water air heat pumps), and water resources (rainwater, household grey water, water vapours), storing and delivering them wherever and whenever needed inside the IRTG system

Fig. 23 External view of the building with the photovoltaic plants and the edible food garden

is a hybrid between a 'home-away-from-home' and 'office-away-from-the-office' for people who live and work in Amsterdam from 5 days to 3 months. In the Green Key Gold-certified hotel, the light and airy entrance is located at the roof level where the roof as a garden and galvanized steel greenhouse are filled with edible greenery to greet visitors and residents.

Facilitating global living and working for the travelling professional, Zoku believes in offering its residents greener spaces to better experience the seasons in an urban environment.

Fig. 24 View of the balconies with vegetation

Fig. 25 Internal view of the roof with the waterholes

Original 1960 Building Architect: Arthur Staal

Architect: Mulderblauw architecten

Green-Roof Design: Concrete Architectural Associates & De Dakdokters

Green Roof and Greenhouse Plant Design, Installation, and Maintenance: HRBS.

Modular Green-Roof System: HRBS.**Greenhouse Box Construction:** Maurice Kassenbouw

Construction: Van Rossum Consulting Engineers BV (Figs. 26, 27, 28, and 29).

Described as a complete ecosystem, Zoku is a truly mixed-use building with an urban greenhouse paradise and its series of outdoor micro-roof gardens for respite and relaxation. Guests can enjoy a cup of coffee or a glass of wine on the roof garden, beautifully lit at night while gazing out over the picturesque Amsterdam skyline.

Figs. 26 and 27 Greenhouse boxes

Fig. 28 Axonometric view of the rooftop with the greenhouse boxes

Fig. 29 A different kind of plants is cultivated in front of the micro-roof gardens

Fig. 30 Dynamic façade detail

Fig. 31 ICTA-ICP building, general view

Greenhouse Research Center ICTA-ICP UAB/H Arquitectes + DATAAE—Cerdanyola, Spain, 2014 (Figs. 30 and 31)

The ICTA-ICP building, located in the UAB Campus (Universitat Autònoma de Barcelona), is a research centre in environmental sciences and palaeontology. The building has an isolated volume of five floors of 40 × 40 m^2 and two basements. On the ground floor we have the hall, bar, classrooms, meeting rooms, and administration area; the next three floors hold the offices and laboratories; on the roof, there is a big rooftop farm mixed together with other spaces. The building has been designed to be an adaptable and flexible infrastructure able to be changed in the use (Marius et al. 2019).

The building has been designed to host three types of climates associated with different intensities of use:

Climate A: in-between spaces that are exclusively acclimatized/heated by passive and bioclimatic systemsClimate B: offices that combine natural ventilation with radiant and semi-passive systems

Climate C: laboratories and classrooms that have more hermetic and conventional functioning

Each type of climate has its associated systems. The behaviour of the building is monitored and controlled by an automatic computer system that processes and manages an important set of information to optimize both comfort and energy consumption.

Fig. 32 Internal view of the rooftop greenhouse

The system has been programmed in favour of the maximum passive comfort of the building and to minimize the use of non-renewable energy sources. The building reacts and adapts constantly with openings, managing to use all the natural possibilities offered by the environment; therefore the comfort perception is much more real, less artificial than usual (Marius et al. 2019).

A mineral material with a lot of thermal inertia and long service life has been chosen for the structure combined with low-environmental-impact materials for the secondary partitions. The use of organic or recycled materials and dry constructive systems has been a priority.

The building optimizes the whole water cycle by reducing the demand and consumption through the reuse of rainwater, grey water, yellow water, and waste water (Figs. 32 and 33).

Lufa Farms, Montreal, Quebec

Lufa Farms is an agricultural and technological company located in the Ahuntsic-Cartierville neighbourhood of Montreal, Quebec. The company, founded in 2009, has installed commercial greenhouses on the roofs of several warehouses in greater Montreal, beginning with a 31,000 square feet greenhouse over their headquarters in 2011.

Lufa Farms breaks ground on their fourth greenhouse, making Montreal home to the world's largest rooftop farm.

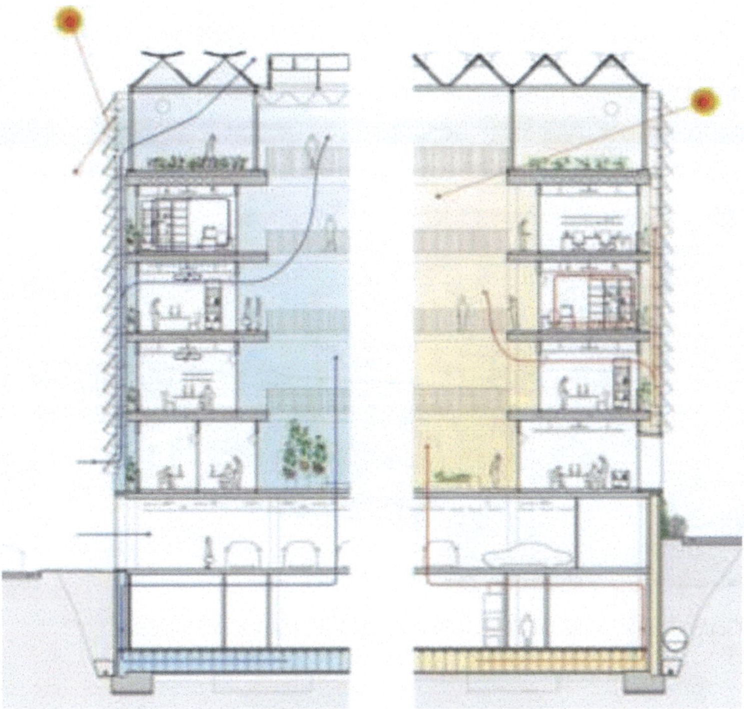

Fig. 33 Detailed section with the scheme of the bioclimatic and sustainable functions

The greenhouse measures 163,800 square feet, making it the largest rooftop farm in the world (Figs. 34 and 35).

This greenhouse repurposes an existing industrial rooftop to further its vision of growing food where people live being sustainable.

The greenhouse is built with double-paned glass and two sets of energy-saving screens for improved insulation, and its integration with the building below provides additional thermal benefit to both structures. The greenhouse will also capture rainwater to be used in the closed-loop irrigation system, as well as offset waste with an on-site composting system.

Lufa Farms' mission is to create a better food system by growing food sustainably on city rooftops and collaborating with hundreds of farmers and food makers, to provide customers with fresh, local, responsible food via their online marketplace (Arosemena 2012).

Lufa Farms was started in Montreal in 2010 and operates two greenhouses that harvest a total of 190 metric tons of products each year, feeding 3000 people. They hope to expand to other cities (Figs. 36, 37, and 38).

Fig. 34 General view of the building with the roof greenhouses in Montreal

Fig. 35 Internal view

Figs. 36 and 37 Lufa Farms—Laval rooftop greenhouse. Year built: 2013. Size: 43,000 square feet Production: Tomatoes and eggplants

Fig. 38 The greenhouse maximizes energy savings and creates an optimal environment for crops to thrive. Hydroponic system for plant growing

Viparis Project: Rooftop Farm Greenhouse in Paris, France

1. The rooftop view over one of the most densely populated cities in Europe: It is an unlikely place to provide fresh food and green spaces but the high-rise Porte de Versailles in Paris is where you will find the world's largest urban farm (https://www.lonelyplanet.com/articles/agripolis-urban-farm-paris).

Paris is cutting back on its carbon footprint by growing its own products in the city. Porte de Versailles in the 15th arrondissement, a sprawling cultural complex just 15 min from the Eiffel Tower, is the site of the world's largest urban farm. By spring 2020, the rooftop of the six-storey building will be transformed into a green roof. Le Perchoir, Paris's renowned chain of rooftop venues, will open a bar and

Figs. 39 and 40 Aerial view of the Viparis project with the rooftop greenhouses

Fig. 41 The Romainville district in Paris with the vertical farm design

restaurant on the panoramic terrace, with a farm-to-fork menu that includes products grown on-site. The rooftop farm will provide food to restaurants, hotels, and supermarkets in southern Paris ©Viparis.

Occupying 14,000 square metres of space (about the size of two football pitches), the farm will be home to more than 20 market gardens providing more than 2000 pounds of fruit and vegetables a day per season, from about 30 different varieties of plants. The plants will grow vertically through aeroponic farming, a method that uses a solution of nutrient-rich mist and rainwater. Using only 10% of the water needed in traditional agricultural settings, it is a more sustainable method of farming. Taking things to the next level, the farm will offer a range of services related to

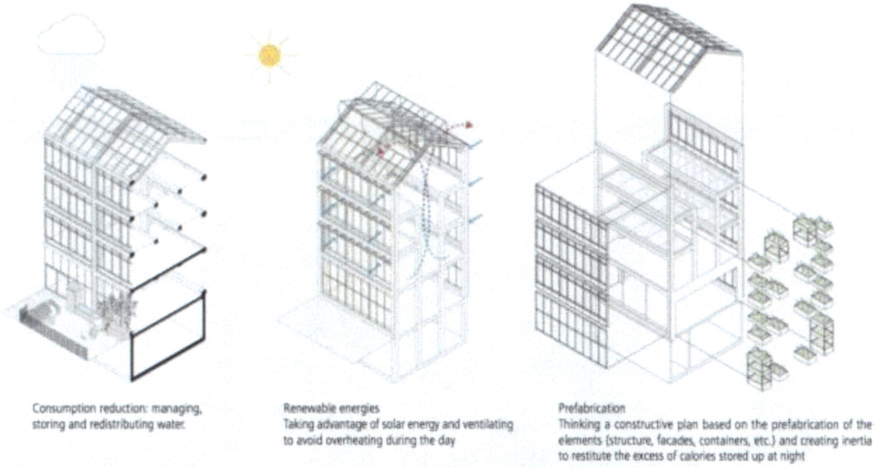

Fig. 42 Detailed section of the bioclimatic design and plant distribution

Fig. 43 Internal view of the farm

urban agriculture: educational tours and team-building workshops for companies. Local residents will be able to secure about 140 vegetable plots, effectively turning the garden into a community space. Locals can lease vegetable plots and Viparis is teamed up with two companies specializing in urban agriculture, Agripolis and Cultures en Ville; they bring the rooftop farm to life. The urban rooftop farm will open towards spring 2020 (Figs. 39 and 40).

Fig. 44 General view of the building (photo Ilimelgo)

Vertical Farming Project in Romainville District, Paris (Fig. 41)

Firm Ilimelgo Architects reimagines the Romainville district with an urban agriculture, a vertical farm complex, in the Parisian suburb. The project integrates the production of green products into the city through a 1000 square metre greenhouse that maximizes sunlight and natural ventilation. Recognizing the developing world's diminishing agricultural space, the project aims to meet the growing demands for crop cultivation in urban environments.

Conscious of this need, Romainville has been supporting sustainable and forward-thinking alternatives to small plot-based agriculture. Though they have implemented many rooftop and allotment gardens in the past decade, the vertical farm represents a comprehensive dedication to sustainability, education, and local economic participation. The building is split into two wings to aid crop growth, taking advantage of sunlight and limiting shade. Organic building materials such as straw bale and wood fibre insulation add to the sustainability of the project.

The form of the vertical farm, a rectangular prism with a triangular roofline, is a reference to the existing architecture in the area (Figs. 42, 43, and 44).

The upper floors feature spaces for bio-intensive farming using culture containers. Specially irrigated to provide healthy environments for specific crops, the containers also allow for a flexible and dynamic organization of space. The facility houses a mushroom farm, orchards, a henhouse, and laboratories that experiment in seed germination.

Conclusions

The projects highlighted are examples of productive rooftop gardens, with great attention to structure, climate, access, and design combined with creativity and a well-thought-out maintenance scheme.

Those that were once grey roofs are now productive and green spaces with plants, food, and social activities. Growing food offers a special experience in the rooftop garden while insulating the building and hosting biodiversity.

Urban farming is an emerging technology aiming to increase crop production per unit area of land in response to heightened pressure on agricultural production using protected horticulture systems such as glasshouses and controlled environment facilities in combination with different levels of growth surface, and indoor and outdoor surfaces.

Greenhouse farming necessitates a combined technical approach for many different factors, including lighting, special growing system, energy efficiency, construction, and site selection.

The technical and economic optimization of roof farming requires further attention with additional research into maximizing productivity and reducing system costs being required (Wilkinson and Dixon 2016).

The rooftop gardens produce food for hospitals, markets, universities, research centres, and communities giving great contribution to sustainability, social networking, social inclusion, community engagement, and education.

Several researches have found out that food production on buildings creates successful social connections, leads to better prospects, and promotes self-sufficiency of food (Saporito 2017; Montero et al. 2017). Urban rooftop farming is gaining importance as a way to further develop local food production. To assess the sustainability of implementing urban rooftop farming projects, a multidisciplinary methodological scheme was designed to approach such a complex system. The method combines four disciplines as follows: (a) qualitative research, to evaluate the qualitative potential by evaluating the perceptions of different stakeholders; (b) geographic information systems (GIS), to quantify the potential roofs for implementing URF; (c) life cycle assessment (LCA), to quantify the environmental impacts of URF forms; and (d) life cycle costing (LCC), to quantify the economic costs of URF forms. Results highlighted the potential of urban rooftop farming in both qualitative and quantitative terms and the potential benefits of different urban rooftop farming types. Urban rooftop farming can contribute to sustainability; available and feasible spaces can be found in retail and industrial parks to deploy commercial, urban rooftop farming activities through rooftop greenhouses (Montero et al. 2017).

Links

1. https://www.tandfonline.com/doi/full/10.1080/14620316.2019.1574214#_i17
2. http://www.aesoptorino2015.it/content/download/458/2441/version/1/file/33_nasr_impagi nato_A.pdf
3. https://link.springer.com/article/10.1007/s13593-015-0331-0#CR44
4. http://www.fairmont.com/
5. The Eagle Street rooftop farm. http://rooftopfarms.org/
6. The Brooklyn Grange rooftop farms of New York (USA). http://brooklyngrangefarm.com/
7. https://www.lonelyplanet.com/articles/agripolis-urban-farm-paris
8. https://books.google.it/books?id=QtpCDwAAQBAJ&pg=PA249&lpg=PA249&dq=edible+food+rooftop&source=bl&ots=kz3hKUo-YI&sig=ACfU3U36MJcFi7hTwz71MmJ3eIg4C9Oc7Q&hl=it&sa=X&ved=2ahUKEwjgqYzv4NjmAhXD5KQKHUn8D0AQ6AEwEXoECAoQAQ#v=onepage&q=edible%20food%20roofto p&f=false
9. https://www.greenhousegrower.com/crops/vegetables/gotham-greens-takes-locally-grown-produce-to-a-whole-new-level/
10. https://www.archdaily.com/636587/research-center-icta-icp-uab-h-arquitectes-dataae/5567b7d2e58ececc6c0000d5-research-center-icta-icp-uab-h-arquitectes-dataae-photo
11. http://www.soliculture.com/lumo-technology/

References

Arosemena, G. (2012). *Urban agriculture: Spaces of cultivation for a sustainable city*. Editorial Gustavo Gili (WD).

De Filippi, F. & Saporito, E. (2018). Urban regeneration through urban farming. Turin and Paris: two case studies. Franco Angeli. https://doi.org/10.3280/TR2018-087020

Marius, M., Nikolic, J., Lile, R., Beiu, R., & Popa, M. (2019). Integrated rooftop greenhouses and green skyline cities. In *Conference: International multidisciplinary scientific geoconference surveying geology and mining ecology management*, SGEM, Vienna.

Montero, J. I., Baeza, E., Muñoz, P., & Stanghellini, C. (2017). Technology for rooftop greenhouses. In *Rooftop urban agriculture*. https://doi.org/10.1007/978-3-319-57720-3_6

Orsini, F. & Dubbeling, M. (2017). *Rooftop urban agriculture*. Springer.

Saporito, E. (2017). OrtiAlti as urban regeneration devices: An action-research study on rooftop farming in Turin. *Future of Food: Journal on Food, Agriculture and Society, 5*(1), 59–69.

Wilkinson, S., & Dixon, T. (2016). *Green roof retrofit building urban resilience*. Hoboken: Wiley.

Urban Green Coverage: Importance of Green Roofs and Urban Farming Policies in Enhancing Liveability in Buildings and Cities—Global and Regional Outlook

Mohsen Aboulnaga and Heba Fouad

Introduction

Green roofs have great effect on climate, construction, and social and economic factors and they have great benefits. Green roofs assist in mitigating urban heat island effect in cities. According to Environmental Protection Agency (EPA) in the USA, using green roofs can lower temperatures to 4.5 °C lower than those of conventional roofs. It can also decrease citywide ambient temperatures (General Services Administration 2011; Santamouris 2014). Also, green roofs can reduce building energy use by 0.7% compared to conventional roofs, reducing peak electricity demand and leading to an annual savings of US $0.23 per square foot of the roof's surface. These temperature reduction and energy efficiency benefits are key contributors to the growing popularity of green roofs in the USA (General Services Administration 2011; Sailor et al. 2011). In 2016, the North American green roof industry was estimated to grow by more than 10% over 2015, continuing industry's growth trend over the past decade. Nearly 900 green roof projects have been reported (more than four million square feet in 40 US states and 6 Canadian provinces) (Green Roofs for Healthy Cities 2017). Green roofs are also known as living roofs or eco-roofs, and are innovative and effective strategies to improve the built environment and mitigate UHIE in big cities. Advantages of green roofs are listed in (Patnaik et al. 2018). There are many benefits that green roofs can offer. Figure 1 illustrates the benefits of green roofs.

Green roofs globally are governed by policies and laws that are enacted by many countries and governments to enhance the use of green roofs in cities in order to

M. Aboulnaga (✉)
Department of Architecture, Faculty of Engineering, Cairo University, Giza, Egypt

H. Fouad
Environmental Design and Energy Efficiency Programme, Department of Architecture, Faculty of Engineering, Cairo University, Giza, Egypt

A. Sayigh, A. Trombadore (eds.), *The Importance of Greenery in Sustainable Buildings*, Innovative Renewable Energy,
https://doi.org/10.1007/978-3-030-68556-0_7

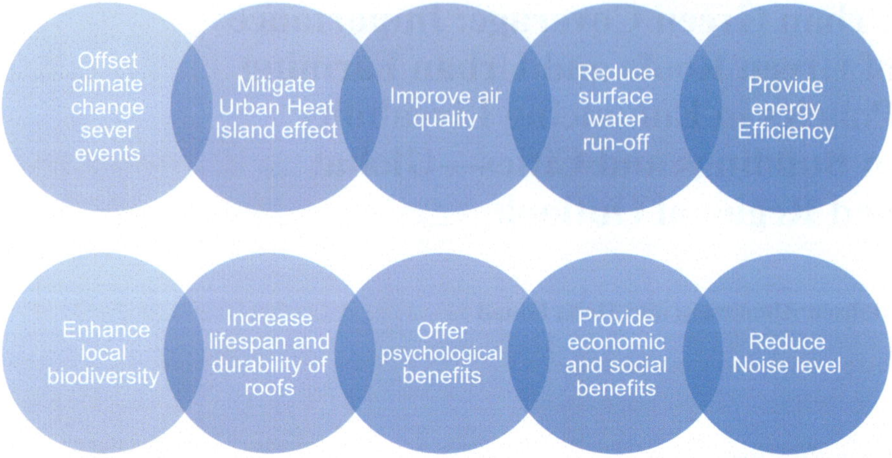

Fig. 1 Benefits of green coverage in cities. (**a**) Green walls in Paris, France. (**b**) Green roofs, EWHA University, Seoul, SK

a. Green walls in Paris France   b. Green roofs, EWHA University, Seoul, SK

Fig. 2 Urban green coverage in cities. *Images source: Authors*

obtain the best advantage of their climatic, construction, social, economic and aesthetical benefits. Figure 2 presents green urban coverage in Paris, France, and in Seoul, South Korea.

Policies and Laws Governing the Design and Implementation of Green Roofs

The design and implementation of green roofs are subject to concerned authorities (municipalities) that grant permits and approvals. The implementation of green roofs should refer to these authorities' required information before initializing any project containing green roofs to follow the policies and be free of defects.

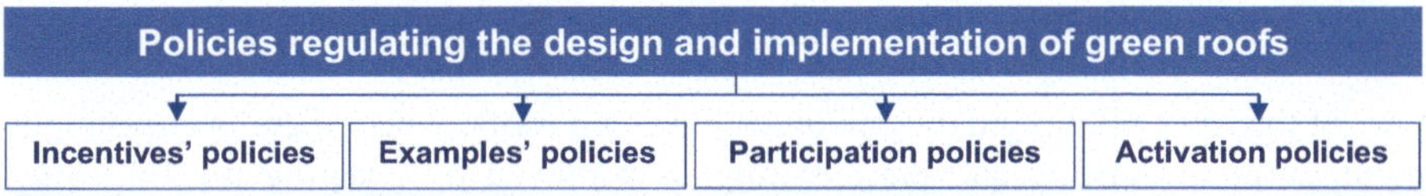

Fig. 3 Policies' types for implementing green roofs in Europe

The policies and legislation that regulate green roofs are more than similar policies concerned with green walls and green facades in some countries and cities around the world. Analysing and studying the experience and challenges in the implementation of policies that support and encourage green roofs are important to get various advantages. The general policies that govern green roofs can be divided into four categories (incentives, best practice (examples), participation, activation) (Carter and Fowler 2008) as shown in Fig. 3. The first policy was developed in the United Kingdom in 2009 by the experience centre in affecting behaviour. Twelve options were put under four policies: incentives, examples, participation and activation categories. In January 2016, San Francesco was the first city in the USA to require between 15 and 30 of the new buildings' roof to include green roofs and solar panels or a blend of both (https://www.nationalgeographic.com/news/2016/10/san-francisco-green-roof-law 2019).

Incentives' Policies

It is a model of the most important regulating policies of green roofs that include the stimulation and encouragement of users to build green roofs by:

- Providing discount fees and local taxes for buildings with green roofs
- Directing financial support for green roofs (Carter and Fowler 2008)
- Encouraging fast response to get approval statements

Examples' Policies

It is an important model of green roof regulating policies that clear out the common responsibility and sharing examples to be followed in future practice. Also, inferring strategies and policies regulating green roofs are through the following two points:

- Location of green roofs that provide models and suggestions to construct and implement such green coverage (green roofs)
- Coordinating strategies and policies on governmental level (Growing Green Guide 2014a)

Participation Policies

One of the most vital regulating policies for green roofs, where there is encouragement towards working and participating, is achieved by enacting the following three points:

- Promotion and encouragement for general discussion around green roofs
- Working with construction industry groups to merge walls, roof and facades in the principal guidelines
- Setting criteria for encouraging organizations and civil associations to use green roofs (Carter and Fowler 2008)

Activation Policies

The last policy type is considered one of the models regulating green roofs that enhance the activation of suggestions and their implementation to grant the ability to operate and build agreements, providing infrastructure or applicable alternatives and education and training through the following three points:

- Assuring the flexibility and availability of the approval statement process
- Providing data and support to residents and companies alike
- Constructing an experimental space for policy tools (Carter and Fowler 2008)

It was imperative to assess and display green roof policies around the world, regionally and locally, as well as to review the enhancement methods used in each country in promoting and implementing urban coverage (green roofs and green walls). Figure 4 illustrates the cities and continents where the green roof policies were developed and implemented. In this study, 50 cities in six continents (North America and South America, Europe, Africa, Asia and Australia) were selected to

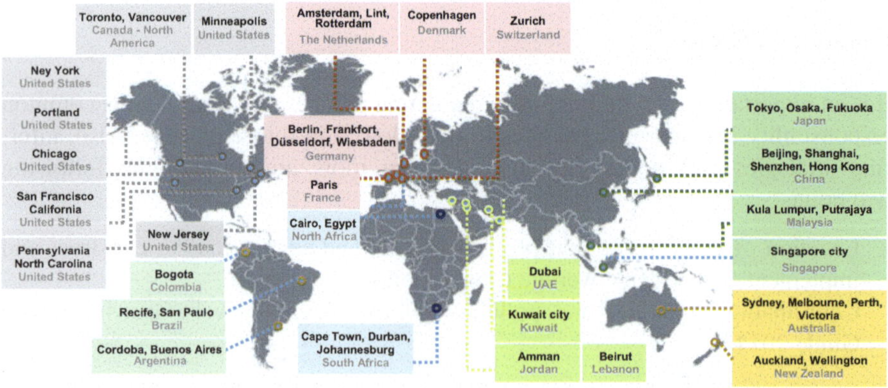

Fig. 4 Cities selected to study the policies regulating green roofs at international, regional and local levels. (*Source: Authors*)

review and present the various policies governing the design, development and execution of green roofs in these cities.

Policies and Legislation Governing Green Roofs Internationally

Multiple countries considered the enhancement of green roofs and enacted governing policies for implementing green roofs. These are either to reduce carbon dioxide (CO2) emissions or to provide benefits in terms of construction, health, psychological, economic, social and aesthetic benefits. The concern of these countries differs in the extent to which they are interested in the idea of green roofs. The following section displays policies in European Union countries.

Policies Governing Green Roofs in European Countries

European countries are those nations that follow the European Union policies. Figure 5 presents the countries that follow the EU policies or laws for green roofs and that which are not following the EU laws.

Policies Regulating Green Roofs in European Union Countries

European Union countries (namely Germany, the Netherlands, France, Denmark, England, Italy) strongly value sustainability and have signed Lisbon Agreement II, which aims at human dignity, freedom, democracy, equality and rule of law. The treaty is concerned with the following main points:

- Legal commitments for sustainable development
- Benefits of public safety and sustainability (Ngan 2004a)
- Strategies for adaptation of climate change
- Attention to the benefits of green roofs for many of the active parties that must be involved and in cooperation

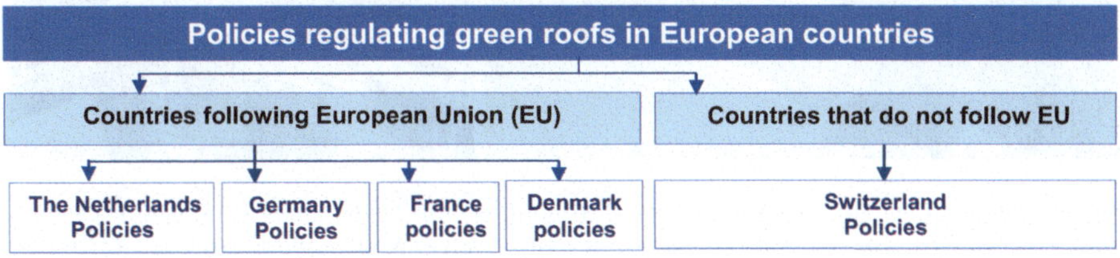

Fig. 5 Countries following policies regulating green roofs

For the last point (green roofs), e.g., it should have the following: (a) reduction in water discharge at the time of floods is the responsibility of water boards and a decrease in fine particles; (b) decrease in the impact of heat islands is the responsibility of municipality; and (c) low energy cost is the responsibility of users due to green surfaces that help and contribute directly to the achievement of Sustainable Development Goals—SDGs (Ngan 2004a).

Policies Governing Green Roofs in the Netherlands

The Netherlands established a target policy to achieve a 50% reduction in CO2 emissions compared to the level of 1990 and also to make the city of Rotterdam a 100% adaptive city to climate change and heat islands by 2030 (van Genuchten et al. 2019). This would assist in making the city and nation alike able to deal with any future severe climate events. The main reason of using green roofs is to increase the ability to retain storm water. Since 2008 to mid-2012, more than 100,000 square metres of green roofs were achieved; nevertheless, the year 2011 is the best till now. Also, an innovative initiative was implemented in the city of Rotterdam, where the largest green roof—DakAkker rooftop urban farm—was built in the city. The innovative green roof is the largest rooftop farm in the city and it was built on the roof of an abundant office building in Rotterdam. Such roof holds activities and city's residents can eat and enjoy life under clean air. The rooftop farm also provides food supply as well as climate action (van Genuchten et al. 2019). Figure 6 shows the rooftop and activities in the city of Rotterdam.

Although the policy for green roofs was achieved, Rotterdam city's policy was not fully active except only for over 5 years. There were only 32 projects implemented in this regard. Nevertheless, the target was reduced from 800,000 to 290,000 m^2 (European green capital award 2016). In addition, several policies that govern green roofs in the Netherlands promote the following specific issues:

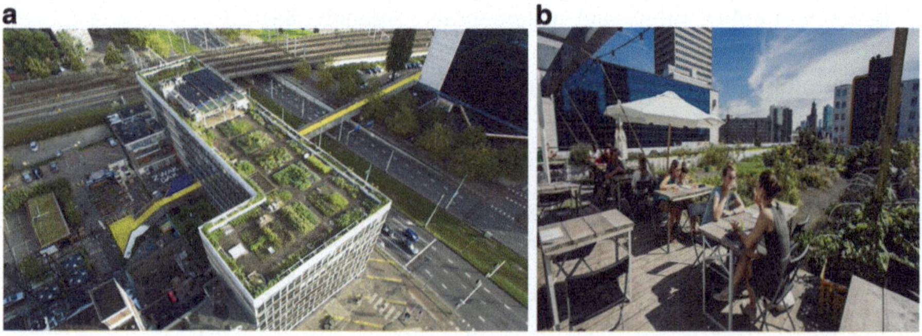

Fig. 6 Largest urban green roof in the city of Rotterdam—the Netherlands. (**a**) DakAkker largest rooftop urban farm. (**b**) Activities on the green roof. (*Source: Sustainability 2019 11, 3310; doi:* https://doi.org/10.3390/su11123310—www.mdpi/journal/sustainability. *Image sources:* http://www.foodurbanism.org/dakakker-urban-rooftop-farm/; https://en.rotterdampartners.nl/venues/op-het-dak-dakakker/)

- The government promotes roof cultivation and active public participation.
- Financing projects through taxes used by municipalities and water councils to pay their costs.
- Financing of green roof subsidies of €30/m^2.
- Making an ongoing effort by municipalities to reach out for participants in construction, agriculture or special care with green roofs to strengthen these policies in the country.

Although there are policies for green roofs in the Netherlands, there is little knowledge about:

- Green roof performance assessment—more research should be carried out on the ideal plant species suitable for climatic conditions.
- Prediction on the impacts of climate change (CC); if such impacts are greater than expected, it may be necessary to raise the policy objectives to address CC.
- Knowledge of the water systems—there are some gaps, but in general, there is enough knowledge.

A private residence in Lint, the Netherlands (https://archello.com/project/living-green-family-home n.d.), is an example of one of the buildings where the indirect policy was implemented and it shows the benefits of green roofs as heat insulation and increased air quality (Fig. 7).

Policies Regulating Green Roofs in Germany

Germany has a long history of policies when it to comes to supporting the development of green roofs. It began back in 1960s when researchers commenced on cultivating roofs with some plants. A research led to interest in green roofs and the formation of a research complex for the development of nature views and design within Germany in 1975, which assisted in strengthening green roofs and innovation in green roof design in Germany. In 1982, the first version of the guidelines for planning, construction and maintenance of green roofs was established (Carter and Fowler 2008). It was then updated seven times up to 2010. Nevertheless, the German

Fig. 7 Green roof of a private residence in Lint—the Netherlands. (*Source:* https://archello.com/project/living-green-family-home)

Government enacted the following regulations to promote green roofs (Carter and Fowler 2008):

- Green roofs have been included in the Building Code since 1984.
- Berlin approved a support programme for green roofs from 1983 to 1997, which paid to the population about 50% of the costs of green roof buildings and thus resulting in about 63,500 m^2 of green roofs.
- German Government continues to provide grants for the construction of green roofs in certain areas, as an example for the promotion of green roofs.
- In 2002, 103 cities were identified to provide direct and indirect incentives to build their green roof, whereas 28 cities had green roof construction requirements in local development plans.
- There are subsidies granted by the state to promote green roofs ranging from 10 to 50% of the initial costs of the green surface (Ngan 2004b).
- The German Government provides direct support to about 50% of cities for building owners to install green roofs on their buildings' roofs (Ngan 2004b).

An example is the green roofs of the Studio S in Wiesbaden, Germany (https://ar.smarthomemaking.com/20-green-walled-buildingsinspiring-your-own-plant-cladded-home n.d.), which has a green insulating cover that increases air quality and provides an attractive view for psychological comfort as shown in Figs. 8 and 9.

Other examples of iconic green roofs in Germany are the Potsdamer Platz in the heart of Berlin. It features elaborate yet naturalistic storm water retention systems that are designed to minimize burden on Berlin's existing water infrastructure, including its many green roofs. The storm water management, harvesting and recycling system is fed by the city's 21 in. of annual rainwater. The green roofs retain and then discharge water to the large on-site buffer pond (Piano Lake) which has five underground storage tanks (https://www.greenroofs.com/projects/potsdamer-platz/ 2019). Also, the Potsdamer Platz controls an estimated 23,000 m^3/year of potable water and its cooling capacity reduces summer temperatures by

Fig. 8 Green roof of Studio S in Wiesbaden, Germany. (*Source: **(a)*** https://ar.smarthomemaking.com/20-green-walled-buildings-inspiring-your-own-plant-cladded-home. ***(b)*** https://inhabitat.com/3-rooms-3-trees-and-a-meadow-adorn-the-roof-of-house-s-in-germany/house-s-roger-christ-2/)

Fig. 9 Green roofs in Berlin, Germany. (*Source:* http://www.archi-house.co.uk/wp-content/uploads/2016/02/berlin-green-roof-small.jpg)

approximately 2 °C resulting in energy savings according to Landscape Architects Network, 2015.

In addition, rainwater from the mostly green rooftops at Potsdamer Platz is collected year-round in three large underground cisterns. The roofs of the 19 buildings have a total surface area of 50,000 m^2 and partially supply the water areas such as the Piano Lake and the water and wastewater network (toilet flushing). In total, around 20 million litres of drinking water is saved through such a system yearly (https://www.greenroofs.com/projects/potsdamer-platz/ 2019). Figure 10 shows the Potsdamer Platz with the building's green roofs.

In the city of Frankfort, another example of green roofs' enacted policy is the Städel Museum, which was founded in 1815 and showcases an exceptional collection of European art. It was expanded in 2012 by the architectural firm Schneider + Schumacher with new space that has roof garden (https://www.archdaily.com/266833/stodel-museum-schneider-schumacher 2019) as illustrated in Fig. 11. In addition, Düsseldorf, Germany, has another large building—the Kö-Bogen retail and office complex with a large green roof. It has a green roof that is intended to offer some design continuity with the area's historic parks as shown in Fig. 12.

Fig. 10 Potsdamer Platz's green roofs, Berlin. (*Source:* https://www.greenroofs.com/projects/potsdamer-platz/)

a b c

Fig. 11 Green roof garden of Städel Museum in Frankfort, Germany. (**a**) The green roof after renovation. (**b**) Green roof with light at night. (**c**) The museum's green roof is used as a social hum for visitors. (*Image source: (a) https://*www.apollo-magazine.com/a-grand-old-museum-at-the-cutting-edge/. *(b)* https://www.architecturaldigest.com/gallery/green-roof-living-roof-designs. *(c)* https://miesarch.com/work/506)

Fig. 12 Green roofs of Kö-Bogen retail and office complex, Düsseldorf, Germany. (*Image source:* https://www.theguardian.com/cities/gallery/2018/aug/17/urban-oases-green-roofs-around-the-world-in-pictures)

Policies Governing Green Roofs in Copenhagen, Denmark

Green roofs represent a view of Copenhagen city through its spread over all educational and public buildings such as green roofs in the famous residential building 8 (Carter and Fowler 2008). Figure 13 presents the large green roof resulting from Danish enacted policy. According to local plans, 200,000 m^2 of green roofs will be built (ENVS—Department of Environmental Science 2010). All suitable new roofs with a slope of less than 30° will become green roofs at no cost to users (https://www.researchgate.net/publication/293196881_An_International_Review_of_Current_Practices_and_Future_Trends_Green_Roof_Policies#downloadCitation n.d.). The obligation to use green roofs is part of the city's goals to be more sustainable and achieve a carbon reduction mechanism, in addition to governing grants and tax exemptions to convert existing roofs to be green (Carter and Fowler 2008).

Also, Copenhagen's Amager Bakke waste-to-energy plant has been developed in 2018 with a 16,000 m^2 rooftop that includes large green roofs, a park and a ski area. The rooftop garden provides outdoor activities for citizens and visitors (https://www.architectsjournal.co.uk/buildings/bigs-copenhill-waste-to-energy-plant-cum-ski-slope-opens-in-copenhagen/10044785.article 2019). Figure 14 illustrates the iconic Amager Bakke waste-to-energy rooftop garden in Copenhagen, Denmark.

Policies Regulating Green Roofs in France

The French Government set a law in March 2015 stating that all new buildings in commercial areas should be partially covered by either green roofs or solar panels; however, they have no legislation or policies in place to regulate the construction of

Fig. 13 Building 8's green roofs in Copenhagen, Denmark. (***Source:*** *ENVS—Department of Environmental Science, Biodiversity, Buildings, Climate and Air Quality, Social Inclusion, Water—Green-Roof Programme—Copenhagen, Denmark, Aarhus University, 2010, and* https://www.greenroofs.com/projects/8-house-8-tallet/)

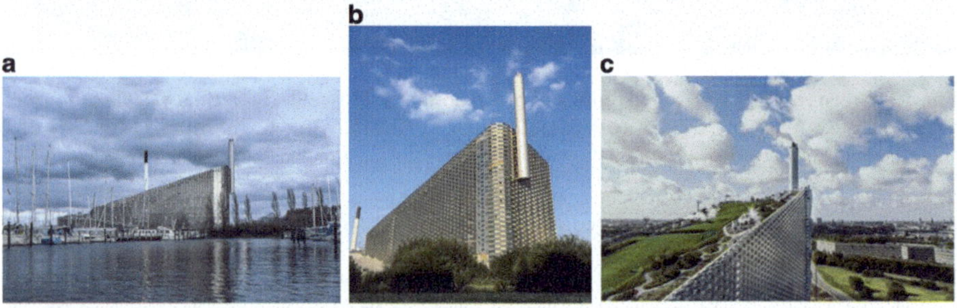

Fig. 14 Copenhagen's iconic waste-to-energy plant with a large top green roof. (**a**) General view. (**b**) Exterior of the plant. (**c**) The top green roof. (*Images sources: (**a**).* https://www.archdaily.com/894058/new-photographs-explore-bigs-waste-to-energy-plant-as-ski-slope-roof-is-installed/5af1b00bf197cc611900031a-new-photographs-explore-bigs-waste-to-energy-plant-as-ski-slope-roof-is-installed-photo. *(**b**)* https://archinect.com/news/article/150044555/big-s-copenhagen-waste-to-energy-plant-is-finally-getting-the-promised-ski-slope-and-rooftop-park. *(**c**)* https://www.archdaily.com/925966/copenhill-the-story-of-bigs-iconic-waste-to-energy-plant/5d9a804a284dd1ffa40000a2-copenhill-the-story-of-bigs-iconic-waste-to-energy-plant-photo)

green roofs (https://www.researchgate.net/publication/293196881_An_International_Review_of_Current_Practices_and_Future_Trends_Green_Roof_Policies#downloadCitation n.d.). In August 2019, Paris started building the largest green roof ever in the world. It is considered the biggest urban rooftop farm (https://www.greenroofs.com/2019/08/28/worlds-largest-urban-farm-to-open-on-a-paris-rooftop n.d.). It is under construction in the south-west of the city and this urban oasis will span approximately over 14,000 m^2 (150,695 sq. ft.), also making it the largest urban farm in Europe (Fig. 15). With the plan to grow more than 30 different plant species, the site will produce around 1000 kg of fruits and vegetables daily in high season. It will be tended by around 20 gardeners using entirely organic methods (https://www.greenroofs.com/2019/08/28/worlds-largest-urban-farm-to-open-on-a-paris-rooftop n.d.). Also, École des Ponts ParisTech is a school built under large green roofs in Paris (Fig. 16).

Fig. 15 The largest urban rooftop farm in Paris, France. (*Image source:* https://www.greenroofs.com/2019/08/28/worlds-largest-urban-farm-to-open-on-a-paris-rooftop/)

Fig. 16 École des Ponts ParisTech, school built under green roofs in Paris, France. (*Images source:* https://www.treehugger.com/green-architecture/green-roofs-are-changing-architecture-heres-whole-school-built-under-undulating-green-roof.html)

Policies Regulating Green Roofs in Non-European Union Countries

There is a disparity between countries interested in green converge/surface policies to regulate green roofs and those which do not put much interest. These can be presented in the following section.

A. Green Roof Regulations and Policies in Zurich, Switzerland

The main objective in Switzerland was to increase biodiversity as well as reduce heat stress and partially compensate for the loss of other green areas through construction activities. The federal government set mandatory conditions to reach this goal encompassing:

- Growing green surfaces on all flat surfaces when building new housing projects or renovating old spaces since 1991 (https://www.researchgate.net/publication/293196881_An_International_Review_of_Current_Practices_and_Future_Trends_Green_Roof_Policies#downloadCitation n.d.).
- There are some recommended plants such as herbs and flowers.
- The growth medium layer (soil) with minimum thickness of 10 cm with a minimum water retention capacity of 45 L/m^2 (4.5 cm) (General Services Administration 2011).
- Government provides training on how to fix green roofs and how to care for it.

Box 1 Green Roof Polices in European Union Countries
By presenting the policies regulating green roofs of European and non-European countries, it can be stated that there are clear policies in European Union countries except France, while non-European countries do not have a clear policy regulating green roofs except Switzerland, which started to pay attention to green roofs in terms of biodiversity.

Policies Governing Green Roofs in North American Countries

The USA has developed and enacted many policies to build urban green surfaces including green roofs. The policy framework is implemented at two levels: federal and municipal. The following section highlights and discusses cities in the USA and Canada, where green roofs' policies were developed and enacted. Figure 17 depicts the cities in North America that enacted green roof policies.

Green Roof Regulatory Policies in Chicago, Illinois

The direct support policy was developed where the government provided a Green Roof Grants Programme (GRGP) for 2005 to finance more than 20 green areas of residential and small projects. These total amounts have been awarded through a selection process based on project location, vision and environmental benefit. Also, the Chicago Department of the Environment (CDE) extended this programme in 2006

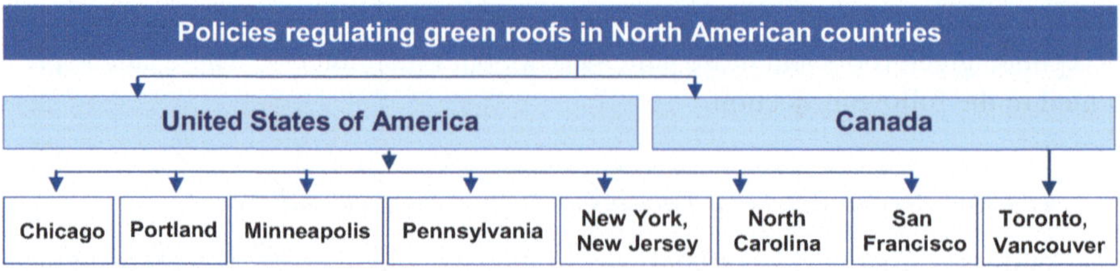

Fig. 17 Cities in North America with policies regulating green roofs

Fig. 18 Green roofs of Peggy Museum of Nature in Chicago—Illinois, USA. (*Source:* http://wendycitychicago.com/chicago-green-roofs/;https://perkinswill.com/project/peggy-notebaert-nature-museum/)

and 2007, covering an additional 52 projects, not only government, but also private foundations that provided small green roof grant programmes supporting 7 green roof projects covering 20% of the cost of green roof construction (Liptan 2003). The Home Depot Foundation provides funding for education and research on green surfaces (https://www.researchgate.net/publication/293196881_An_International_Review_of_Current_Practices_and_Future_Trends_Green_Roof_Policies#downloadCitation n.d.). In addition, the Peggy Nature Museum is one of Chicago's green roof models, which features five different, 2400 square feet green roofs (Liptan 2003) as shown in Fig. 18. The new green roof is built over 2400 square feet that demonstrates green roof technology that features wetland, extensive, semi-intensive and intensive systems. Also, a limestone wall allows for water from the upper green roof to drip down the facade, irrigate the bluff vegetation and stream into a wetland garden, where it overflows into a cistern in the underground. The captured water is recycled and reused throughout the museum landscape (Conservation design Forum 2019).

Green Roof Policies in Portland, Oregon

An indirect subsidy policy was used where Oregon State reduced 35% of storm water fees, and if 70% of the surface was covered with vegetation, the owner could obtain total credit all over the city (Carter and Fowler 2008). The Green Building Policy Guidelines (GBPG) for the city of Portland allowed for all new city-owned facilities to use green roof (Liptan 2003). An example of this policy is the City Building, which is one of Portland's green roof models (Steier 2018). Its roof surfaces have been renovated and converted from a traditional roof to a green roof, to take advantage of the environmental and structural benefits of the green roof and reduce urban heat island effect (UHIE) in the city. Figure 19 presents the renovated green roof in the City Building in Portland.

Fig. 19 City Building's green roof in Portland—Oregon, USA. (*Images source:* https://www.greenroofs.com/projects/the-portland-building/)

Green Roof Policies in San Francisco, California

San Francisco municipality has issued an internal regulation that includes green roofs. The government also started to clarify that the roof of the building is one of the city's environmentally untapped resources and that it needs to formulate strategies on how to exploit it and develop regulatory policies to access clean, green air roofs. San Francisco also joins growing cities that do not have green roof policies (https://ecourbanhub.com/san-francisco-green-roofs-legislation-buildings-wiener/ n.d.). The state policy enacted in 2016 by San Francisco Planning Department (SFPD) requires 30% of roof space of new buildings to be built as green roofs or a combination of green roofs and solar panels (https://ecourbanhub.com/san-francisco-green-roofs-legislation-buildings-wiener/ n.d.). Examples of green roof policy are California State University San Marcos and California Academy of Sciences, San Francisco, that was designed by Renzo Piano as illustrated in Figs. 20 and 21. Figure 21 shows the green roof of the student union in California State University.

Also, the City College is one of San Francisco's green roof models. Green roofs have been grown from local plants, and many other sustainable elements in the LEED Platinum project that was designed to reach clean air. Another policy was enacted to encourage rooftop gardens on new San Francisco buildings (https://www.proudgreenbuilding.com/news/bill-to-encourage-rooftop-gardens-on-new-san-francisco-buildings/ n.d.). Figures 22, 23, 24, and 25 show these green roof examples.

Green Roof Policies in Minneapolis, Minnesota

An indirect support policy has been used where the city provides the owner and building users with up to 100% of rainwater charges for rainwater management involving green roof systems (Earth Pledge 2005). An example is the City Center in

Fig. 20 Green roof of California Academy of Sciences, San Francisco—California, USA. (*Images source:* https://www.greenroofs.com/projects/city-college-san-francisco/; http://www.syserco.com/projects/city-college-of-san-francisco-conlon-hall)

Fig. 21 Green rooftop of California State University San Marcos, California, USA. (*Images source:* https://www.greenroofs.com/projects/city-college-san-francisco/; http://www.syserco.com/projects/city-college-of-san-francisco-conlon-hall)

Fig. 22 Rooftop garden on a residential building in San Francisco—California, USA. (*Images source:* https://www.proudgreenbuilding.com/news/bill-to-encourage-rooftop-gardens-on-new-san-francisco-buildings/)

Fig. 23 The largest rooftop garden in Silicon Valley, San Francisco—California, USA. (*Images source: https*://www.pinterest.co.uk/pin/666392694642341O5/)

Fig. 24 Large green roofs on Apple new office building in Silicon Valley, California, USA. (*Image source:* https://www.dezeen.com/2015/10/07/apple-silicon-valley-campus-hok-california-green-roof/)

Fig. 25 Microsoft mountain campus with green roofs in Silicon Valley, California, USA. (*Image source:* https://www.bizjournals.com/sanjose/news/2016/01/21/microsoft-submits-plans-for-major-mountain-view.html)

Fig. 26 Green roofs in Minneapolis—Minnesota, USA. (**a**) Minneapolis City Center. (**b**) Minneapolis Airport—Terminal 2. (*Image source:* https://www.greenroofs.com/projects/city-of-minneapolis-target-center-arena/; https://www.pinterest.com/pin/610097080750960366/?lp=true)

Minneapolis, which is the fifth largest wide green surface and the first green roof installed on a nucleus in North America. It is considered a nucleus to promote plant diversity and environmental adaptability and to express the mix of local plants as shown in Fig. 26.

Green Roof Policies in Pennsylvania, New Jersey and North Carolina

Pennsylvania, New York and New Jersey as well as North Carolina set standards for storm and storm water management for future development using green surfaces to reduce floods and storms (https://www.livingroofsinc.com/projects 2019). Philadelphia Library is an example of Pennsylvania's green roofs as shown in Fig. 27 (https://archpaper.com/2015/06/gallery-barclays-centers-green-roof-sedum-installation-50-complete/ 2019). Also the green roofs in Brooklyn—New York—and North Carolina are shown in Figs. 28 and 29, respectively.

Fig. 27 The green roof of Philadelphia Library, Central Parkway, Pennsylvania, USA. (*Image source:* https://www.greenroofs.com/projects/free-library-of-philadelphia-parkway-central/)

Fig. 28 Barclays Center's green roof, Brooklyn, Lower Manhattan—New York, USA. (Image source: https://archpaper.com/2015/06/gallery-barclays-centers-green-roof-sedum-installation-50-complete/)

Fig. 29 Green roofs in North Carolina, USA. (**a**) AC Marriott Hotel in Chapel Hill. (**b**) Blue Ridge Community College-Health Sciences Center, Hendersonville. (*Images source:* https://www.livingroofsinc.com/projects)

Green Roof Regulatory Policies in Toronto and Vancouver, Canada

Toronto, the capital of Ontario, is the first city in the world to implement an internal system covering all buildings over 2000 m^2, and the government has made a mandatory requirement for surface cultivation at rates ranging between 20 and 60% of the surface area (Ngan 2004a). Figure 30 shows the green roof examples in Toronto where about 36,000 square feet of green roof is composed of interlocking and gravel pathways with concrete and wooden benches surrounding the twin towers, bringing the entire project to 150,000 square feet, as well as the Skyluxe green roofing in Greater Toronto Area. Figure 30 also shows green roofs of the iconic Convention Centre in Vancouver, which uses extensive and semi-intensive green roof inclined types to manage rainwater run-off effectively (https://www.greenroofs.com/projects/nathan-phillips-square-toronto-city-hall-podium-green-roof/ 2019).

Policies Regulating Green Roofs in South America

There are several countries in South America that have started developing policies and laws for green roofs. These are mainly Brazil, Argentina and Colombia. The following section portrays the enacted policies that regulate the implementation of green roofs.

Green Roof Regulatory Policies in Recife, Brazil

The Recife Municipal Council set a law regulating green roofs on 16th December 2014. This law requires buildings with more than four floors to have their roofs covered with native vegetation. The law also applies to any commercial building with more than 400 m^2. An example of this law is Torre Charles Darwin, which is the first 35-floor green roof project. The building will be covering 2.8 million square feet as shown in Fig. 31. Recife in Brazil is the first city in South America to convert the roofs of new or

Fig. 30 Green roofs in Toronto and Vancouver, Canada. (**a**) Green roofs of City Hall podium in Toronto. (**b**) The Convention Centre green roof—waterfront at Canada Place in Vancouver. (*Images source:* https://www.greenroofs.com/projects/nathan-phillips-square-toronto-city-hall-podium-green-roof/; https://www.skyluxe-roofing.com/green-flat-roof/; https://www.goodnet.org/articles/5-impressive-green-roofs-from-across-globe. https://torontoimages.wordpress.com/2011/07/15/green-roof-toronto-city-hall/)

existing buildings into green roofs (https://greenroofsaustralasia.com.au/news/inspiredcopenhagenmunicipal-council-recife-brazil-approves-green-roof-law n.d.). Recife is one of the two Urban-LEDS model cities in Brazil. As much as 46% of Recife's total area is green, 60% of which is protected under conservation laws (https://greenroofsaustralasia.com.au/news/inspiredcopenhagenmunicipal-council-recife-brazil-approves-green-roof-law n.d.).

An Urban Afforestation Plan (UFP) aims at preserving and increasing this unique environment. In addition, another example, outside of São Paulo in Brazil, is the green roof of a private house with extended green roof, which was developed by Studio MK27 firm (HE MIAOMIAO 2011). The roof is a horizontal type (https://greenroofsaustralasia.com.au/news/inspired-copenhagen-municipal-council-recife--brazil-approves-green-roof-law 2019) as shown in Fig. 32.

Green Roof Policies in Córdoba, Argentina

In July 2016, the city of Córdoba rehabilitated the green roofs of existing buildings supported by an incentive programme. The new policy aims to limit the rising levels of air pollution and CO2 absorption by achieving a network of green roofs, but the

Fig. 31 Green roofs of Torre Charles Darwin in Recife, Brazil. (*Images source:* https://www.skyscrapercity.com/showthread.php?t=1354523)

Fig. 32 Green roofs of Studio MK27 outside São Paulo, Brazil. (*Images source:* https://worldarchitecture.org/architecture-news/ehfvg/studio-mk27-completes-planar-house-with-extended-green-roof-outside-of-so-paulo.html)

new regulation covers only the central areas of the city, but there is a clause allowing more living in the future (World Architectural Community 2019). An example of green roof policies in Córdoba is illustrated in Fig. 33.

In Buenos Aires, green roofs were installed across schools and government buildings, a significant proportion of the 1400 ha of green roof that is believed to be enough to generate a positive environmental impact across the city of some

Fig. 33 Green roof in Córdoba, Argentina. (*Images source:* https://twitter.com/dtdchange/status/774182841311649792)

Fig. 34 Green roofs in Buenos Aires, Argentina. (*Images source:* https://www.lafargeholcim-foundation.org/projects/green-roofs-buenos-aires-argentina)

20,000 ha (http://wendycitychicago.com/chicago-green-roofs/ n.d.). Schools are the perfect starting point for the project because of its impact on the neighbourhood. Green roofs have become an outdoor classroom for learning about ecology and agriculture as well as teaching and learning media for pupils in Argentina (https://www.lafargeholcim-foundation.org/projects/green-roofs-buenos-aires-argentina 2019). The green roof also improves insulation and retains up to 90% of rainfall to mediate against potential flooding from run-off (Fig. 34).

Fig. 35 Green roof in Santalaia Building in Bogotá, Colombia. (*Image source:* https://www.greenroofs.com/projects/santalai. *Image source:* https://www.euronews.com/2017/07/09/colombia-bogota-vertical-gardens-hint-at-greener-future)

Fig. 36 Vertical gardens (green walls) of Hotel B3, Bogotá, Colombia. (*Image source:* https://tropicalcommons.co/en/2018/09/04/vertical-gardens-and-green-roofs-of-colombia/)

Policies Regulating Green Roofs in Bogotá, Colombia

There is an initiative to develop a new policy at the World Green Infrastructure Congress in 2016. An example is shown where green roofs and large green walls of 360 m^2 were installed in 2012 in Santalaia, a high-end multifamily residential building, which is situated on the eastern edge of the densely populated Colombian capital of Bogotá (https://www.greenroofs.com/projects/city-college-sanFrancisco/ n.d.). Figure 35 shows these large green walls and green roof in Bogotá. According to Groncol (a Colombia-based company) the Colombian capital has approximately 32,000 m^2 of green roofs. Bogotá has more than 1100 m^2 of vertical gardens implemented in hotels, universities, shopping centres, business buildings and restaurants.

Figure 36 illustrates the vertical garden of Hotel B3, which was installed in 2012. It was the first hydroponic wall of this magnitude in Latin America. The project consists of a wall that covers the entire western facade of the hotel. It has more than 360 m^2 of vertical garden (https://www.greenroofs.com/projects/santalai 2019).

It is clear from the above review of policies and laws developed to regulate urban green coverage mainly green roofs that policies and laws differ from the federal level to the municipal level. In line with the federal policies and legal basis, a variety of sub-policies have been developed at the municipal level. Table 1 lists the difference between the federal level and the municipal level.

Table 1 Policies regulating green roofs—difference between federal and municipal levels

Local government (municipal) level	Federal government level
• Direct financial support: – Direct support is provided through green roof grant programmes offering a lump-sum payment • Indirect financial support: – Offer 35% reduction in storm water charges and/or if 70% of the surface is covered with green plants, the owner can get a total subsidy of 100% of the rainwater management fee that includes green roof systems • Rational in developmental policies: – The US Government is incorporating green roofs as a tool in its development laws such as Green Building Guidelines (GBG), Land Use Plan (LUP) and Rain Water Management Standards (RWMS) • Other tools: – Demonstration projects, competitions, media coverage and performance evaluations are ways to promote green roofs and improve public awareness. Because of the success of pilot projects, Portland started to adopt a series of policy mechanisms to promote environmental roofs throughout the city	• The Federal Water Pollution Control Act (FWPCA) or the Federal Clean Water Act (FCWA) and rainwater management that can be supported by green roofs required. Since 2003, all new federal buildings were required to obtain a LEED certification, and were encouraged to achieve LEED Silver by the Public Services Department with implementation of green roofs • The Ministry of Defence has requested all new construction projects to meet LEED silver requirements, which include: – Mitigate heat islands to reduce their effects, requires covering 50% of the car parking space, and planting surfaces to operate contrary to sunlight – Reduce water and energy use, to promote fluid restoration, and increase open space, storm water intake, quality control and water-saving natural views – Green roofs: Use green roofs in any building to meet 15 LEED requirements, one of the best ways to reduce rainfall by 18% and reduce pollutants

Source: HE MIAOMIAO, Promoting Green Roofs in China: A Comparison of Green Roof Policies in the (U.S.) and China, Fulfillment of the Architectural Design, University of Florida, 2011

Box 2 Green Roof Policies in North American Countries

By presenting some of the countries of North American continent and some of the states of the United States, we reach the interest of most North American countries to use green surfaces and know their benefits and there are many policies, laws and initiatives to promote in North America.

Policies Regulating Green Roofs in Asian Countries

The policies regulating the green roofs in Asia differ from country to country due to the different climatic conditions of each country. Policies and laws regulating green surfaces in some Asian countries, mainly in China, Japan, Singapore and Malaysia (Fig. 37), are presented in the following section.

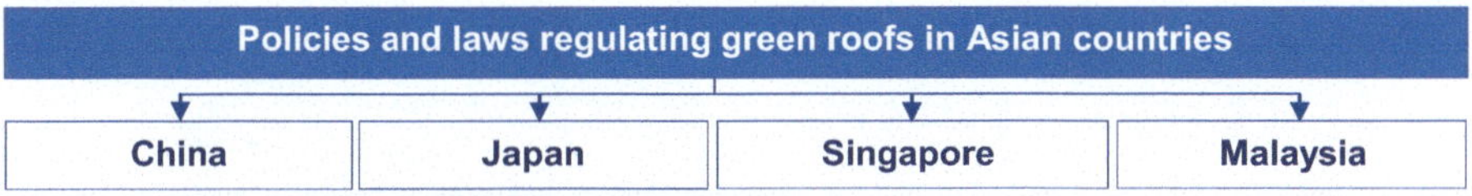

Fig. 37 Countries enacted green roof policies in Asia

Policies and Laws Governing Green Roofs in China

China took a step forward towards environmental protection acts and has begun to promote green roofs in recent years. This has been manifested in several cities that established and enacted policies to promote green roofs since 2002. The first green roof policy was initiated in Shanghai 2006. Between 2009 and 2011, about 300,000 green roofs were built which are forming about 3.2 million square feet of green roofs (https://tropicalcommons.co/en/2018/09/04/vertical-gardens-and-green-roofs-of-colombia/ 2019). Shanghai's Congress also approved Green Roof Regulations (GRR) on May 1, 2007. In addition, the law (2010) stated the following main articles as shown in Fig. 38. Examples of large Greenland and green roofs in Shanghai are illustrated in Figs. 39–44. One of the major developments is the Shanghai Greenland Centre, which is located above a two-line subway station and forms the city's largest urban park (approximately 20,000 m^2). By merging human activities with nature, the centre offers various commercial facilities through a roof with cascading terraces and slopes topped with trees and other types of greenery (ICLEI n.d.) as shown in Fig. 40. Shanghai planned to install and plant 400,000 m^2 of rooftop gardens in 2016 alone, an area roughly the size of Vatican City (https://www.archdaily.com/905876/shanghai-greenland-center-nikken-sekkei/5bec3ed108a5e5767c00005a-shanghai-greenland-center-nikken-sekkei-photo 2019).

In Beijing, Beijing Urban Plan (BUP) 2004–2008 stated that the plan would cover 30% of the roofs of high-rise buildings and 60% of low-rise buildings by green roofs. In this regard, such sound plans led to immense environmental impacts: According to the UNEP (United Nations Environment Programme), 80% reduction in CO_2 levels in Beijing is estimated should green roofing reach a rate of 70%. In addition, Gardens and Forest Department (GFD) in Beijing has approved the Green Roof Specifications Manual (GRSM), which provides basic requirements, type of roof, plant design, plant selection, and green roof technology for buildings in Beijing area (SLR 2007). Furthermore, Beijing Municipal Government has set a target of 430 million square feet of green roofs where the government provides financial support at 50% of government-subsidized costs (SLR 2007). Figures 45–50 present iconic green roofs installed in Beijing, China. In another province Shenzhen, the local government issued a plan in 1999 to build green roofs. However, the government began promoting green roofs in November 2005. The green roof area in Shenzhen exceeded 1 million square metres in 2008 (https://www.archdaily.com/788409/shenye-tairan-building-zhubo-design 2019). The Garden and Forest Technology Institute (GFTI) is assisting the government for seeking green roof

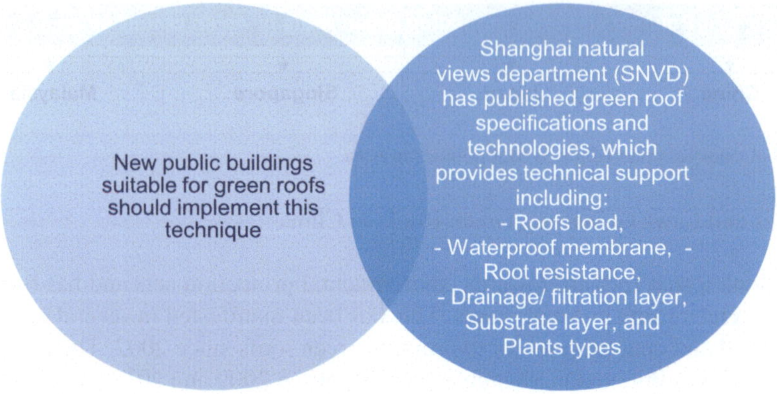

Fig. 38 Shanghai 2010 law articles for green roofs—China

Fig. 39 Green roofs of Giant Interactive Headquarters—SWA Shanghai, China. (*Images source: https://www.swagroup.com/projects/giant-interactive-headquarters/*)

technology. Figures 51–52 present some examples of green roofs in Shenzhen, China (http://www.roofchina.com/article/show_article.php?id=459 2019).

In Hong Kong, green roofs are found in some new commercial and residential buildings. Also green spaces and natural views have been widely integrated into the planning of new cities and new development areas, and green roofs become the only option for densely populated urban areas. Nevertheless, there is no clear policy framework for green roofs (SLR 2007). Examples of green roofs are illustrated in Figs. 53–56. One of the residential examples is Choi Tak Residential Building

Fig. 40 Greenland Center, largest urban park in Shanghai, China. (*Images source: https://www.archdaily.com/905876/shanghai-greenland-center-nikken-sekkei/5bec3ed108a5e5767c00005a-shanghai-greenland-center-nikken-sekkei-photo*)

Fig. 41 Green roofs of Novartis new campus, Shanghai. (*Images source: https://www.designboom.com/architecture/kengo-kuma-novartis-building-shanghai-features-folding-green-roof-china-06-06-2017/*)

Fig. 42 Green roofs Shanghai's Natural History Museum. (*Images source:https://cdn.archpaper.com/wp-content/uploads/2015/04/01_Shanghai-Natural-History-Museum.jpg*)

Fig. 43 Green roofs of MVRDV Park-Topped Community Center for Shanghai Neighborhood, China. (*Images source: https://www.archdaily.com/881800/mvrdv-create-park-topped-community-center-for-shanghai-neighborhood/*)

Fig. 44 Green roofs of Hongqiao Flower Building in Shanghai, China. (*Images source: https://www.mvrdv.nl/projects/234/hongqiao-flower-building*)

Fig. 45 Green roofs of Beijing's Chemical Mansion and Sunshine Mansion, China. (*Images source: https://www.treehugger.com/corporate-responsibility/chinas-learning-to-love-green--roofs.html*)

Fig. 46 Green roof of Temporary Guanganmen Green Technology Exhibit in Beijing, China. (*Images source: https://inhabitat.com/green-technology-showroom-by-vector-architects/*)

Fig. 47 Green roofs of Beijing University of Chinese Medicine. (*Images source: https://www.chinadaily.com.cn/beijing/2013-07/26/content_16848344.htm*)

Fig. 48 Green roofs of Beijing Civic Centre, China. (*Images source: https://inhabitat.com/giant-ski-slope-to-top-green-roofed-civic-center-in-beijing/*)

Fig. 49 Green roof of Beijing airport. (*Images source: https://www.topchinatravel.com/beijing/how-to-spend-a-6-hour-layover-in-beijing.htm*)

Fig. 50 Green roofs of Garden School in Beijing, China. (*Images source: https://architizer.com/blog/inspiration/collections/green-roof-high-schools/*)

Fig. 51 Cascading green roofs in Shenzhen, China. (*Images source: https://www.archdaily.com/788409/shenye-tairan-building-zhubo-design*)

Fig. 52 Green roofs in Shenzhen, China. (*Images source: http://www.roofchina.com/article/show_article.php?id=459*)

Fig. 53 Green-roofed West Kowloon Terminus in Hong Kong, China. (*Images source: https://inhabitat.com/green-roofed-west-kowloon-terminus-will-link-hong-kong-with-chinas-high-speed-rail/*)

Fig. 54 Green roofs of residential buildings in Hong Kong, China. (*Images source: https://www.sempergreen.com/us/references/housing-authority*)

Fig. 55 Roof garden of a Zero-Carbon Building, surrounded by urban woodlands in Hong Kong, China. (*Images source: https://www.lowandbonar.com/what-we-do/xeroflor/*)

Fig. 56 UrbanFarm: Rooftop of Bank of America Tower, Central Hong Kong, China. (*Images source: https://gogreenhongkong.com/2016/01/16/rooftop-farming/*)

(Fig. 56), which has a green roof of 220 m^2 was installed on top of this residential building that is developed by Kong Housing Authority in Hong Kong (https://www.sempergreen.com/us/references/housing-authority 2019).

In Fig. 55, a green roof of 750 m^2 is installed over the top of the roof of Fu Shan Estate. In this green roof, the Sedum mix blankets are particularly developed and tailor-made for Hong Kong's climate (https://www.sempergreen.com/us/references/fu-shan-estate 2019).

Policies governing green roofs in Japan

Established in November 1990, the Urban Green Natural Views Development Organization (known as "Urban Green Technology") is responsible for the promotion of green spaces and green roofs in Japan. It has published and developed some important guidelines on roofs and wall greening with the support of the Ministry of Land, Infrastructure and Transport (MoLIT) and the Ministry of the Environment (MoE) of Japan providing financial incentives since 2002. It also coordinates research and development, conferences, and incentive schemes on urban greening where for more widespread green roofs, the lower the cost of energy consumption (Dunnett & Kingsbury 2008).

Tokyo has introduced policies that require green roofs installed on 20% of all new flat roof surfaces on public buildings exceeding 250 m^2, and 10% of all flat roofs on private buildings exceeding 1000 m^2 (Eidt 2014).

Japan has green roof suppliers; companies prepare and provide planning guides and information to practitioners. Local provincial governments have also developed their requirements and standards, as well as incentive schemes. This helps to ensure the quality of work for green roof systems and promote its development (Dunnett & Kingsbury 2008).

Despite the fact that Japanese policy on climate change has been criticized for concentrating more on mitigation than adaptation at the local level, Fukuoka City has considered adaptation to the effects of environmental change within its green space and urban plans since the late 1990s (Mabona et al. 2019).

The 2018 Climate Change Adaptation Act (CCAA) now mandates municipal governments to form Local Climate Change Action Plans (LCCAP). Yet up until this point, few Japanese cities have had specific laws and plans to address adaptation action, and there is limited evidence of practical adaptation actions at the local level (Hijioka et al. 2016).

In Fukuoka, by contrast, locally based researchers have undertaken applied research on the relationship between urban greening, the built environment and urban thermal environments for several decades (Hui 2010). Fukuoka City Government also released in 2016 a climate change countermeasures action plan to integrate climate and environmental plans and link mitigation with adaptation, continuing actions (Hui 2010).

The ACROS Fukuoka Prefectural Hall is considered Japan's premier building designed over 15 terraced green roofs forming 100,000 m^2 in the city centre. It is a new solution to common urban problem. Figures 57–65 illustrates the examples of green roofs in Japan (Hui 2010; https://inhabitat.com/japans-namba-parks-has-an-8-level-roof-garden-with-waterfalls/; https://www.theguardian.com/cities/gallery/2018/aug/17/urban-oases-green-roofs-around-the-world-in-pictures n.d.; https://resources.realestate.co.jp/living/japan-green-roof-buildings/ 2019).

In Osaka, Namba Parks (a retail and office complex) have an eight-level rooftop garden next to the Namba Train Station. This rooftop garden with waterfalls was developed after the closure of Osaka Stadium (Fig. 58). It presents a great redevelopment of green roofs in Osaka (Hui 2010; https://resources.realestate.co.jp/living/japan-green-roof-buildings/ 2019).

At the residential level in Japan, the green roof installed in Ville Ronde is considered as a natural buffer and the blurring of the natural boundaries between architecture and natural views of Japanese coast (Hui 2010) as shown in Fig. 66.

Policies Regulating Green Roofs in Singapore

Singapore government has agreed to reach 500,000 m^2 of Sky Greenery by 2030. This will be done using financial support and incentive schemes for roofs. This will be done using financial support and the incentive system for roofs. However, there

Fig. 57 Cascading green roofs of ACROS Fukuoka Prefectural Hall in Fukuoka. (*Images source: https://metaefficient.com/architecture-and-building/amazing-green-building-the-acros--fukuoka.html*)

Fig. 58 Green roofs of Namba Parks in Osaka, Japan. (*Images source: https://inhabitat.com/japans-namba-parks-has-an-8-level-roof-garden-with-waterfalls/*)

Fig. 59 Green roofs of top floor of Keio University in Tokyo, Japan. (*Images source: http://www.archivitamins.com/keio-university-roof-garden-michel-desvigne/?utm_source=feedburner&utm_medium=feed&utm_campaign=Feed:+archivitamins/JUHL+(Archivitamins)#.XeJ7gpMzZVl*)

Fig. 60 Green roof in Tokyo, Japan. (*Images source: http://www.asiagreenbuildings.com/6520/world-green-building-council-and-ifc-join-forces-for-promoting-sustainable-construction/*)

is no known legislation or policies to regulate the construction of green roofs (https://www.researchgate.net/publication/293196881_An_International_Review_of_Current_Practices_and_Future_Trends_Green_Roof_Policies#downloadCitation n.d.). Examples of Green roofs in Singapore are depicted in Figs. 67–70.

The clean and green Singapore 2015 campaign has taken momentum, even before the campaign commenced; few attempts have taken place to foster rooftop gardens on buildings' roofs. In addition, Singapore promoted 'sustainable urbanism

Fig. 61 Chofu City Gymnasium in Tokyo. (*Images source: https://thetokyofilesmedia.files.wordpress.com/2016/05/chofu-city-gymnasium-green-roof-tokyo-japan-3.jpg*)

after year 2014, where Singapore's landscape replacement policy (SLRP) requires that any greenery lost due to development must be replaced with publicly accessible greenery in equal area within the new building (https://www.designboom.com/architecture/woha-kampung-admiralty-singapore-10-30-2018/ 2019).

An iconic example is Maria Bay Sands, where a rooftop garden of 1.2 hectare (3 acre) tropical oasis is installed on the top of the towers in an example of SLRP execution (https://www.greenroofs.com/projects/marina-bay-sands-integrated-resort-skypark/ 2019). The leafy and grassy gardens include 250 trees and 650 plants as depicted in Fig. 67.

This has been also manifested in high-rise office building 'Robinson Tower' green roof and top garden in Singapore as shown in Fig. 68. Another example of this policy is promoting green roof in community, which is manifested in the rooftop garden and green roofs of block 38 in Jurong East, Singapore (https://resources.realestate.co.jp/living/japan-green-roof-buildings/ 2019).

Also, the sky garden of PARKROYAL hotel is anointer implementation of the SLRP enforcement. In addition, the iconic green roof of the School of Art, Design and Media, Nanyang Technological University, is an example of urban greenery in Singapore as shown in Fig. 69. This mix-used green building which is also a hybrid building incorporates a huge area of greenery (https://www.designboom.com/architecture/woha-kampung-admiralty-singapore-10-30-2018/ 2019).

It provides more than 100% of its footprint in a series of layered levels that have created embracing biodiversity as shown in Figs. 71, 72.

Fig. 62 Tokyo's Green Towers. (*Image source: https://www.archdaily.com/785775/ingenhoven-architects-reveal-plans-for-green-tower-in-tokyo/571547b4e58ece8f3800000e-ingenhoven-architects-reveal-plans-for-green-tower-in-tokyo-photo*)

Policies Regulating Green Roofs in Malaysia

Despite of the significant benefits of green roofs in Malaysia, so far green roofs are not widely spread when compared to Europe and Asian cities (Japan, Singapore and China). There are no regulated green roof policies in Malaysia, although they are listed in the Environmental Building Index (BEI). The Green Buildings index (GBI) indicates that if a green roof is installed, the building gets 2 points. In addition, Malaysian Carbon Reduction and Environmental Sustainability Tool (MCREST) includes ways to reduce carbon emissions and achieve environmental sustainability. This tool is considered a part of the government assessment.

In contrast, a survey on green roofs' users in high-density residential buildings revealed:

- Users did not realize the economic and climate benefits compared to the social, aesthetic and psychological benefits of green roofs,
- Maintenance officials were unaware of the economic benefits of green roofs since some of the problems took place after operation, and

Fig. 63 Rooftop gardens in Plaza Omotesando Harajuku. (*Image source: https://www.timeout.com/tokyo/things-to-do/best-rooftop-gardens-in-tokyo*)

- Architects and implementers of green roofs were concerned with climate benefits such as reducing temperature and reaching thermal comfort, especially in the use of large green roofs (Ismail et al. 2018).

There are some obstacles in using green roofs in Malaysia such as the lack of local studies on green roofs, insufficient awareness and shortage of experts (Ismail et al. 2018). Nevertheless, stakeholders should benefit from Singapore's advanced experience of green roofs since both countries have the same climatic conditions. Hence, the government started to encourage users to use green roofs. Figures 73 and 74 show examples of green roofs in Kula Lumpur, whereas Fig. 75 presents Heriot-Watt University campus green roof in Putrajaya, a smart and green city in Malaysia. Also, Klang Valley's vertical garden is shown in Fig. 76.

Policies Regulating Green Roofs in Australia

In Australia, green roof policies have been enacted since 2002. Also, there has been an increase in green roofs in Australia over the last 10 years. The city of Sydney and city of Melbourne have launched policies to regulate green roof and green wall in 2012 and 2015, respectively (Growing Green Guide 2014b). These policies are aligned with 2030 and 2040 sustainability targets of both cities.

Fig. 64 Mori Garden, first floor park in Roppongi district, Tokyo, Japan. (*Images source:* http://www.asiagreenbuildings.com/6520/world-green-building-council-and-ifc-join-forces-for-promoting-sustainable-construction/)

Fig. 65 Tokyo's Shinjuku Gyeon roof Garden. (*Images source: https://www.travelandleisure.com/attractions/festivals/tokyo-japan-cherry-blossom-peak-season-sakura?*)

Fig. 66 Villa Ronde in Japan. (*Images source: https://www.archdaily.com/134110/villa-ronde-ciel-rouge/501448d428ba0d5b4900079d-villa-ronde-ciel-rouge-photo & https://www.archdaily.com/134110/villa-ronde-ciel-rouge*)

Fig. 67 Green roof of Marina Bay Sands, in Marina waterfront bay in Singapore. (*Images source:* https://www.greenroofs.com/projects/marina-bay-sands-integrated-resort-skypark/; https://mcf-consultoria.com.br/blog/?p=658)

Sydney has established the Green Roofs and Walls Policy Implementation Plans (GRWPIP), whereas Melbourne has the Growing Green Guide 2014 (GGG) (Growing Green Guide 2014b).

Since enacting of this policy in Sydney, green roofs had increased by 23%, while Melbourne has also reported an increase in green roofs and walls, but the percentage is not known (Ismail et al. 2018). Nevertheless, in Melbourne local government area 28 green roof and wall projects were developed in 2016 and beyond. In contrast, Sydney local government area has 123 green roof and wall projects that started in 2016. Based on the policies in the modelled four scenarios, Table 2 lists the increase of GR based on policies of the four case studies modelled (Growing Green Guide 2014b). Figures 77, 78, and 79 illustrate green roofs developed in Sydney, Australia. In terms of greenery high-rise building, the One Central Park, Chippendale, in Sydney utilizes hydroponics and heliostat technologies to cultivate plants around the periphery of the building at all levels as shown in Fig. 77. This solution offers shading and reduces cooling energy and costs (http://imap.vic.gov.au/uploads/

Fig. 68 Robinson Tower's Green roof and top garden in Singapore. (a) Roof garden, (b) double-skin facades, atrium garden. (*Images source:* https://www.designboom.com/architecture/kpf-robinson-tower-singapore-sustainable-urbanism-07-31-2019/)

Fig. 69 Sky gardens and green roof of PARKROYAL on Pickering in Singapore. (*Images source: https://fcl.ethz.ch/research/high-density-cities/dense-and-green.html*)

Fig. 70 Green roof of the School of Art, Design and Media (ADM) at Nanyang Technological University in Singapore. (*Images source: http://news.ntu.edu.sg/pages/newsdetail.aspx?URL=http://news.ntu.edu.sg/news/Pages/NR2019_Jan 21.aspx&Guid=96eb2d18-367e-4f08-a96f-6053da41c852&Category=News+Releases*)

Fig. 71 Rooftop garden and green roof of an integrated mixed-used development—Kampung Admiralty' in Singapore. (*Images source: https://www.designboom.com/architecture/woha-kampung-admiralty-singapore-10-30-2018/*)

Fig. 72 Rooftop garden of block 38 in Jurong East, Singapore. (*Images source: http://www.asiagreenbuildings.com/wp-content/uploads/2014/11/DAS9f5yCxmezqq1jmYCoF3NOo1_1280.jpg*)

Growing%20Green%20Guide/Policy%20Options%20Paper%20-%20Green%20Roofs,%20Walls%20and%20Facades.pdf 2019; https://www.architectureanddesign.com.au/news/one-central-park-sydney-named-the-world-s-best-tal 2019).

It is clear that two drivers are behind green roof and green wall development in Australia; these are the environmental and people benefits: the environmental added values are establishing green roofs, walls and facades that create a microclimate system in cities which lead to the reduction of 3–4 °C, enhancing people's well-being, and creating fauna habitat and production of oxygen. For the people benefits:

Fig. 73 Green roof of Cantilever House, Malaysia. (*Images source:* https://inhabitat.com/green-roofed-cantilever-house-floats-above-the-malaysian-rainforest/cantilever-house-by-design-unit-sdn-bhd-3/)

Fig. 74 Green roof of a new building in Kuala Lumpur, Malaysia. (*Image source:* https://siouxcityjournal.com/lifestyles/home-and-garden/here-s-how-malaysia-is-figh)

the exposure to green nature has a positive effect on people's quality of life. The initiative will improve the emotional, psychological and physical well-being of people residing in high-density city areas. Another added value is the prospective increase of building value and forming of flora tourism (https://www.fma.com.au/news/green-roofs-walls-and-facades 2019). Also, the Green Roof Melbourne (GRM) is one of the programmes run by the Melbourne Committee; it was announced in 2010 and described as the largest green roof project in Australia (https://www.melbourne.vic.gov.au/building-and-development/sustainable-building/Pages/rooftop-project.aspx 2019).

Figures 80, 81, and 82 present green roofs, green walls and vertical gardens' examples in Melbourne. These examples include the residential tower (the Freshwater Place) in Melbourne that was to be completed in 2020.

Fig. 75 Green roof of Heriot-Watt University campus in Putrajaya, Malaysia. (*Images source: https://www.hw.ac.uk/about/work/malaysia.htm*)

Fig. 76 Klang valley vertical garden in Malaysia. (*Image source: https://www.havehalalwilltravel.com/secret-attractions-klang-valley-malaysia ting-the-climate-crisis-with/article_b02b1175-62d4-5135-8d06-7c45fbc2cc28.html*)

Table 2 Outcomes of GR projects in Sydney, according to modelling of four scenarios

Total number of projects at the end of period	Base level of projects	Annual growth rate	Short term (5 years to 2022)	Medium term (10 years to 2027)	Long term (15 years to 2032)
Scenario 4—voluntary heavy (Singapore)	123	29.8%	453	1668	>3245
Scenario 3—voluntary medium (Rotterdam)	123	17.1%	271	595	1310
Scenario 2—voluntary light (London)	123	12.4%	220	395	707
Scenario 1—mandatory (Toronto)	123	9.6%	194	307	485
Scenario 4—voluntary heavy (Singapore)	123	29.8%	453	1668	>3245
Scenario 3—voluntary medium (Rotterdam)	123	17.1%	271	595	1310

Source: Expanding the Living Architecture in Australia Project Get the data

Fig. 77 Vertical gardens of One Central Park, Chippendale, a greenery high-rise building in Sydney, Australia. (*Images source:* https://urbannext.net/one-central-park/; https://fineartamerica.com/featured/one-central-park-chippendale-sydney-martin-berry.html)

In addition, a half-acre roof garden (level 10) was developed in CH2 commercial building that houses Melbourne City Council.

Comparing the building that achieved maximum Six Green Star rating, the upgraded CH2's emission is 64% less. The CH2 building introduced an innovative concept of original natural vegetated state. This is accomplished by using a roof garden, which also serves as a break-out and recreation space for staff. The northern

Fig. 78 Green roof of the new VOCO hotel in Sydney, Australia. (*Images source:* https://www.traveltalkmag.com.au/hotels-and-resorts/sydneys-newest-hotel-boasts-outdoor-green-waterfalls)

facade also incorporates planter boxes situated east and west of each northern balcony (Fig. 80). The greenery, when compared to the existing Council House, has contributed to the efficiency of the building. Figure 81 shows the estimated reduction in energy, water and emissions (https://www.wikiwand.com/en/Council_House_2 2019).

Another example is the Condor Tower (2005) that has a 75 m^2 garden on the fourth floor. Since 2008, city councils and influential business groups in Australia have been active in promoting the benefits of green roofs. Furthermore, ecologist Lloyd Goodman and other experts in a ground-breaking experiment developed in June 2014 Tildencia plants that were installed on four floors (92, 91, 65 and 56) of the Eureka Tower, which is considered one of the world's tallest rooftop garden in Melbourne's planned skyscraper (https://www.australiandesignreview.com/news/worlds-tallest-rooftop-garden-planned-for-eureka-tower/ 2019). Figure 82 depicts Eureka Tower's tallest roof garden.

In Victoria, the Victorian Desalination Project contains a "living tissue" of 98,000 native Australian plants on a surface area of over 26,000 m^2, to be a roof of the desalination plant, designed to blend the building with natural views and provide acoustic protection, corrosion resistance, thermal control and low maintenance (https://www.archdaily.com/876189/the-victorian-desalination-project-and--ecological-reserve-aspect-studios 2019). Figure 83 shows the green roofs of Victorian Desalination Project in Victoria, Australia (http://www.livingroofs.org.nz/policy/ 2018).

Fig. 79 Rooftop garden of high-rise office tower in Sydney, Australia. (Images source: https://i2.wp.com/www.urbangardensweb.com/wp-content/uploads/2013/03/bligh-street-solar-rooftop-architectus.jpg)

Policies and Laws Regulating Green Roofs in New Zealand

New Zealand promoted green roofs and fences in the central city through demonstration projects, incentives and best practice guides. There is no direct legal requirement to provide a green roof in New Zealand. There may be an indirect need to mitigate the impact on the utility environment or protected species and improve rainwater (https://www.greenroofs.com/2019/08/28/worlds-largest-urban-farm-to--open-on-a-paris-rooftop n.d.). The Waitakere City Council (WCC) in the west of Auckland, which became Auckland Council in 2011, has approved a set of information on green roofs, from which a living green roof model was achieved in the Henderson Council building. New Zealand is one of the laggards in the use of green roofs compared to Germany and the USA (https://www.greenroofs.com/2019/08/28/worlds-largest-urban-farm-to-open-on-a-paris-rooftop n.d.), so new and existing roofs should be considered as a real opportunity to adapt to climate change. Examples of green roofs are presented in Figs. 84, 85, and 86. In New Zealand, green roofs can make commercial buildings more efficient and environmentally friendly such as NZI Centre building (Fig. 85).

Auckland International Airport has taken further steps in promoting green roof in the form of a sculptural multifunctional space surrounded by woven mesh screens and topped with a living green roof. The "Te Kaitaka" is known as The Cloak (https://inhabitat.com/the-cloak-fearon-hay-architects-install-a-maori-inspired-green-roof-at-auckland-

Fig. 80 Green garden and green facades, CH2—Melbourne City Council, Australia. (*Images source:* https://rpc6yg.wordpress.com/2011/10/25/the-sustainable-case-study-council-house-2-ch2-part-i/; https://inhabitat.com/ch2-australias-greenest-building/ch22; https://www.arch2o.com/ch2-melbourne-city-council; https://realty.economictimes.indiatimes.com/slide-shows/10-greenest-building-around-the-world/51365306 house-2-designinc/arch2o-ch2-melbourne-city-council-house-2-designinc-11/; https://www.ronstantensilearch.com/council-house-2-greening-systems/; http://www.solaripedia.com/13/174/1743/council_house_2_roof.html)

Box 3 Green Roof Policies in Australasia

From the above, we conclude that there are some initiatives and projects to use and implement green roofs and Melbourne which is the first city to support green roofs and walls.

New Zealand has no direct legal requirement or regulatory policies for green.

international-airport/te-kaitaka-the-cloak-fearon-hay-architects-2/ 2019). It helps in reducing cooling loads and offers a unique aesthetic; it was inspired by the fine weaving of Māori cloaks as illustrated in Fig. 86.

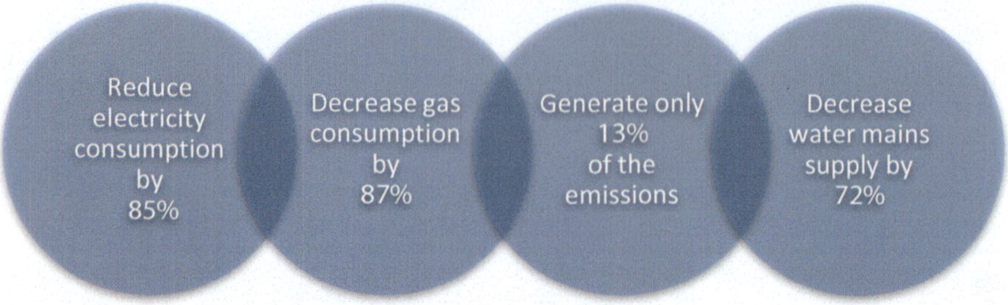

Fig. 81 Reduction in CH2 performance resulting from greenery and other technologies. (*Source:* https://www.wikiwand.com/en/Council_House_2)

Fig. 82 Eureka Tower in Melbourne. (*Image source:* https://www.wikiwand.com/en/List_of_tallest_buildings_in_Melbourne)

Fig. 83 Victorian Desalination Plant and Ecological Reserve, https://www.archdaily.com/876189/the-victorian-desalination-project-and-ecological-reserve-aspect-studios

Fig. 84 Green roof of former Waitakere City Council in Auckland, New Zealand. (*Image source:* https://www.rnz.co.nz/news/national/353597/sale-of-57-million-former-waitakere-council-hq-on-the-cards)

Fig. 85 Green roof of NZI Centre in Auckland, New Zealand. (*Images source:* https://www.colliers.co.nz/news/2018/green%20roofs%20could%20bring%20hidden%20costs/)

Policies Regulating Green Roofs in South Africa

According to research work published in 2016, it indicates that green roof systems are a new concept in South Africa (Labuschagne and Zulch 2016). The implementation of green roofs may require a different approach from other cities around the world. The government provides incentives to encourage the development of green roof systems that contributes to increasing green roofs. Adequate income from

Fig. 86 Innovative green roof at Auckland International Airport, Auckland, New Zealand. (*Image source:* https://inhabitat.com/the-cloak-fearon-hay-architects-install-a-maori-inspired-green-roof-at-auckland-international-airport/te-kaitaka-the-cloak-fearon-hay-architects-2/)

construction in the private sector can be obtained without incentives; 57.35% of respondents felt that green roof systems are unknown or popular in South Africa, as they have never seen a green roof system in South Africa (Labuschagne 2016). Nonetheless, there are three types of green roof systems in the South African provinces (Gauteng—Western Cape—KwaZulu-Natal, Kaitburn Cape). It has been pointed out that the semi-intensive green roof system is most appropriate for South African conditions and the intensive green roof systems are not feasible at all (http://scholar.ufs.ac.za:8080/xmlui/bitstream/handle/11660/4698/LabuschagnePH.pdf;jsessionid=1EC25D96BCB40D67C15F694DA2ADD00B?sequence=1 2019).

Although there is little knowledge of green roof, the required materials for building green roof systems are available and there is demand for green roof systems in South Africa, specifically Johannesburg, where there are public green gardens, but due to the following problems, green roofs could be of great added value for the city:

- Citizens avoid parks because of the increased crime rate, so green roofs provide a safe natural environment.
- High unemployment in South Africa, so green roofs create new jobs.
- Insufficient energy supplies in South Africa, so there is a need to increase isolated buildings in order to provide electricity and improve air quality, which is provided by the green roofs, specifically in the capital Johannesburg because of the presence of bad air (http://scholar.ufs.ac.za:8080/xmlui/bitstream/handle/11660/4698/LabuschagnePH.pdf;jsessionid=1EC25D96BCB40D67C15F694DA2ADD00B?sequence=1 2019).

Green roof policies were introduced in Cape Town and Durban in 2009. Cape Town's Management of Urban Storm Water Impacts Policy (CTMUMSIP) was developed in 2009, which states about water-sensitive urban design (WSUD). It provides an opportunity for green roofs to be absorbed into the policy. The policy describes such type of design as practices that encourage biodiversity, amenity and aesthetics (Management of Urban Stormwater Impacts Policy, Roads and Stormwater Department 2009). It also states that the city may introduce incentives to encourage these types of design. Cape Town Smart Building Handbook (CTSBH) has been launched in 2012, in which a pilot garden roof was developed on a city building as a model (City of Cape Town 2012). In addition, other projects show support for green roofs (World Architectural Community 2019). These embrace two intensive garden roofs implemented in city-owned buildings (2000 and 2006). Another innovative project includes the implementation of green walls within informal settlements to promote a greater quality of living for the lower income residents of Cape Town.

When it comes to green roof development, Durban is considered miles ahead than Cape Town. The city of Durban has been running a Municipal Climate Protection Programme since 2006 that entails a series of phases including extensive local research and plan formulation. As part of this programme, the Green Roof Pilot Project (GRPP) was launched in 2008 where eight green roofs were developed on a city building, each featuring different plants, growing mediums and green roof techniques (Booysen 2014). Figures 87, 88, 89, and 90 show examples of green roofs in South African cities (Cape Town, Johannesburg and Durban).

Policies and Legislation Governing Green Roofs: Regional Outlook

Around 75% of buildings and infrastructure in the MENA region are directly exposed to climate change, heat island effect and natural disasters such as droughts, floods, hot days and storms resulting in floods (nd n.d.). However, there are few

Fig. 87 Roof garden—Western Cape's Environment Department, Cape Town, South Africa. (*Image source:* http://www.carolizejansen.com/DEADP_Daktuin2.html)

Fig. 88 Green roof at SANRAL, South Africa. (*Images source:* https://i.pinimg.com/originals/f3/11/eb/f311eb5fe90e23 7081cf62a38733358e.jpg)

Fig. 89 Green roof at DBSA, South Africa. (*Image source:* https://i.pinimg.com/originals/ee/f7/06/eef706f0df50be6e33ae7acfcc29b122.jpg)

Fig. 90 Green roof of Master Builders Centre in Durban, South Africa. (*Image source:* https://www.greenroofs.com/projects/master-builders-association-durban-westville/)

policies to promote the use of green roofs in the MENA region and no guidance and targets for development.

Some green building systems are being developed in the MENA region, which partially encourage the presence of green roofs on green buildings, but there are no policies regarding the design and installation of green roofs. However, currently there are only a few examples of green infrastructure, especially green roofs in the MENA region. The use of green roofs needs to encourage the users and establishing of policies to regulate these green roofs. In addition, planners and policy and decision makers should approve an integrated policy that adopts green infrastructure strategies to provide opportunities to reduce the environmental impacts of dense urban areas, where green roofs can play an important role.

Municipalities should also use different tools, and provide direct or indirect incentives that have already been implemented in Europe, North America and Asia, to encourage and promote the implementation of green roofs in the MENA region, yet to gain the climate, structural and economic benefits.

Green roof policies and programmes are unique in each country due to different climatic conditions. Resident policies can be reviewed and identified in the following points. In 2019, the United Nations Framework Convention on Climate Change (UNFCCC) indicated that there is high potential for green roofs in Egypt (https://www.emaratalyoum.com/local-section/2009-10-01-1.1924 2019; https://unfccc.int/climate-action/momentum-for-change/activity-database/

momentum-for-change-the-future-of-green-roofs-in-egypt 2019). The following section presents green roofs in some countries in the MENA region such as Kuwait, UAE, Jordan and Lebanon as well as Egypt (Fig. 91).

Policies and Laws Regulating Green Roofs in Kuwait

Kuwait Council began in 2009 to work on accelerating the transformation of Kuwait into a sustainable country, focusing on the benefits to the country by using green roofs in reducing temperatures and managing rainwater and drainage. Citizens play a major role in spreading the idea of green roof cultivation due to the fact that they are the direct beneficiaries through awareness campaign on the importance of such cultivation, besides the instructions by experts and specialists in the field, e.g. the conditions set by the government to implement green roofs, which are described in the following points (Researchgate.net n.d.):

- Roof plants should be exposed to direct sunlight for at least 4–5 h a day, so dense shadow places (from buildings or trees) should be avoided, yet use of ornamental plants is permitted.
- Places which are exposed to strong winds should be protected.
- Plants should be close to a water source (about 15–20 m), so as not to load users.
- Green roofs should not be sprayed or any unauthorized chemical pesticides should not be used, where recently it has been observed that some non-specialists have adopted improper policies on establishing roof systems and how to plant.
- NGOs play an active role through encouraging and activating the role of citizens in providing technical and financial support to spread the idea of roof cultivation among urban residents (Researchgate.net n.d.). Figures 92 and 93 show green roof and green wall examples in Kuwait.

Policies and Laws Governing Green Roofs in the UAE

UAE's government aims to convert all roofs of buildings and houses in Dubai into green roofs. Dubai Municipality has identified the buildings, which will be applied to the green roof system. Dubai Municipality has also identified the new buildings under construction and under license issuance to be applied first (https://www.greenroofs.com/2018/03/05/greenroofs-com-project-of-the-week-dubai-opera-garden-greenroof-and-walls/ 2019). According to Dubai Licensing Department

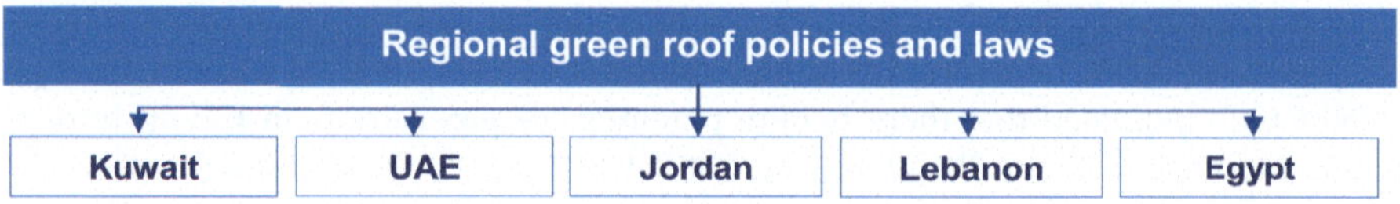

Fig. 91 Cities in the MENA region where green roofs are developed and enacted

Fig. 92 Green roof of Al Shaheed Park—Phase 2 in Kuwait. (*Image source:* https://248am.com/mark/information/al-shaheed-park-phase-2-now-open/; https://trends.archiexpo.es/zinco-gmbh/project-66390-258977.html)

Fig. 93 Green roof in Kuwait. (*Images source:* https://i.pinimg.com/originals/1b/74/5b/1b745beeedfb20ddfc537dd831f5d625.jpg)

(DLD), the application of green roofs will be mandatory for all types of buildings except for industrial buildings with metal roofs, with a minimum of 30% of the total roof area and optionally on facades (https://www.greenroofs.com/2018/03/05/greenroofs-com-project-of-the-week-dubai-opera-garden-greenroof-and--walls/ 2019).

The government also provides incentives to motivate buildings' owners to implement the project by allowing the construction of a lounge on the roof of residential villas with an area equivalent to 25% of the cultivated area of the roof to increase the construction area or by allowing the establishment of swimming pools or lounges on the roof of public and multilevel buildings with an area equivalent to 25% of the cultivated roof area. Nevertheless, the UAE Green Roofs Guidelines (UAE-GRG) specify the following (https://www.greenroofs.com/2018/03/05/greenroofs-com-project-of-the-week-dubai-opera-garden-greenroof-and-walls/ 2019):

- There is no specific type of plants required, but it is preferable to use local plants and ornamental plants.
- The selection of plants with low water requirements and low cost of planting and maintenance.

Dubai—UAE—is one of the first sustainable cities in the Middle East to serve as a model for future cities to absorb rapid urbanization, reduce energy use and water consumption, and lower air pollution. Examples of green roofs and green walls resulting from green roof policies and initiatives are illustrated in Figs. 94, 95, and 96. Dubai Silicon Oasis Authority (DSOA) is one of these leading examples. DSOA increases green area by 30% compared to the 2014 level. The areas of green roofs have increased from 1700 m^2 in 2014 to 2800 m^2 in 2016 (https://www.greenroofs.com/2018/03/05/greenroofs-com-project-of-the-week-dubai-opera-garden-greenroof-and-walls/ 2019) as shown in Fig. 94.

Another example of green roof is the Dubai Opera Garden. The green roof has 32,292 square feet of green area and cultivated vegetated terraces. It also has approximately 500 square feet green walls as depicted in Fig. 95. This utility house has been facing a challenge of weight restraints as originally it was not planned to include green roof, which will serve also as a venue place. The installed green roof is a heavy type (between 30 and 50 cm) of local sweet soil, which brings the weight up to 750 kg/m^2 (https://www.greenroofs.com/2018/03/05/greenroofs-com-project-of-the-week-dubai-opera-garden-greenroof-and-walls/ 2019).

The first rooftop garden in Dubai was installed in one of the high-rise buildings in the main thoroughfare—Sheikh Zayed Road. Figure 96 presents the urban garden on the roof of the high-rise car parking annexed building (https://whatson.ae/2016/02/dubais-first-rooftop-garden-middle-tecom 2019; https://www.thenational.ae/lifestyle/food/dubai-s-first-rooftop-garden-in-pictures-1.194393?videoId=5712466005001 2019).

Policies and Laws Regulating Green Roofs in Amman, Jordan

In 2006, the Greater Amman Municipality (GAM) established the Urban Agriculture Bureau (UAB) and began its formal programme as part of the Urban Agriculture Initiative (UAI) and the Food Security Initiative (FSI). Through this initiative, the UAI became an integral part of the municipal agenda and more than 400 green roofs were established in Amman, but unfortunately, UAO was integrated with the park management in 2015 (nd n.d.).

Fig. 94 Rooftop green roof of Dubai Silicon Oasis Authority's building, UAE. (*Images source:* http://www.hillsandfort.com/landscape/gallery.php?portId=15&page=projects)

Fig. 95 Dubai Opera Garden and green roof and green walls, UAE. (*Images source:* https://www.greenroofs.com/2018/03/05/greenroofs-com-project-of-the-week-dubai-opera-garden-greenroof-and-walls/)

Fig. 96 Dubai rooftop garden and green roof. (*Image source:* https://whatson.ae/2016/02/dubais-first-rooftop-garden-middle-tecom/; https://www.thenational.ae/lifestyle/food/dubai-s-first-rooftop-garden-in-pictures-1.194393?videoId=5712466005001)

Also, the GAM joined a regional initiative related to urban agriculture, where 100 houses were selected to establish some urban agricultural activities, including green roofs. Green roofs were introduced through individual initiatives in 2009 for food security, green space, storm water management and energy efficiency.

Amman Green City Committee (AGCC) appointed specialized committee in 2015 to increase Amman's green area from 2.5% to 5% in 2020 (https://whatson.

ae/2016/02/dubais-first-rooftop-garden-middle-tecom 2019). There are sites to raise awareness of the use of the green roofs in Amman, but there is no clear policy regulating the green roofs at the national or municipal levels in Amman and this is not integrated into the policies of other sectors such as water and energy. Nonetheless, there were some incentives from the UAI from 2006 to 2009. Later, green roofs were combined with the Green Building Assessment (GBA) within the Jordan Green Building Council (JGBC) (http://www.cedro-undp.org/Projects/details/19 2019). Figures 97, 98, and 99 show traditional green roof and green wall examples.

Policies and Laws Regulating Green Roofs in Beirut, Lebanon

It seems that Lebanon has no evidence of green roof policies, but in 2006, the Central Bank of Lebanon (CBL) installed the large green roof. This project was an initiative supported by UNDP and was funded by the European Union and the Netherlands (https://www.egypttoday.com/Article/1/52769/In-pics-Egypt%E2%80%99s-rooftops-going-green 2019). Figure 100 presents the CBL green roof in Beirut.

Policies and Legislations Governing Green Roofs in Egypt

Before 2018, green roofs in Egypt were mainly depending on the donors' initiatives such as the Food and Agriculture Organization—German Initiative (GIZ)—and private initiatives and have no broad public impact that seeks to strengthen the green roofs (Steier 2018). Nevertheless, in 2018, three governmental buildings in Egypt including the city of Hurghada decided to join the state's project of going green, and used their rooftops for agriculture (https://www.egypttoday.com/Article/1/75486/Egypt-starts-green-roofing-initiative-in-government-buildings 2019).

The Governorate of Red Sea stated that three green buildings are located in the city of Hurghada, including green roofs (Fig. 99). On September 29, 2019, Egyptian Ministry of Environment launched green roof initiative to increase the flora area nationwide. An example in Cairo is planting the Egyptian Environmental Affairs Agency, EEAA (https://www.egypttoday.com/Article/1/75486/Egypt-starts-green-roofing-initiative-in-government-buildings 2019).

According to the Director of Agriculture in Governorate of Cairo, the capital does not have agricultural areas and if we can plant 10% of the houses' roofs it will be a great achievement and will work to face high prices, in addition to the aesthetic appearance of each house, pointing out that the municipality provides all support to its citizens. However, green roof initiatives are still less widespread (https://www.emaratalyoum.com/local-section/2009-10-01-1.1924 2019).

In Egypt, less soil green roof systems (such as a wooden table or potted system—a mobile aquatic green roof system) are used, and fish farming is tested next to plants in a closed cycle of plant nutrient solution to eliminate the need for roof insulation layers and provide a source of fresh, pesticide-free healthy food. The

Fig. 97 Urban agriculture green of a roof in a house in Amman, Jordan. (*Image source:* https://cityfarmer.info/urban-agriculture-greens-roofs-and-homes-in-amman-jordan/)

Fig. 98 Green wall covering a building in Jabal Weibdeh, Amman—Jordan. (*Image source: ttps://*i.pinimg.com/originals/d6/d3/6a/d6d36aca8e320f70662c6e82cb65c33e.jpg)

Ministry of Agriculture has also produced and distributed a technical manual to educate people, help people produce vegetables and raise public awareness.

It could be concluded that many entities seek to build green roofs in order to provide fresh vegetables and fruits to gain money without paying attention to the

Fig. 99 Lebanon Central Bank Building's green roof, Beirut. (*Images source:* https://www.architecturelab.net/central-bank-of-lebanon-roof-garden-beirut-green-studios/; https://cdn.architecturelab.net/wp-content/uploads/2014/12/91.jpg)

Fig. 100 Green roof on top of a governmental building in Egypt. (*Images source:* https://www.egypttoday.com/Article/1/52769/In-pics-Egypt%E2%80%99s-rooftops-going-green)

environmental benefits such as the temperature reduction, carbon dioxide emission mitigation and percentage of savings in energy consumption.

There are some initiatives to use green roofs, but there are no clear policies or laws regulating the cultivation of the green roofs in Egypt. Some initiatives resulted from some private companies to promote green roofs and green walls (https://schaduf.com/schaduf-projects/tag-sultan/ 2019) as shown in Fig. 100.

Obstacles to Strengthening Green Roofs in Egypt

Although green roofs lead to many benefits, there are many obstacles to the promotion of this technology in Egypt as shown in Table 3.

Proposals for the Development of Policies Governing Green Roofs

There are six stages for the development of the green roof policy (GRP) that have been tried before in Melbourne and have proved to be highly successful (Ngan

Table 3 Obstacles to enhancing green roofs in Egypt

Lack of research	Lack of awareness	Cost
– The lack of technical research for the development of green roofs in Egypt when refitting green roofs, low-cost technology for maintenance and light weight are major obstacles to the development of green roofs in Egypt	– Although there are many NGOs working to train citizens on how to create a green roof – Indicators stated in the environmental reports on the importance of green roofs, most of the roofs of buildings are still traditional and this is a sign of lack of awareness – More information is needed to provide citizens with not only the social and economic costs and benefits, but also the environmental potential that can reduce the impact of thermal islands – Use the roofs for various activities such as washing and drying or for holding social events/for storage	– The cost of the full life cycle must be analysed. Intensive and expanded green roofs may have a higher initial cost than traditional roofs, but given the full life cycle, maintenance costs are lower than traditional roofs with or without energy saving – Rainwater management, energy conservation and improved air quality should be considered

Source: https://unfccc.int/climate-action/momentum-for-change/activity-database/momentum-for-change-the-future-of-green-roofs-in-egypt

2004a), and Melbourne is considered one of the most advanced cities in the use of green roof. Hence, countries wishing to promote green roof policies should follow the methods listed in Figs. 101 and 102.

Conclusions

By identifying the world's green roof policies, it has been found that green roofs are widely used in Europe, North America and Asia. Some countries have introduced policies and laws that encourage the use of this technology in Europe. Cities like Munich (Germany), Copenhagen (Denmark) and Paris (France) have laws requiring the installation of green roofs for certain types of buildings, while in North America, Toronto and Vancouver in Canada were the first cities in the world to implement an internal system containing all buildings over 2000 m^2 between 20 and 60% of green roofs.

Similarly, the American cities such as Portland, Minneapolis and Chicago as well as San Francisco and North Carolina have already incorporated green roofs into urban planning laws and there are many cities around the world that have internal regulations for green roofs. Many other cities, namely London (guidelines on

Fig. 101 Green walls of a private building in Cairo, Egypt. (*Images source:* https://schaduf.com/schaduf-projects/tag-sultan/)

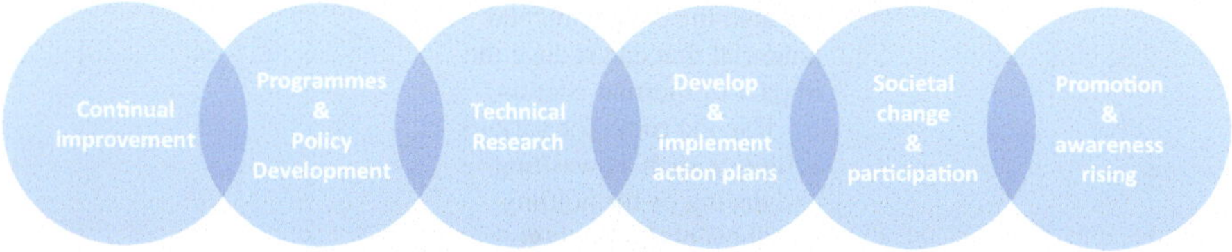

Fig. 102 Methods of policy development, design and implementation of green roofs

living room surfaces), Paris, Berlin, Shanghai and Beijing, have direct and indirect policies that include green roofs.

In addition, there are countries that developed and used green surface in Australasia such as Sydney, Melbourne, Berth and Victoria in Australia as well as Auckland and Willington in New Zealand.

In South Africa, some policies were developed, but not in large numbers. These were noticed in Durban, Cape Town and Johannesburg; nevertheless, more policies and laws are needed to govern the cultivation of green roofs and urban farming.

In the MENA region, the development of green roof policies is at a slow pace, but recently regional countries such as Kuwait, UAE, Jordan and Egypt have adopted many initiatives for cultivating green roofs. These countries seek to achieve green roofs, but it seems without policies to implement them, so they are not advanced in roof cultivation. Egypt recently made efforts, but it is needed to raise awareness of the use of green roofs and urban farming, especially in Cairo, not only in economic terms, but also in terms of environmental, construction, health, social, aesthetic and psychological.

In synopsis, Tables 4, 5, 6 and 7 present a brief summary of green roof policies in European Union countries and internationally such as North America, South America and Asia as well as Australia, New Zealand and South Africa. Also, Table 7 lists regional countries, including Kuwait, UAE, Jordan, Lebanon and Egypt.

Table 4 Summary of policies regulating green roofs in European Union countries

Policies and legislations governing green roofs internationally		
Some of European Union countries		
Location	Policies	Summary
Austria		
Linz	Mandatory requirement + incentives	Government commits the new building owner to establish green roofs where it will grant the cost and the user will be paying 5% of the cost
Denmark		
Copenhagen	Achieving sustainability goals +	– Part of the city's goals is to be more sustainable and achieve a carbon reduction system – Commitment to green roof in buildings, and there are grants and tax exemptions provided by the government to convert the roofs of existing buildings to green
Germany		
Berlin Munich Cologne Linz Stuttgart	Mandatory requirement Mandatory requirement + incentives Storm water reducing storm water charges Mandatory requirement	Mandatory targets to increase permeability and "green" surfaces, through planning regulations – Using a wide range of green roof techniques – There are grants for the use of green roofs in urban areas and discounts on storm water charges – Commitment to the use of green roofs in the appropriate flat surfaces, and as a result green – There are a range of financial incentives to reduce and control storm water run-off – Green roofs are promoted as a means to manage storm water – Wide green roofs on buildings are used as a means of reducing rainwater charges – There is a plan to reduce storm water charges on the user of green roofs. Commitment to use the green roof leads to creation of a large amount of green roofs in the city
England		
London	Guidelines on living room surfaces	The London Plan policy expects that provision of intensive, extensive or recreational roof space (or a combination of these) should be provided on all new developments Recommended standards are provided including suggestions for the area of roof space covered, depth of substrate, etc. Outside of government, the green roof industry along with academics has developed the GRO Code—best practice green roof installation
Switzerland		
Basel	Mandatory requirement	It is required amendment of the Building and Construction Law and the renovation of all new roofs and their conversion into green roofs, with the provision of incentives and tools

Source: Green Roofs, Walls & Facades Policy Options Background Paper, February 2014

Table 5 Policies and legislation governing green roofs internationally

Policies and legislation governing green roofs internationally		
USA		
Location	Policies	Summary
USA		
Austin Boston Chicago Los Angeles Milwaukee Minneapolis New York Philadelphia Portland Washington San Francisco Seattle	Increased incentives Meet LEED certification Plans for Grant Program Meet LEED CR Grant + tax exemption Offering indirect Offering tax exemption Tax exemption Offering grant + tax Meet LEED CR Increased incentives Achieving GB Requirements + grants	There are various incentive programmes, and it is proposed to increase grants and incentives, and to award bonuses to the green roof which cover 100% of the roof of the building. New projects must use the green roof system to be LEED certified and there are points awarded to retain rainstorm water on-site – Formulate policies to promote the use of green surfaces – Promote public education to spread the knowledge of green roofs – A series of seminars on "green roofs for homeowners" – Large projects must use green roof system to be LEED certified – Offering tax exemptions for green roofs per square foot price – The government works on reducing storm water fees for efficient storm water management when using green roofs – Offering 25% of green roof costs, from January 2009 to March 2013 – Buildings using green roofs covering more than 50% of their areas will be having a yearly tax exemption – There are up to 25% tax exemptions for green roof users that cover 50% of the roof of the building – Offering grants to reduce rainwater run-off from the use of green roofs – All buildings owned by the city are required to cover their roofs with 70% green roof – A special discount programme is offered to green roof users – New projects must use green roof system to be LEED certified – There are grants for the use of green roofs and sustainable buildings – New buildings must use green roofs to obtain a green worker grade to obtain a building permit – Promote innovative designs such as green roofs and green walls through grants for use – A range of bonuses and incentives are provided to users of green roofs

Source: Green Roofs, Walls & Facades Policy Options Background Paper, February 2014

Table 6 Policies and legislation governing green roofs internationally

	Policies and legislation governing green roofs internationally		
	Location	Policies	Summary
	Canada		
North America	City of Richmond City of Port Coquitlam Toronto Vancouver	Building by-law storm water Green roof specific by-law Green roof-specific by-law + financial incentives LEED certification requirement	By-laws requiring new buildings with over 2000 m2 of floor area to meet storm water management objectives. The by-law called "*Green Roofs and Other Options Involving Industrial & Office Buildings Outside the City Centre*" does not mandate green roofs The by-law was introduced in 2008, but as of 2011 only 2 or 3 projects had been completed, none of which installed green roofs Green roof requirement on new construction over 5000 m2 within the city's general zoning It is the first North American city to pass by-laws requiring a green roof on new commercial, institutional and residential developments with a gross floor area of 2000 m2 or above. Requirements range from 20% of assessed available roof space for 2000 m2 to 60% for 20,000 m2 or larger buildings. Applies to industrial building plans submitted after April 30, 2012, with lesser requirements. Also used is the financial Leadership in Energy and Environmental Design (LEED) programme that provides third-party verification of green buildings. Policy and by-laws have adopted requirements for LEED gold certification, which does not require, but encourages, green roofs and green walls, and has led to several large-scale green roof installations
South America	**Brazil**		
	Recife	Green roof-specific by-law	Buildings with a height of more than four floors, its roofs will be covered by local plants. It also applies to commercial buildings with an area of more than 400 m2
	Argentina		
	Córdoba	Financial incentives	In July 2016, the city of Córdoba rehabilitated the green roofs of existing buildings supported by the incentive programme
	New Zealand		
	New Zealand		
	Wellington	Government recommendations	The Government of New Zealand is promoting green roofs and fences in the central city through demonstration projects
Africa new	**South Africa**		
	Johannesburg	Government recommendation	The municipality noted that the semi-intensive green roof system is most suitable for South African conditions

Source: Green Roofs, Walls & Facades Policy Options Background Paper, February 2014

Table 7 Policies and legislation governing green roofs internationally, regionally, locally

	Local	Regional	International
	Location	Policies	Summary
Asia	**Singapore**		
	Singapore	Planning system incentives	– Rooftop greenery is promoted by not including certain portions of the areas used for greenery in the calculation of a building's gross floor area
	Japan		
	Tokyo	Government recommendations	– A 25% coverage for new buildings was mandated in 2001; this led to an increase from approx. 52,500 m2 of roof coverage to approx. 102,500 m2
	China		
	Shenzhen Beijing Shanghai	Government recommendations	– There are pilot projects with green roof over 5000 m2 – Identifies the basic requirements, roof's type, plant design, plant selection and green roof technology of facilities in the Beijing area – Provides technical support for green roofs
Asia	**Malaysia**		
	Kuala Lumpur **Kuwait** **UAE** **Jordan**	Analytical studies Government recommendations Mandatory condition + incentives Guidelines incentives by the urban agriculture	– There are some studies conducted to enhance green roofs in Malaysia, such as that conducted in Klang Valley through a questionnaire for green roof users of intensive residential buildings – Technical guides and financial incentives are provided to support the idea of green roof cultivation among urban residents – Applying green roofs is mandatory on all types of buildings except for industrial buildings with metal roofs. The government provides incentives to motivate building owners to implement green roofs – Integration of green roofs with green building assessment in the Jordan Green Building Council, where it was one of its targets to increase the green area in Amman from 2.5% to 5% in 2020
Egypt	**Cairo**	Government recommendations	– There are some government and NGO initiatives that seek to strengthen the green roofs, but with the aim of benefiting from the economic and aesthetic returns only and not looking at the environmental, structural, social, etc.

Source: Green Roofs, Walls & Facades Policy Options Background Paper, February 2014

References

Booysen, K. (2014). Our future cities: Co-creating more inclusive and visionary cities; serious about green roofs: Are Cape Town and Durban serious? [Online]. http://futurecapetown.com/2014/07/what-green-roof-incentives-and-policies-are-national-cities-using-part-iii/

Carter, T., & Fowler, L. (2008). Establishing green roof infrastructure through environmental policy instruments. *Environmental Management, 42*(1), 151–164.

City of Cape Town smart building handbook: A guide to green building in Cape Town, Published by the City of Cape Town, June 2012, ISBN: 978-0-9870321-0-2. https://grthardware.co.za/wp-content/uploads/2019/07/smart-building-green-building-handbook_2012-06.pdf

Conservation design Forum. Retrieved November 28, 2019., from https://www.cdfinc.com/Project?project_id=31

Dunnett, N., & Kingsbury, N. (2008). *Planting green roofs and living walls*. Portland, OR: Timber Press.

Eidt, J. Landscape urbanism: Green roofs, community farms in Japan. *Sustainability*, 2014, Wilder Utopia. Retrieved December 8, 2019, from https://www.wilderutopia.com/sustainability/landscape-urbanism-green-roofs-community-farms-in-japan/

ENVS—Department of Environmental Science. (2010). *Biodiversity, buildings, climate and air quality, social inclusion, water - green-roof programme*. Copenhagen: Aarhus University.

European green capital award 2018 technical assessment synopsis report April 2016.

General Services Administration. 2011. The benefits and challenges of green roofs on public and commercial buildings. Exit.

Green Roofs for Healthy Cities. 2017. 2016 annual green roof industry survey: executive summary.

Growing Green Guide, Green Roofs, Walls & Facades Policy Options Background Paper, February 2014a.

Growing Green Guide. (2014b). https://theconversation.com/australian-cities-are-lagging-behind-in-greening-up-their-buildings-97088/

HE MIAOMIAO, Promoting Green roofs in CHINA: A comparison of Green roof policies in the (U.S.) and CHINA, fulfilment of the architectural design University Florida, 2011.

Retrieved December 1, 2019, from http://imap.vic.gov.au/uploads/Growing%20Green%20Guide/Policy%20Options%20Paper%20-%20Green%20Roofs,%20Walls%20and%20Facades.pdf

Retrieved December 1, 2019, from http://scholar.ufs.ac.za:8080/xmlui/bitstream/handle/11660/4698/LabuschagnePH.pdf;jsessionid=1EC25D96BCB40D67C15F694DA2ADD00B?sequence=1

http://wendycitychicago.com/chicago-green-roofs/, n.d.

Retrieved December 5, 2019, from http://www.cedro-undp.org/Projects/details/19

http://www.livingroofs.org.nz/policy/, 2018

Retrieved November 30, 2019, from http://www.roofchina.com/article/show_article.php?id=459

https://ar.smarthomemaking.com/20-green-walled-buildingsinspiring-your-own-plant-cladded-home, n.d.

https://archello.com/project/living-green-family-home, n.d.

Retrieved November 30, 2019, from https://archpaper.com/2015/06/gallery-barclays-centers-green-roof-sedum-installation-50-complete/

https://ecourbanhub.com/san-francisco-green-roofs-legislation-buildings-wiener/, n.d.

https://greenroofsaustralasia.com.au/news/inspiredcopenhagenmunicipal-council-recife-brazil-approves-green-roof-law, n.d.

Retrieved November 30, 2019, from https://greenroofsaustralasia.com.au/news/inspired-copenhagen-municipal-council-recife-brazil-approves-green-roof-law

https://inhabitat.com/japans-namba-parks-has-an-8-level-roof-garden-with-waterfalls/; https://www.theguardian.com/cities/gallery/2018/aug/17/urban-oases-green-roofs-around-the-world-in-pictures, n.d.

Retrieved December 1, 2019, from https://inhabitat.com/the-cloak-fearon-hay-architects-install-a-maori-inspired-green-roof-at-auckland-international-airport/te-kaitaka-the-cloak-fearon-hay-architects-2/
Retrieved December 8, 2019, from https://resources.realestate.co.jp/living/japan-green-roof-buildings/
Retrieved December 5, 2019, from https://schaduf.com/schaduf-projects/tag-sultan/
Retrieved November 30, 2019, from https://tropicalcommons.co/en/2018/09/04/vertical-gardens-and-green-roofs-of-colombia/
Retrieved December 3, 2019, from https://unfccc.int/climate-action/momentum-for-change/activity-database/momentum-for-change-the-future-of-green-roofs-in-egypt
Retrieved December 7, 2019, from https://whatson.ae/2016/02/dubais-first-rooftop-garden-middle-tecom, [online]
Retrieved November 29, 2019, from https://www.archdaily.com/266833/stodel-museum-schneider-schumacher
Retrieved November 30, 2019, from https://www.archdaily.com/788409/shenye-tairan-building-zhubo-design
Retrieved December 1, 2019, from https://www.archdaily.com/876189/the-victorian-desalination-project-and-ecological-reserve-aspect-studios
Retrieved November 30, 2019, from https://www.archdaily.com/905876/shanghai-greenland-center-nikken-sekkei/5bec3ed108a5e5767c00005a-shanghai-greenland-center-nikken--sekkei-photo
Retrieved November 29, 2019, from https://www.architectsjournal.co.uk/buildings/bigs-copenhill-waste-to-energy-plant-cum-ski-slope-opens-in-copenhagen/10044785.article
Retrieved December 1, 2019, from https://www.architectureanddesign.com.au/news/one-central-park-sydney-named-the-world-s-best-tal
Retrieved December 1, 2019, from https://www.australiandesignreview.com/news/worlds-tallest-rooftop-garden-planned-for-eureka-tower/
Retrieved November 30, 2019, from https://www.chinadialogue.net/article/show/single/en/8904-Why-we-re-still-up-in-the-air-about-green-roofs
Retrieved December 1, 2019, from https://www.designboom.com/architecture/woha-kampung-admiralty-singapore-10-30-2018/
Retrieved December 5, 2019, from https://www.egypttoday.com/Article/1/52769/In-pics-Egypt%E2%80%99s-rooftops-going-green
Retrieved December 5, 2019, from https://www.egypttoday.com/Article/1/75486/Egypt-starts-green-roofing-initiative-in-government-buildings
Retrieved December 2, 2019, from https://www.emaratalyoum.com/local-section/2009-10-01-1.1924
Retrieved December 1, 2019, from https://www.fma.com.au/news/green-roofs-walls-and-facades
Retrieved December 3, 2019, from https://www.greenroofs.com/2018/03/05/greenroofs-com-project-of-the-week-dubai-opera-garden-greenroof-and-walls/
https://www.greenroofs.com/2019/08/28/worlds-largest-urban-farm-to-open-on-a-paris-rooftop. n.d.
https://www.greenroofs.com/projects/city-college-sanFrancisco/, n.d.
Retrieved December 1, 2019, from https://www.greenroofs.com/projects/marina-bay-sands-integrated-resort-skypark/
Retrieved November 30, 2019, from https://www.greenroofs.com/projects/nathan-phillips-square-toronto-city-hall-podium-green-roof/
Retrieved November 29, 2019, from https://www.greenroofs.com/projects/potsdamer-platz/
Retrieved November 30, 2019, from https://www.greenroofs.com/projects/santalai
Retrieved November 30, 2019, from https://www.lafargeholcim-foundation.org/projects/green-roofs-buenos-aires-argentina
Retrieved November 30, 2019, from https://www.livingroofsinc.com/projects
Retrieved December 1, 2019, from https://www.melbourne.vic.gov.au/building-and-development/sustainable-building/Pages/rooftop-project.aspx

Retrieved November 29, 2019, from https://www.nationalgeographic.com/news/2016/10/san-francisco-green-roof-law

https://www.proudgreenbuilding.com/news/bill-to-encourage-rooftop-gardens-on-new-san-francisco-buildings/, n.d.

https://www.researchgate.net/publication/293196881_An_International_Review_of_Current_Practices_and_Future_Trends_Green_Roof_Policies#downloadCitation, n.d.

Retrieved November 30, 2019, from https://www.sempergreen.com/us/references/fu-shan-estate

Retrieved November 30, 2019, from https://www.sempergreen.com/us/references/housing-authority

Retrieved December 7, 2019, from https://www.thenational.ae/lifestyle/food/dubai-s-first-rooftop-garden-in-pictures-1.194393?videoId=5712466005001, [online]

Retrieved December 1, 2019, from https://www.wikiwand.com/en/Council_House_2

Hui, S. C. M. (2010). *Development of technical guidelines for green roof systems in Hong Kong*. Retrieved November 30, 2019, from https://core.ac.uk/download/pdf/37962912.pdf

ICLEI (n.d.) Local Government for Sustainability.

Ismail, W. Z. W., Abdullah, M. N., Hashim, H., & Rani, W. S. A. An overview of green roof development in Malaysia and a way forward, 27 September 2018.

Labuschagne, P. H. (2016). Green rooftop systems: A South African perspective. Master Thesis, [online]. http://scholar.ufs.ac.za:8080/xmlui/bitstream/handle/11660/4698/LabuschagnePH.pdf;jsessionid=1EC25D96BCB40D67C15F694DA2ADD00B?sequence=1

Labuschagne, P., & Zulch, B. (2016). Green rooftop systems: A South African perspective. *Energy Procedia, 96*, 710–716, Elsevier, ScienceDirect. https://doi.org/10.1016/j.egypro.2016.09.131. https://www.sciencedirect.com/science/article/pii/S1876610216307706

Liptan, P. (2003). *Planning, zoning and financial incentives for eco roofs in Portland, Oregon, first north American green roof conference: Greening rooftops for sustainable communities*. Toronto: The Cardinal Group.

L. Mabona, K. Kondob, H. Kanekiyob, Y. Hayabuchic, A. Yamaguchib, 2019. Fukuoka: Adapting to climate change through urban green space and the built environment. *Cities, ScienceDirect—Elsevier, 93*, 273–285. https://doi.org/10.1016/j.cities.2019.05.007. https://www.sciencedirect.com/science/article/pii/S0264275118312125#

Management of Urban Stormwater Impacts Policy, Roads and Stormwater Department—Catchment, Stormwater and River Management Branch, Version 1.1, C 58/05/09, 27 May 2009, [online]. https://www.westerncape.gov.za/assets/departments/transport-public-works/Documents/management_of_urban_stormwater_impacts_policy.pdf

nd (n.d.)نوف إكس 2102 » 021 . المؤتمر العالمي لألسطح الخضراء الخامس

Ngan, G. (2004a). Green roof policies: Tools for encouraging sustainable design.

Ngan, G., (2004b). Green roof policies: Tools for encouraging sustainable design. http://www.coolrooftoolkit.org/wpcontent/uploads/2012/04/Green-Roof-Policy-report-Goya-Ngan.pdf.

Earth Pledge. (2005). *Green roofs: Ecological design and construction*. Atglen, PA: Schiffer Publishing Ltd..

Patnaik, B. et al. 2018. Impact of Green Roofs on Urban Living. *International Journal of Current Engineering and Technology, 8*(6). Retrieved November 26, 2019, from https://www.researchgate.net/publication/330185352_Impact_of_Green_Roofs_on_Urban_Living

Researchgate.net/publication/314497533_Water_Energy_and_Rooftops_Integrating_Green_Roof_Systems_into_Building_Policies_in_the_Arab_Region (n.d.)

Sailor, D. J., Elley, T. B., & Gibson, M. (2011). Exploring the building energy impacts of green roof design decisions—A modeling study of buildings in four distinct climates. *Journal of Building Physics, 35*(4), 372–391. EXIT.

Santamouris, M. (2014). Cooling the cities – A review of reflective and green roof mitigation technologies to fight heat island and improve comfort in urban environments. *Solar Energy, 103*, 682–703.

SLR. (2007, 51). Shanghai landscaping regulations. Retrieved 1020, 2011, from Shanghai Government http://www.shanghai.gov.cn/shanghai/node2314/node3124/node3177/node3186/userobject6ai1770.html

Steier, G., (2018). *Advancing food integrity: GMO regulation, agroecology, and urban agriculture* (1st ed.). CRC Press, Taylor and Francis Group, ISBN: 13: 978-1-138-30525-0. Retrieved December 8, 2019, from https://www.routledge.com/Advancing--Food-Integrity-GMO-Regulation-Agroecology-and-Urban-Agriculture/Steier/p/book/9781138305250

van Genuchten, E., Gonzalez, A. C., & Mulder, I. (2019). Open Innovation strategies for sustainable urban living. *Sustainability, 11*, 3310. Retrieved November 10, 2019, from https://doi.org/10.3390/su11123310. www.mdpi/journal/sustaiablity.

World Architectural Community. Retrieved November 30, 2019., from https://worldarchitecture.org/

Food in the City: Enabling Integrated Solutions for Urban Agriculture and Food Production in Buildings

Michele D'Ostuni, Antonella Trombadore, and Francesco Orsini

Introduction

This contribution starts from the consideration that urbanization, which has brought half of the world population living in the cities, appears to be irreversible and, as a consequence, the number of megalopolis is constantly increasing. The always growing demand of food in great urban areas makes alimentation one of the biggest problems which is being carried on by the uncontrolled development of cities. The aim of this chapter is to underline the role of buildings' design towards the realization of circular urban agriculture projects.

Recently, the combination of factors like industrialized agriculture and global food chain, together with worldwide issues like urbanization and migration, brought the debate over topics such as food security and safety to the attention of the European Union. As a consequence, countries and big municipalities like Amsterdam, London, Brussels, and Vienna decided to act by promoting projects of urban agriculture. In this scenario, a group composed of professors and researchers from the University of Florence (architecture) and Bologna (agricultural sciences) joined forces to design a vertical farm in Amsterdam.

The "Living Tower" focused on the retrofitting action of one of the existing towers of the former Bijlmerbajes prison complex in Amsterdam. The project won the first prize at the EcoGreenTech award in the category "*Studies, Researches, Patents, Thesis & Prototypes concerning the green hi-tech.*"

M. D'Ostuni · A. Trombadore (✉)
DiDA Dipartimento di Architettura, University of Florence, Florence, Italy
e-mail: antonella.trombadore@unifi.it

F. Orsini
Research Centre on Urban Environment for Agriculture and Biodiversity (ResCUE-AB), Bologna University, Alma Mater Studiorum, Bologna, Italy

A. Sayigh, A. Trombadore (eds.), *The Importance of Greenery in Sustainable Buildings*, Innovative Renewable Energy,
https://doi.org/10.1007/978-3-030-68556-0_8

The effort of interdisciplinary team was oriented to demonstrate that it is possible to imagine the architectonic environment as a living organism, part of a new food production chain, sponsoring zero-kilometer products. Buildings' sustainable comfort is here improved while their energy consumption is reduced. Thus, the goals we wanted to achieve with our proposal were to promote a healthy lifestyle and raise awareness on how to environmentally behave by keeping the ecological footprint as low as possible.

Enabling the Environment for Sustainable Urban Agriculture (UA) Projects in Continental Europe

Recently, a combination of factors created the favoring conditions for the development of urban agriculture projects in Europe, bringing the debate on the matter to an international level. Accordingly, main elements that shall be considered when approaching UA projects include:

- Urbanization
- Migration and social inclusion
- Health and global food chain
- Political background
- Economic development

Urbanization

Population growth is nowadays a re-known phenomenon investing in every part of the world. If it is true that developing countries will be the ones facing a bigger urbanization trend, European nations have already reached high level of urban population, with picks of 70/80% inhabitants living in big cities and megacities such as Brussels, Amsterdam, and London. While urban areas become larger and the income of their residents increases, a growth in the food demand is also experienced (Miccoli et al. 2016). As a consequence, the world's agricultural production will have to respond to this increased demand by increasing production from between 70 and 100% of current volume by 2050; this points out that it is expected that the area of arable land will not be able to grow by more than 12% compared to today. Historically, there has always been a link between the development of organized agriculture and the process of urbanization. As an example, in the rise of urbanization during the industrial era, urban agriculture has emerged as part of a countermovement to protect the population from social dislocation or as a form of coping strategy. In this context, the integration of food production systems in urban areas appears to be one possible solution to meet the increasing demand of food caused by the same urbanization (Keeffe 2016) (Fig. 1).

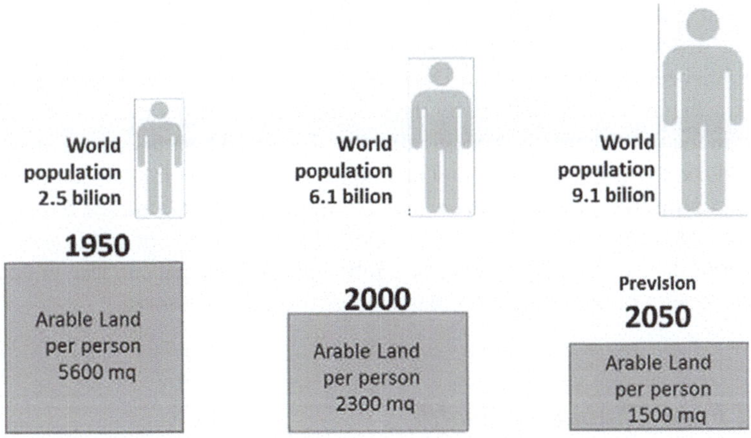

Fig. 1 Arable land for person, 1950–2050 (source Perfetti (2010) (Perfetti, P., 2010. The New Millennium Risk Game Explodes: the Capture of Territories has just Started. Energia, Ambiente e innovazione 2. Proctor, W., Drechsler, M., 2006. Deliberative multi-criteria evaluation. Environ. Plan 24, 169–190))

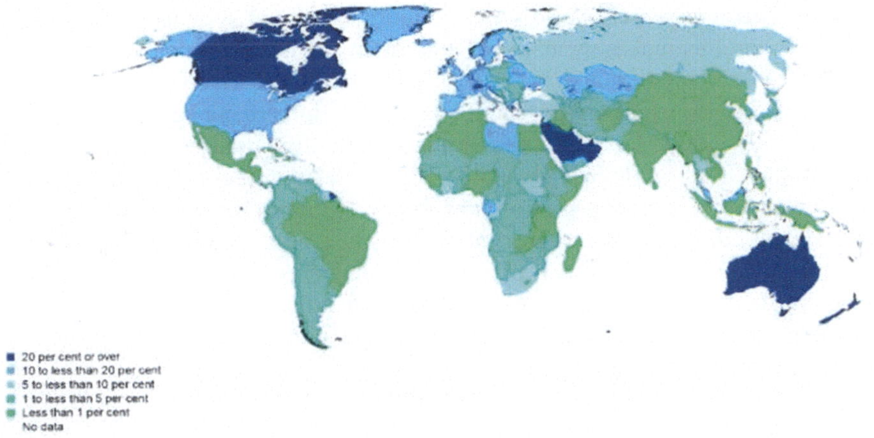

Fig. 2 UN Department of Economic and Social Affairs—International migrants as a percentage of total population, 2019, in International migration wallchart 2019

Migration and Social Inclusion

Migration flows from poor or disadvantaged parts of the world to Europe have grown constantly: In Europe, instead of growing by 2%, the size of the population would have fallen by 1% in the absence of a net inflow of migrants (UN International Migration Report 2017), and incoming flow is thought to increase in the next 30 years (Fig. 2).

In this scenario, what worries the most is how to integrate such mixed groups of people of different cultural backgrounds, religions, and cultures. In an overview study published in 2013 (Moulin-Doos 2016) it was found how migrants gained respect for themselves and others by developing a sense of their own worth through working in the urban gardens. Specifically, the associated benefits were greater in the female migrant population: the gardens helped to restore their socioeconomic role, which is central to a person's self-respect (*case study: The Gottingen Experience—Intercultural Gardens in Germany*). Moreover, the introduction of arable areas within the city borders is expected to create new job opportunities that can also educate and employ a portion of the population that is now struggling to work.

Global Food Chain and Health

At the European level, a 2013 report by the International Federation of Red Cross and Red Crescent Societies (IFRC) on the humanitarian impacts of the recent economic crisis indicated that in 22 European countries, the number of people dependent on food aid increased on average by 75% between 2009 and 2012. There has been an increasing separation between places of food production and those of consumption. Urban areas rely heavily on a multitude of food systems to meet their food needs and this makes them vulnerable to any crisis in the food supply chain (Steel 2008). It should be noted that a 2015 analysis of total food consumption in the Netherlands and the estimated production of food within urban boundaries demonstrate that only 0.0018% of the consumption is currently produced within the city. One corollary of this commentary is that cities will have to consider the issue of food security, including strategies on how to develop more localized food production systems. European cities make great efforts to feed themselves, and environmental costs of food systems are becoming more and more unsustainable. If we take London as an example, it has been calculated that it needs "around 150 times its own footprint just to feed itself." Today, in Europe, give our food for granted is a mistake we cannot afford anymore.

Economic Considerations

It is possible, through the analysis of the case studies and historical references, to trace four dimensions of the economical sphere of UA:

1. During times of crisis urban agriculture has made important contributions to food production.
2. The peri-urban fringe of cities has been identified as the location of larger agricultural activities, where "significant scope exists for up-scaled social and public enterprises …." One local authority example from the UK highlights a market garden strategy aimed at generating 1200 jobs catalyzed through urban agriculture interventions on municipal land (Zeunert 2016).

3. Zeunert argues that evidence from the potential use of several well-known green spaces of various sizes in European cities indicates the potential of such areas to generate significant economic returns, if urban agriculture is executed at 25% (or a more substantial 50%) of their area.
4. Potential opportunities for the development of small-scale rural entrepreneurs: A key finding (Lohrberg et al. 2015), based on an analysis of more than 100 case studies of urban agriculture enterprises in Europe over a period of 3 years, was how they are the "hidden champions" of an urban green development strategy (Lorleberg 2016). The common business strategies in urban agriculture include (a) cost reduction, (b) differentiation, (c) diversification, (d) shared economy, (e) experiments, and (f) experience.

Political Background

The EU Commission recognizes that "city farms" could have a positive impact on the environment, though this depends on the farming practices adopted (McEldowney 2017). Although the EU rural development policy over the period 2007 to 2013 did not include any specific support for city farms, urban farms could be subject to support in the framework of that policy but only insofar as they were located on land fulfilling the respective eligibility criteria established by the member states. These could potentially include aid for activities such as modernization of agricultural holdings; development of new products, processes, and technologies in the agricultural and food sector; or participation in food quality schemes, as well as other forms of aid, such as agro-environmental measures.

In the absence of defined European laws and funding opportunities on UA, whether largely accepted and recognized, some member states decided to develop national plans and invest in local initiatives in order to promote farming within the city borders: municipalities like Amsterdam (Gementee Amsterdam 2016), London (Mayor of London 2018), and Brussels (PRDD 2016) have adopted strategic plans to promote and facilitate food production within the city to implement a new local circular economy (Fig. 3).

Fig. 3 National food strategic plan covers

"Could You Design an Ultimate Urban Greenhouse" Student Challenge by W.U.R.

The Netherlands is a small, densely populated country, with approximately 450 ab. per km2. Nevertheless, it results to be world's number two food exporter by value, second only to the United States (Viviano 2017). This has become possible thanks to decades of investments and research in "precision farming" and the high concentration of greenhouses throughout the national land. In a 2016 document produced by the Gemeente (municipality) of Amsterdam and TNO in collaboration with FABRICations, entitled "*CIRCULAR AMSTERDAM—A vision and active agenda for the city and metropolitan area*," they inserted urban food production as one of the pillars for achieving circular economy goals within the municipality area. In this scenario, it must not come as a surprise that in 2018 the Dutch "*Wageningen University & Research*" promoted its first international student challenge—*Could you design an ultimate urban greenhouse*—with the aim "to contribute to innovative technologies and out of the box concepts to explore and unleash the potential of urban agriculture."

The challenge selected scenario of intervention was the new Bajes Kwartier designed by OMA, LOLA Landscape, and FABRICations (Bajes Kwartier Sustainability Masterplan—Amsterdam n.d.). As they reported in the handbook *The Challenge*, the challenge is focusing on the significant more than symbolic production of food. The Challenge is NOT the same as the assignments that are commissioned to FABRICations (designer of the tower), OMA (designer of the building program), and LOLA (designer of the total layout and public space who is contracted by the commercial developer).

So you are free to deviate from their plans and we encourage you to do so. However it is not necessary that food production is the main source of income for the tower or that it is being produced on a commercial scale. The deliverable was the project of a vertical farming within the neighborhood explained in a 50-page dossier.

Introduction to the Project of the "Living Tower": The Role of Architecture in the Implementation of UA Projects

*Urban agriculture can be considered a relatively new approach by which planners, engineers, architects, and agronomists are trying to shape the cities of the future enhancing circularity, promoting more resilient urban spaces. Urban agriculture involves four different crossing spheres of the urban living (*Van Tuijl et al. 2018*): food production and citizens' health, social impact, environmental aspects, and economic development.*

For this reason, the first approach to the project of the "Living Tower" was to start investigating the relationship between architecture and urban agriculture. For this purpose 20 other projects were analyzed and categorized into six macro-categories representing six different urban scales:

1. Neighborhoods
2. Buildings
3. Urban installations
4. Rooftop greenhouses
5. Facades
6. Interiors and design products

The six categories represent the urban scale of the selected projects. This choice was made in order to make it easier to compare projects of similar scale and goals. Second step was to identify four main features that could define every project. These include:

1. Function (social, educational, food production, promotional)
2. Type of project (renovation or new construction)
3. Food technology (applied to the project)
4. Project status (built, about to be built, project concept)

The comparative analysis of the case studies showed how architectonic choices such as morphology, structure, and materials can positively implement circularity through the integration of food production systems. See the following sheets (Fig. 4).

Food Production

For the hereby presented project of an advanced vertical farm in the city of Amsterdam examples of high-technology food production systems integrated in buildings and cities were primarily studied and evaluated. Coherent with the research, the projects selected represent the state of the art of the integration within cities of advanced food production technologies. In this phase goals and technologies of urban agriculture projects together with the circularity approaches previously listed were addressed and compared. The aim was to identify a pattern of good practices where food production could implement and complete circular goals of the projects.

Social Embedding and Urban Transformation

Most of the analyzed projects tend to remark the importance of urban agriculture as a boost for social integration. It was noticed that this is considered a fundamental goal in transformation projects such as the new rooftop greenhouses, as well as the Bijlmerbajes Kwartier in Amsterdam. The pillar of the social embedding is the integration of new groups of people in an already consolidated social environment,

ReGen Villages, Almere (NL) - EFFEKT

INTRODUCTION

The ReGen Villages project was first introduced to the United Nations community in the form of a Global Sustainability Development Report authored in 2015. Titled *"RegenVillages – Integrated village designs for thriving regenerative communities"*, this document outlines plans for developing symbiotic resource systems at neighborhood scales and with closed loops. The specific concept outlined in this report was aspirational and has since proven to be highly desirable by the general public through its global mass media exposure at the 2016 Venice Biennale for Architecture.

Today, **the operational goal of ReGen Villages is focused on establishing a perpetual state of regenerative resource units for community consumption, that under ideal conditions, require no external inputs for water or electricity**. In this respect, system capacities would support approximately 150+ residential units and their respective occupants. While some dependency will exist on off-site sources for certain forms of nutrition (40% estimated), the system yields being planned and experimentally prototyped suggest an overall technological capability of generating greater than 100% of any single community's demonstrated demand for food, water, electricity, and for compostable waste.

The first ReGen Villages pilot community will be developed in the Oosterwold organic farm district of Almere, Netherlands, which is approximately a 30-minute from Amsterdam. The district designated by the Dutch government is intended to showcase off-grid capable housing and responsible neighborhood development planning that considers conservation and stewardship of natural areas and resources. The ReGen Villages master plan maximizes whole ecosystem thinking to establish a prescribed balance between built and open spaces. Planning for resiliency is key for developing neighborhood-scale services that establish internal capacities for resource unit self-reliance despite external connectivity to municipal lifeline systems.

In conclusion, self-reliant and resilient communities such as ReGen Villages introduce new benefits to community members while reducing burdens for governments at all levels, alleviates demand for healthcare systems, reduce stress on banks, and decreases risk for insurance companies and pension funds alike. Above all, the ReGen Village concept establishes just and equitable residential housing solutions for increasing personal well-being while reducing exposure to impacts of climate-induced threats to lifeline resource systems.[4]

DETAILS

Client: Tesla & Regen Villages Holding B.V

Year: 2016

Starting construction: 2018

Design: EFFEKT

Project area surface: 15 500 m^2

Total built area: 135000 m^2

Residential units: 203

Budget (euro): 25 millions

House prices (euro): from 300 to 700 thousands

Fig. 4 Examples of state-of-the-art project sheets

creating new job opportunities within the production process, giving a specific education to people in need to learn a job.

Improve Biodiversity

We grew to imagine cities as grey entities, the most detached reality from nature. Cities and nature have been perceived for more than a century as two separated worlds. The emergency of climate change and the documented health risks connected to the urban reality brought architects, planners, and agronomists to start thinking on how to reduce the green/natural gap in our cities. The majority of the selected projects aimed to improve biodiversity through the building components or, in case of the urban-scale projects, even through the integration of trees and wider green spaces. Some ideas like the Algae Canopy Facade studied by the EcoLogicStudio tend to promote breathing organisms with innovative perspiring materials and the integration of algae as urban oxygen producers.

Land Saving

In soil-based industrial agriculture, soil health is the most important foundation of a healthy farm ecosystem. Nowadays, common farming techniques employed in industrial crop production, such as overuse of synthetic fertilizers and intensive monoculture, can result in soil degradation over time, causing a cascade of problems that, in the end, result in further need for agrochemicals and increased impact on climate change. Some urban agriculture projects, like rooftop greenhouses and indoor vertical farming systems (e.g., GrowX and SkyGreens), want to limit soil degradation by bringing intensive production within cities.

It is important to stress that indoor crops are much more productive than traditional soil-based ones: as we see in the Gotham Greens "1500 m2 (0.14 ha) of greenhouses produce yields equivalent to over 40 ha of conventional field farming." Gotham Greens' methods yield 20–30 times more products per hectare than field production while eliminating any use of arable land. This process, if expanded to city scale, can impede the land-grabbing phenomenon and the abandonment of newly poor soil for new ones.

Reduce the Food Miles

"Food mile" is a unit used to measure the distance that a food product travels from where it is produced to where it is sold or consumed: "I came up with the term 'food miles' to try to help consumers engage with an important aspect of the struggle over the future of food—where their food come from, and how," said Professor Tim Lang who invented the term in the early 1990s.

Food miles are calculated based on the distance traveled by each food ingredient and the associated amount of carbon dioxide that is released due to the transport means used. Although the environmental assessment of a specific food process does not provide an absolute indicator of sustainability, it may be used for comparing alternative food sources and improve sustainability of urban food systems. UA projects aim to generate production in proximity to the final users/customers, eliminating the need for long-distance, refrigerated food transportation. When sustainable growing systems are used, UA projects may allow to dramatically reduce both fossil fuel needs and the associated carbon emissions and air pollution (Sanyé-Mengual et al. 2018).

Educational: Raising Awareness on the Topic

Some of the selected projects share with the food production educational goals. It is the case for example of the "Manhattan School for Children," part of "The Greenhouse Project" promoted by a group of parents and educators to create environmental science laboratories on the rooftops of New York City public schools. The idea of educating children about sustainability and healthy food from an early age is being adopted also in Europe in order to build a more conscious future society (Pennisi et al. 2020).

Other projects like the UrbanFarmers BOX and the Biotope use urban voids to create installation with the aim of raising awareness about urban agriculture, showing that a different "agri-tectural" approach in cities is possible.

Economic Development and Value Creation

Even though UA was always integrated in cities, the rapid innovation experienced by the sector still requires high investments in order to allow for the uptake of sustainable and technology-rich solutions. Among studied case studies, both small and big projects (including those at urban scale) had to struggle for an economic return to comply with the loans or the initial investment. It was noticed that projects that only produce and sell food are at a much higher risk to fail. That is why most of the projects try to create a new business model where food or plant production is the core of a more complex system. When projects are extended at urban scale, the production of food may be integrated in the buildings in order to complete circular flows, creating new lifestyle models and overall increasing the value of the real estate. In Bijlmerbajes for example, food production was limited to one tower only, while the whole neighborhood encompasses a number of related activities and housing services that can comply with different demands, making the diversification of both functionalities and house typologies the core of its business model.

The "Living Tower" Project for the Bajes Kwartier in Amsterdam

The student competition requested for the design of an ultimate system of urban farming, which included the concepts of sustainability, originality, and aesthetics. Accordingly, an interdisciplinary team—the FlorGreen Team—of students and researchers from the Faculty of Architecture of the University of Florence (Antonella Trombadore, Michele D'Ostuni, Beatrice Paludi, Fiona Zimmer) and the Department of Agricultural Sciences and Technologies of the University of Bologna (Francesco Orsini, Giuseppina Pennisi, Mattia Paoletti, Andrea D'Alessandro, Gloria Steffan, Federico Taibi, Francesca di Cesare, Pietro Venturi) was created in order to join forces to propose a holistic project that could integrate food production systems within the spaces of an innovative, green building. The definition of the food production goals and their integration into a circular economy approach combined the knowledge from different fields. The resulting productive model was linked with the four dimensions of urban agriculture for sustainable urban development, which are, namely, (a) social inclusion, (b) food production and citizens' health, (c) environmental aspects, and (d) economic development (Erwin van Tujil et al. 2008).

A New Integrative Approach: Food Production and Global Energy System

Producing food in urban contexts and making the process directly accessible to the consumers represent a great opportunity for training and raising awareness on environmental sustainability issues as well as the general widespread of knowledge and well-being. We believe that the only way to change our future's course is to reach people, meaning that the implementation of healthy food production and education is inevitable. Buildings like the "Living Tower" are important not only for their production and educational role, but also as prototypes that could shape the city of the future: the integration of multiple productive eco-buildings may lead to the creation of more organized food cities, where the energy systems could be interconnected (Fig. 5).

Crops Choice

Since designing a greenhouse we have understood the productive part of the building as the point of origin for most of the architectonic considerations and not only as an additional component in a multifunctional building. Therefore not only 1/3 of our buildings area is dedicated to the productive part, but also additional spaces for research and development of new techniques are designed.

Fig. 5 From the vertical farm to the food urban city

We wanted to create a building which facilitates scientific progress and stimulates further projects to communicate the significance of urban farming in a world which day by day gets more spatially consolidated.

Food Production: Plants and Techniques

The choice of the cultivated crops that the agronomist from the FlorGreen Team made followed two basic principles: (1) local diets and (2) considering of the revenue of each crop.

1. *Tubers*: Potatoes represent an important component of the daily Dutch diet: they are plenty cultivated in the Netherlands, and in economics terms, potato is the main open-field cultivation. For this reason, we decided to cultivate this type of crop inside our tower, choosing more particular species of potatoes (blue-violet and black potatoes) and other tuber crops, especially *Andean tubers*, which have also edible leaves. We wanted to create a niche product and raise the curiosity of Amsterdam's inhabitants. Besides, the requirements of these crops are very similar to the general ones, simplifying the choice and the design of the cultivation systems.

The different species of tubers will be cultivated in three different systems: (1) *raised bed* in wooden pallets; (2) cultivation in *jute bag*; and (3) experimental cultivation with an *aeroponic system* in close-controlled conditions (rhizotron).

2. *Berries*: A good solution to cultivate small fruits is the soilless method. This method is practiced in several farms with excellent results, because it does not require complicated systems and the plants are suitable for growing in pots.

Small fruits are also a very profitable crop: their consumption is increasing and they are a healthy food. In addition, producing these fruits near the place of consumption has great advantages for their short shelf life and delicacy, saving both transport and packaging costs.

The main cultivable species are raspberry (*Rubus idaeus*), blackberry (*Rubus fruticosus*), and bush currant (Ribes spp.).

The cultivation system is *in pots* for all species but there may be small differences for special needs. The pots must have a capacity of 7–8 L and the distance of the vessels arranged in rows must be 1–2 m on the row and 2.5–3 m between rows.

3. *Pharma plants:* The choice of this type of plants has been made both because of their general benefit to human health as medicine and for their possible use as natural cosmetic products. Also, these plants can be a useful tool because of distillation laboratory, which allows to extract essential oils from their various parts (depending on the species and characteristics required). These products of transformation represent a niche production which could create a good market. The species that will be cultivated are:

- *Cannabis sativa L.* (use Indica varieties with high CBD content, female plant) on SOG (sea of green) within an NFT hydroponic system.
- *Calendula officinalis* on raised bed.
- *Curcuma longa* which is a plant native to India and Indonesia: Indonesian cuisine is diffused throughout Amsterdam, a legacy of the colonial period. It is cultivated on raised beds.
- *Rosmarinus officinalis* with a hydroponic system.

4. *Mushrooms*: We know that our typical linear economy cannot be a long-term solution. It is wasteful and inefficient. Mushrooms use our waste materials as inputs for growing food and, with a little extra work, soil as an output. In this sense we propose to create a circular system where the organic waste generated by the activities carried out in the building is then used as a substrate for mushrooms' growth. It is cultivated on substratum.
5. *Sprouts and microgreens*: Cultivating these products is a great opportunity because they are suitable for cultivation without soil and their demand is increasing. Their market is related to *haute* cuisine and therefore profitable. In addition to the economic aspect, these products guarantee healthy and nutritionally rich food, in the case of untreated seeds. There are no treatments or substances needed during the short crop cycle.

The species used for the production of sprouts are divided into cultivated species and wild edible species; their number is very high and their characteristics are very varied. Some exemplary species are *garlic, cabbage, canola, wheat, radish, purslane,* and *sinapis.*

The species used for the production of micro-crops belong to several botanical families, including *Brassicaceae* (cauliflower, cabbage, radish, rocket), *Asteraceae* (lettuce, chicory), *Apiaceae, Amaranthaceae, Amaryllidaceae* (garlic, onion, leek), and *Cucurbitaceae* (melon, cucumber, pumpkin). These are cultivated in germination chambers.

6. *Fish in aquaponic:* The aquaponic system that will be used is a nutrient film technique (NFT) and it will be installed on the ground floor giving a first aesthetically appreciable impact on visitors and guests of the tower. The species chosen for the aquaponic system can be divided into plant species, lettuce (*Lactuca sativa*) and basil (*Ocimum basilicum*), and aquatic species, shrimps (*Macrobrachium rosenbergii*) and white fish (*Oreochromis niloticus* and *Oncorhynchus mykiss*). We have chosen to grow lettuce and basil because they are the species that adapt easily to the aquaponic system and because we want to promote nutritional biodiversity by introducing also leaf vegetables. Lettuce, herbs, and specialty greens (spinach, chives, basil, and watercress) have low to medium nutritional requirements and adapt well to aquaponic systems. The prawns were chosen for an environmental motivation: currently most of the shrimps consumed in Europe derive mainly from the Asian continent; therefore by producing them directly in the place of consumption we can reduce the *food mile* impact. On the other hand, we chose tilapia and trout, because they are high-protein foods and contain high amounts of omega fatty acids and vitamins but also have productive characteristics that allow to reduce energy consumption for the heating necessary for the breeding of these species.

The cultivation system will be organized in two tanks for aquatic species:

- *The tilapia* will be grown during the hottest season because its development requires higher temperatures than those for the trout. Tilapia exhibits optimal growth in water at temperatures of 27–30 °C; as a consequence it will be raised from May until September, when the temperatures in Amsterdam are high enough to allow the production of tilapia without having excessive energy expenditure.
- *The trout* will be grown during the colder season because its development requires much lower temperatures than those required for tilapia. The optimal growth of trout occurs in waters that have temperatures of 10–18 °C; consequently the trout will be cultivated from October until April, that is, when the temperatures of Amsterdam are low enough to allow the production of trout without having excessive energy expenditure (Fig. 6).

Energy and Management Systems

The objective set for the greenhouse management is to develop an automatic system, through the use of new technologies and wireless solutions, for:

- Optimal control of irrigation of plants cultivated in the greenhouse, starting from measurements of soil moisture and main micrometeorological parameters
- Control of the greenhouse microclimate by managing the existing environmental conditioning systems: opening/closing of doors, fan heater, ventilation system, misting system, shading sheets, lighting, etc.

Fig. 6 Fish in aquaponic: the pilot project developed at the environmental research center of the University of Arizona in Tucson

For the heating system, we will use the biodigester of the waste transformer to produce a percentage of the energy required to warm up the indoor climate during the colder season and to cool down during the warmer seasons in all the plant production sections. Considering the plant production surface (1200 m2) and volume (3250 m3), the energy consumption for the acclimatization will be around 1100–1200 MJ/m2 for 1 year, 1.32–1.44 mln of MJ/year for all the plant production surface.

The biogas production will produce also CO2 (35–45%) that can be used for the CO2 supply of the plant production section of the tower.

Production surface	Estimation crop residues	Potential energy produced (waste)
1200 m2	3600–6000 kg/year	180–300 m3/y = 3750–6250 MJ/year
1200 m2	3600–6000 kg/year	180–300 m3/y = 3750–6250 MJ/year

Another important factor is the integration of the plant production and the other sections of the building. Currently there are very few greenhouses in the world that are connected with the nonproductive sections of the building in terms of energy, water resources, and gases such as CO2 and O2. One example is ICTA-RTG: compared to conventional greenhouses, recent evaluations of ICTA-RTG (Toboso-Chavero et al. 2018) have shown interesting peculiarities in terms of greenhouse climate performance. In this way, we will use pipes and conduit to transfer warm air from the greenhouse section to the other sections of the building. We promote a climate management of the greenhouse using the minimal energy requirements.

The productive section of the Living Tower will be equipped with ventilation openings in the inner facade (1.35 × 1.67 m in size), LED lamps (12–30 W/m2) with an efficiency of conversion electricity into PAR of 50–55%, a pipe rail heating system (1.1 m of pipe/m2), and, for each 80 m2 greenhouse area, one air-to-water heat exchanger (OPAC-106) that can be used to heat, cool, and dehumidify the greenhouse air. The bound of the system will be based on the current status and the needs of each specific crop. With this method, we can estimate a minimum energy expenditure for heating and cooling of 740,220 MJ/year (excluded costs for water pump) for the entire surface of the tower for plant production. In order to make the system work, the climate parameters must be controlled within the confined environment. Those are the temperature and relative humidity of the air, the temperature of the soil or of the growing substrate, and the level of radiation to which the plants are exposed.

The electrical command and control system		
Electronic control unit	Extractor control	Cooling system control
Control window opening	Heating program, shading control	Humidification and mist control
Air circulator control	CO2 control	Automatic irrigation control
Air circulator control	CO2 control	Automatic irrigation control

Optimum microclimatic conditions within a greenhouse ensure the suitable environment for plant growth, energy saving, and a safe setting for workers. The proposed system aims to reduce the consumption of water, nutrients, and energy, safeguarding and regulating the daily distribution according to the real needs of the plant and the characteristics of the production system. This can happen thanks to different sensors that measure all the parameters necessary for a good production. Useful sensors to optimize fertigation are brightness sensor, anemometer, rain, camera module, air temperature sensor (indoor-outdoor), air humidity sensor (indoor-outdoor), electric conductivity sensor (EC), soil moisture sensor, shading and thermal screens, and darkening. In order to decrease the incoming radiation and/or the internal temperature a shading/reflective sheeting system is being used, which is placed within the greenhouse about 1 m from the walls and from the ceiling in order to create an inner tube. When the internal temperature rises excessively due to high external radiation level, the reflecting cover returns a part of the radiation to the outside. In this way, the warmer air, remaining in the space formed between the sheet and the walls, can be easily sucked in by the ventilation system. Alternatively, the shading sheets can be applied on superstructures placed on the greenhouse coverings to protect crops from unwanted radiation.

The development of plants is greatly influenced by light in terms of quantity, quality, and durability. In some types of greenhouses there are lighting system auxiliaries with the task of carrying out a dual function: (1) increase the intensity of the radiation inside the PAR (photosynthetically active radiance) in the presence of overcast or low sun on the horizon (to favor the photosynthesis) and (2) lengthen the duration of the day to change the photoperiod.

Embedding the Neighborhood: A Vertical Farm for Social Inclusion

One of the decisive factors for the success of the project and thus the proper performance of the tower is the degree of satisfaction and inclusiveness that the spaces and activities guarantee to those who are supposed to use them the most: the citizens. Our main strategy is to understand and respect the heterogeneity of the users and the plurality of their needs for adapting the building to those in the best way possible.

Being aware of the history of the Bijlmerbajes Kwartier, which served as a prison and temporary refugee camp, we want the tower to be a place of re-inclusion and social work for less well-off categories: political refugees and ex-prisoners but also, e.g., disabled people and people recovering from drug addiction. Through training courses and remunerated internships, both in food production and in other sectors, it is possible to dedicate appropriate management positions to these figures if in conditions of employment difficulty.

Agricultural Program

We aim to increase the value of the district by making the tower a public place of encounter, where knowledge can be passed in different forms, encouraging the residents or visitors to interact with each other but also encouraging to a healthy lifestyle and a conscious, environmentally responsible behavior. Our tower will be a 360° agri-food experience: we will not only produce food, but will also illustrate and teach processes and techniques. In order to encourage local consumption and thus minimize the carbon footprint of the food product, the inhabitants of the neighborhood will have access to the food products with facilitations. But beyond this, educational workshops will be activated, which can be modulated according to the age of the users, including:

- *Farmers' workshop:* The basic techniques of horticulture and gardening will be illustrated, from sowing to harvest, from planting to graft, with particular reference to the correct and sustainable management of inputs.
- *Processing laboratory:* Production inside the tower, such as pharma plants, could be of little interest for daily consumption. Teaching how to handle them could be an incentive for their purchase and direct consumption. Through the use of steam distillers (steam distillation is a technique for extracting essential oils from the plant tissue by transporting them by water vapor) it will also be possible to produce and sell essential oils, giving added value to the final product.

- *Creative recycling workshop:* The waste materials produced by the various activities could be reused, including for teaching purposes, to produce objects for daily use by people and for the tower's rooms, or even for decoration purposes.

Spatial and Functional Integration

Our building is supposed to be a mostly public building beginning from the ground floor, extending to the promenade, inviting people to enter the tower and to take part in various activities (Fig. 7). When designing the interior spaces of our tower, we thought about creating public areas that could be used by the inhabitants of the Bijlmerbajes district, as well as by people living in Amsterdam: we started from the consideration that the Dutch capital is one of the most multicultural cities in Europe, hosting people from all over the world. Students, professionals, refugees, and also tourists are a fundamental part of Amsterdam contemporary reality. As people, also

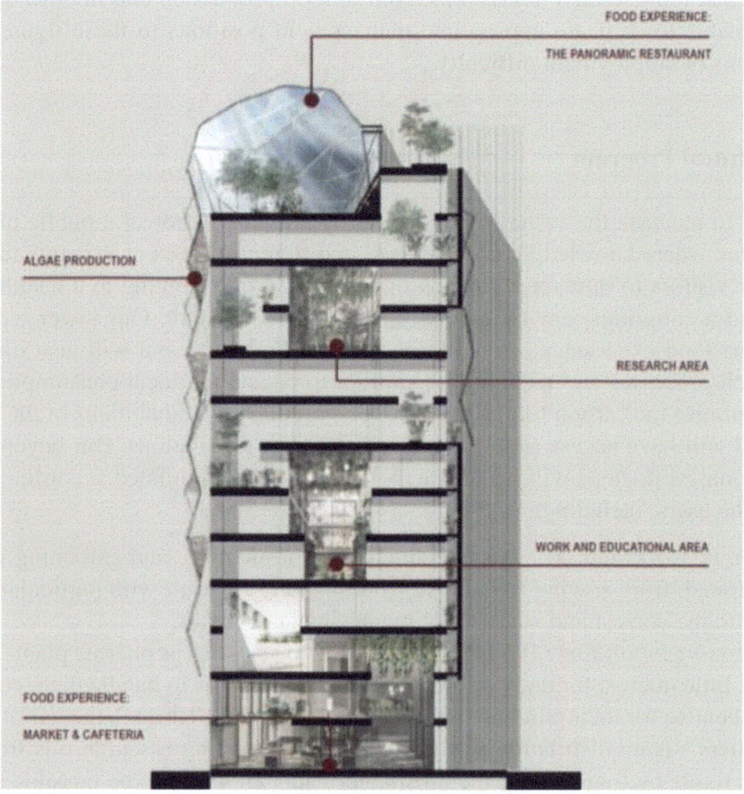

Fig. 7 Building section diagram elaborated by the team to explain the special and multifunctional integration

culture and needs are different, and though integration is growing fast in the Dutch capital, still many things can be done in order to improve acceptance and/or integration of foreign people. Considering this, we decided to make a distinction between spaces that will be used for the sole production (which access will be granted to authorized personnel only, through a private staircase) and spaces that will be given to the citizens:

- Spaces for Interaction*Market (300 sm)*: Amsterdam is well known all over the world for the quantity and quality of the markets that you can find all over the cities: here you can find from flowers to food and from clothes to books while walking through the city. We see the market as a common ground for all cultures and people. For these reasons, at the ground floor of our building we decided to open the whole space and give it all to the citizens, placing a big market where products coming from the production greenhouses can be sold every day to people who want to enjoy fresh and healthy food (Fig. 8).
- *Bar/cafeteria (110 sm):* On the first floor, overlooking the market area, we placed a small cafeteria, where people can consume fresh products. We decided to create a visual connection between the market and the bar, creating a high space that it is in direct relation with the outside. On the opposite side of the kitchenette where coffee and fresh food are being prepared (located in the west area), we put a small expositive greenhouse, where tubers are cultivated with the aeroponic method.

Fig. 8 View of the market km 0 in the ground floor and first floor

- Educational Spaces*Wine tasting and exposition area (75 sm):* As soon as the visitor lands at the third floor, he/she will meet an open area where it is possible to taste wine, and where, in special events, producers can share their experiences and their stories. Wine has been considered for centuries the drink of the Gods, and certainly a good way to meet and get to know people from different backgrounds.
- *Cooking school (60 sm):* At the third floor of our tower, we designed an open space that will be destined to cooking school where people can learn how to cook healthy and good food. We imagine here that classes are given by chefs and scientists of the upper floors, in order to teach not only how to cook, but also how to recognize the best, fresh products and how to choose them, spreading awareness of the benefits of safe food.

Workshop and Courses*Farmers workshop and processing course (120 sm):* On the fifth and sixth floors of our tower we dedicated two spaces that can be used to form people about cultivation and processing techniques. These workshops have an extremely important social value for us: we imagine that they can be joined by curious and interested citizens, but also by people who have difficulties in finding a job. We designed these places mostly for these people; here they can learn an actual subject, practicing in the labs that are placed all over the building (Fig. 9).

Children Area*Child care (95 sm) and creativity room (50 sm):* We dedicated the whole tenth floor to children. The shapes of the space are designed to be enjoy-

Fig. 9 View of the research center for hydroponic production, laboratories, and space for training area

able by children, the partition walls can be easily moved, and colors should be dominant. We thought the child care to be free, so that the mixed group of families living in the neighborhood can share their experiences and get to know each other thanks to the relations that their children are making during the day at the day care. Adjacent to the child care we put a closed space where children can work with elements and materials used for the production in the greenhouses: here they will learn at a young age the beauty of working with food, and the importance of food health and safety.

Expositive greenhouse (120 sm): Part of the children area is an expositive greenhouse, where they can learn, visit, and see how food is produced in the tower where they spend their daytime. Here they can talk with professionals, and get in touch with the plants every day! Watching them growing, and then being harvested.

Restaurant*Restaurant (300 sm):* The last two floors of our building will entirely be occupied by the restaurant, where people can eat the products made in the tower, and watch the chefs preparing them. The big ETFE cupola will make the space attractive and open. Ornamental plants will be surrounding the tables and a big table will be placed at the 12th floor, facing the space beneath, where people can sit randomly together (Fig. 10).

Fig. 10 View of the restaurant and the terrace

Conclusions

The proposal that was made for the Living Tower in Amsterdam provides a wide range of social, environmental, and economic benefits. It is important that the architectonic program could offer attractive activities and adaptable open spaces in order to create a public point of encounter, which should promote inclusion of all neighbors and citizens: when approaching urban agriculture projects, participation from the local community is necessary for the success of the project as people must be seen not only as "cohabitant" of the building, but also as "co-producers (Cockrall-King 2012)." In this sense, architects and their designs must guarantee open spaces for gathering and confrontation.

This model of a socially integrated vertical farm, in combination with new strategies of a total conjunction of architectural and agricultural concepts, enabled the design of this building, which has the potential to become an emblem and tourist attraction. New technologies make the old prison an experimental, preliminary model which can function as a showcase of a new era of sustainable construction in Amsterdam and thereby can be highly interesting for potential investors. In addition, the role that the tower adopts as a commercial hub (bar, restaurant, guided tours, fruit and vegetable market) and research center (experimental labs and start-up areas) contributes strongly to the neighborhood economy and with the added possibility of creating job opportunities. This created framework also functions as a source of attraction for various stakeholders, like universities and retailers. The activities related to agricultural production as laboratories, but also spaces for discussion, study, and recreation, create a stimulating and multidisciplinary environment and increase the social value of the neighborhood by improving the citizen quality of life (Fig. 11).

Producing food in cities, with specific buildings designed and built *ad hoc* for this purpose, might raise a more conscious behavior in citizens' lives and their way of perceiving food. This is why pilot projects are now necessary to be developed, and spread into civil society: even nowadays, a great skepticism can be encountered in people when asked if they would grow or eat food produced in their very houses or living spaces in small-scale vertical farming. That is because they do not know how that food has been produced and even less how to produce it (Jansen et al. 2016). Taking this in mind, it is easy to understand why educational spaces are so important when designing integrated, small/neighborhood-scale vertical farms: communicating how to produce food in a healthy way and involving consumers into the production process are fundamental to make them realize that another food system is possible, shifting the paradigm from horizontal, intensive, soil-disruptive crops to integrated, vertical, and sustainable crops. This is crucial if we want to develop more integrated vertical farming projects in our neighborhoods, promoting the image of an architectonic environment as a living organism, being part of the new food production chain, sponsoring zero-kilometer products, increasing buildings' sustainable comfort while reducing their energy consumption.

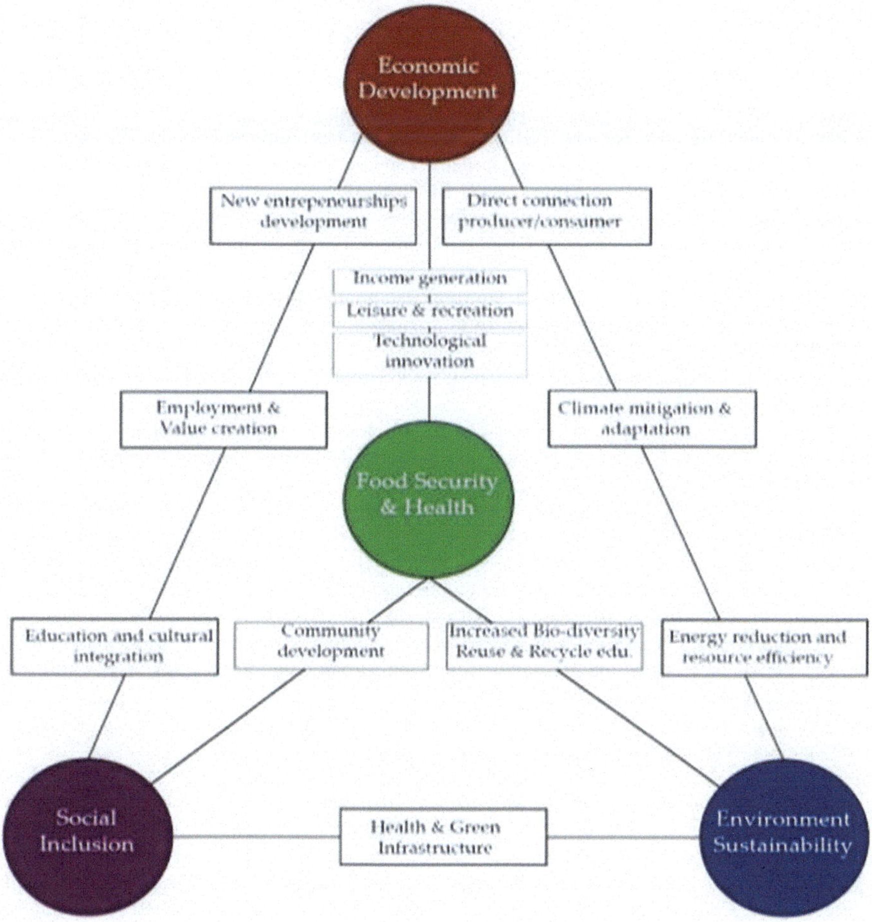

Fig. 11 Diagram elaborated by M. D'Ostuni based on "Opportunities and Challenges of Urban Agriculture for Sustainable City Development," Erwin van Tujil et al. Article published in European Research and Policy, December 2018

Web Sources

Bill & Moira@Cultivariable. *How to grow Ulluco, Oca, Mashua*. Retrieved from https://www.cultivariable.com/

Somerville, C., et al. (2014). FAO fisheries and aquaculture small-scale aquaponic food production.

References

Bajes Kwartier Sustainability Masterplan – Amsterdam. [Online] Available at: https://www.fabrications.nl/portfolio-item/bejeskwartiersustainablemasterplan/ [14]

UN International Migration Report, (2017). Department of Economic and Social Affairs [3]

Claire Moulin-Doos, (2016) Intercultural Gardens: The Use of Space by Migrants and the Practice of Respect. Journal of Urban Affairs 36 (2):197-206 [4]

Erwin Van Tuijl, Gert-Jan Hospers, Leo Van Den Berg, (2018) Opportunities and Challenges of Urban Agriculture for Sustainable City Development. European Spatial Research and Policy 25 (2):5-22 [15]

Esther Sanyé-Mengual, Francesco Orsini, Giorgio Gianquinto, (2018) Revisiting the Sustainability Concept of Urban Food Production from a Stakeholders' Perspective. Sustainability 10 (7):2175 [16]

Frank Lohrberg, Lilli Lička, Lionella Scazzosi, Axel Timpe, (2015). Urban Agriculture Europe, Berlin, JOVIS [7]

W. Lorleberg (2016). Urban agriculture has an economic dimension, Chapter 3, in Urban Agriculture Europe, F. Lohrberg, L. Licka, L. Scazzosi, A. Timpe (eds.), European Cooperation in Science and Technology (COST), Berlin, Jovis [8]

Urban agriculture in Europe: Patterns, challenges and policies, European Parliament December 2017 [9]

G. Keeffe, "Hardware software interface – A strategy for the design of urban agriculture" in R. Roggema (ed.), Sustainable urban agriculture and food planning, Routledge, 2016 [2]

G. Pennisi, F. Magrefi, N. Michelon, G. Bazzocchi, L. Maia, F. Orsini, E. Sanyé-Mengual, G. Gianquinto, (2020) Promoting education and training in urban agriculture building on international projects at the Research Centre on Urban Environment for Agriculture and Biodiversity. Acta Horticulturae (1279):45-52 [17]

Gementee Amsterdam (2016), CIRCULAR AMSTERDAM: A vision and action agenda for the city and metropolitan area [10]

Jansen G., Kanis M., Cila N., Slaats Y., Attitudes Towards Vertical Farming at Home: A User Study. CHI'16 Extended Abstracts, May 07-12, 2016, San Jose, CA, USA [20]

Jennifer Cockrall-King (2012), Food and the city: Urban agriculture and the new food revolution Prometheus books, New York. [19]

J. Zeunert, (2016). Urban agriculture up-scaled: economically and socially productive public green space, Chapter 7, in R. Roggema (ed.), Sustainable urban agriculture and food planning, Routledge studies in food, society and environment, London, Routledge [6]

Mayor of London (2018), THE LONDON FOOD STRATEGY: healthy and sustainable food for London [11]

PRDD: Plan Régional de Développement Durable, 2016 [12]

Viviano, Frank, How The Netherlands feeds the World, National Geographic Magazine (September, 2017) [13]

Saverio Miccoli, Fabrizio Finucci, Rocco Murro, (2016) Feeding the Cities Through Urban Agriculture The Community Esteem Value. Agriculture and Agricultural Science Procedia 8:128-134 [1]

Steel C. (2008), Hungry City: How food shapes our lives, Random House [5]

Susana Toboso-Chavero, Ana Nadal, Anna Petit-Boix, Oriol Pons, Gara Villalba, Xavier Gabarrell, Alejandro Josa, Joan Rieradevall, (2018) Towards Productive Cities: Environmental Assessment of the Food-Energy-Water Nexus of the Urban Roof Mosaic. Journal of Industrial Ecology 23 (4):767-780 [18]

The Role of Natural Factors in Courtyard Houses of Hot-Arid Climate of Iran

Seyedehmamak Salavatian

Introduction

Unbalanced urban population growth and poor standards in cities' development as well as social industrialization in the last decades ignored human physical and mental well-being which ended in the critical situation, especially for developing societies. Due to the raised problems, nowadays the higher prominence of nature and its linkage to the residential areas are agreed in most countries. In this regard, domestic courtyards are considered in the current typologies again. In Iran, a proper connection between the built environment and the nature appears in the traditional architecture; thus, the environmental potentials of natural factors as natural light, water, and vegetation have been utilized in several ways; however this approach has been weakened in contemporary buildings and its merits have been ignored by architects in the recent centuries. Lately, more attention has been paid to this typology in order to eliminate the challenges tied to the dense city housing. Thus, an analytical insight is required to derive their effective features in order to apply them in up-to-date residential designs.

Courtyards are defined as enclosed areas surrounded by buildings or walls which are open to the sky (Almhafdy et al. 2013). The courtyard house is a type of historic dwelling where the main spaces are arranged around a central courtyard. It is one of the world's oldest housing typology that can be traced back to 5000 years ago, which were built in the Middle East (Edwards et al. 2005). According to Haji--Qassemi (1999), a central courtyard in this region is classified into three categories: (a) Narenjestan, a small-sized yard for planting fruit trees; (b) Biruni (the outer yard): mostly decided for guests and strangers; and (c) Andaruni (the inner yard), a private yard allocated to the family.

S. Salavatian (✉)
Department of Architecture, Rasht Branch, Islamic Azad University, Rasht, Iran
e-mail: salavatian@iaurasht.ac.ir

A. Sayigh, A. Trombadore (eds.), *The Importance of Greenery in Sustainable Buildings*, Innovative Renewable Energy,
https://doi.org/10.1007/978-3-030-68556-0_9

Table 1 The major functions of central courtyards in Iranian traditional houses (Memarian and Brown 2000)

Social/cultural	Introversion and protection from strangers' sight
	An appropriate space for holding ceremonies
	Close connection to the nature
Climatic/environmental	As a light source for the boundary spaces
	Providing leeward spaces regarding unpleasant winds
	Cooled spaces due to the shaded areas by high walls
Architectural	As a connector for the major spaces
	Flexibility of space for multiple roles in various situations
	Defining spaces' hierarchy and borders

The most determining aspects of central courtyards in Iranian traditional houses are summarized in Table 1.

Two major strategies are the key factors for the courtyard performance: first, protecting the building from heat gain and second, providing natural ventilation (Agha 2015). During the day, buildings are protected from the direct solar radiation and outside conditions by shielding each other and are preserved regarding undesired heat (Shaheen and Bahjat 2011). At night, courtyard surfaces radiate their stored heat to the sky and release the excess heat; the pressure difference ends in air movement and facilitates space cooling. The light-weight hot air in the courtyard rises to be replaced by cold one due to the wind forces (Almhafdy et al. 2013).

Several researches on the typologies of historical houses and their internal courtyards have been performed in this region. However, most of them are focused on the spatial-physical features, architectural styles, historical aspects, and climatic analysis. The studies on the role of natural factors in courtyards are scarce. This study was carried out in order to determine the key properties of natural sources in central courtyards of hot-arid climate of Iran. This aim was achieved through the investigation of physical features of the most prominent traditional courtyard houses in this climate as well as their comparison and analysis. The ultimate resultants are useful recommendations to be applied in nowadays residences of the region.

Climatic Performance of Central Courtyards

Energy efficiency advantages of applying courtyards in all types of climates, particularly hot-arid climates, are investigated and confirmed in the previous researches (Aldawoud 2008). In hot-arid climate of Iran, most traditional courtyard houses are oriented along the north–south, northeast–southwest, or northwest–southeast directions. The summer and winter living spaces are configured in north and south sides while service spaces at the east façade (receiving west daylight) acting as a buffer zone for heat (Pirnia 2005) (Ghobadian 2006). There is a seasonal movement between summer and winter spaces as a response to climate conditions. Spaces in the southern part—suited for hot seasons—face north to gain minimum radiation and maximum airflow to enhance passive cooling and natural ventilation (Memarian and Sadoughi 2011).

A vast amount of literature asserts that, especially in hot climatic regions, courtyard buildings perform effectively regarding the reduction of cooling loads. The environmental effects of courtyards in hot-arid region of Iran were studied in 2007 (Soflaee and Shokouhian 2007) and relevant design criteria were presented in a number of categories as courtyard orientation, proportions associated to courtyards, properties of natural and man-made bodies, and features of surrounding elevations. In a more recent study by Toe and Kubota (2015), Malaysian vernacular buildings were investigated to apply their passive cooling techniques in modern terraced houses.

The effect of using water body (pond) and water spray in the courtyard was studied in a research by Al-Hemiddi and Megren (2001). They showed that internal courtyards with pools and water spray during hot seasons had a considerable role in cooling the surrounding spaces of the courtyard. Safarzadeh and Bahadori (2005) found that trees, shrubs, and plants are effective elements in improving the thermal comfort of courtyards since they provide proper shading against intensive sun radiation.

Numerous studies have been conducted in the geometric issues of courtyards to optimize their function as a climate modifier. Researches showed that deep and long configurations cause greater reduction in energy consumption since they provide more effective shaded areas (Muhaisen and Gadi 2006). Al-Masri and Abu-Hijleh (2012) examined a number of variant factors as orientation, footprint size, and wall height surrounding the courtyard to find out their thermal effect on the courtyard and its peripheral spaces. In a research by Almahfdy et al. (2013), experimental and simulation methods were combined to examine the design variants of internal courtyards in a tropical climate and the optimum configuration for a better microclimate-modifying performance was suggested. A design guideline for courtyards of hot-arid climate of Iran was suggested by Heidari (2010) in terms of the optimized thermal comfort. In 2015, central courtyards as passive cooling strategies in five major cities of Iran were analyzed in order to optimize new generation of houses with improved models of courtyards (Soflaei et al. 2015). Soflaei et al. (2016) also studied various features of central courtyards in terms of orientation, proportion, and depth to reduce the energy consumption in the surrounding rooms.

A comprehensive literature review illustrates that many studies have determined the effects of physical variants of courtyards on the overall thermal comfort of houses in Iran and other hot-arid climatic regions. However, only a few studies have focused on the impacts of natural factors as water and greenery on the improvement of comfort conditions.

Case Studies

Literature review and survey studies were performed to collect data. The comprehensive literature review implemented the sustainable and climate-responsive architecture of hot-arid region of Iran, particularly the central courtyards to identify their most influential parameters on thermal comfort improvement. The field study was carried out for 20 selected houses to evaluate the role of main natural and environmental factors in the efficiency of courtyards in this climate. The houses under study were selected based on the criteria of the variety in typology, scale, and historical values.

Lastly, the survey-based data were classified and analyzed to propose design guidelines for courtyards as an effective energy-efficient solution for contemporary sustainable housing in this region.

Geography and Climate

This research focuses on hot-arid climate, which covers nearly two-thirds of Iran area. This area receives nearly no rain for at least 6 months annually, and hence is extremely dry and hot. Fluctuations of temperature are too much in summer and winter, during the day and night, which is a unique feature of the region, normally causes great air movement, and induces wind motion especially during the seasons of spring and summer (Dehghan 2011).

Sustainability in Traditional Courtyard Houses

In traditional houses of this region, providing thermal comfort in interior spaces has been considered in various forms and buildings are designed and constructed with a serious view on the environmental aspects. The layout of central courtyards and taking advantage of water and greenery were a major strategy to lower the air dryness and balance the high air temperature. Apart from this, other minor solutions have been applied to assist in this regard. A number of the most frequent climatic techniques are illustrated in Fig. 1.

Introversion

House entries are the only connection to the harsh outside climate. No opening is considered for the outward facades while inward walls benefit vast openings and decorated facades (Fig. 1a).

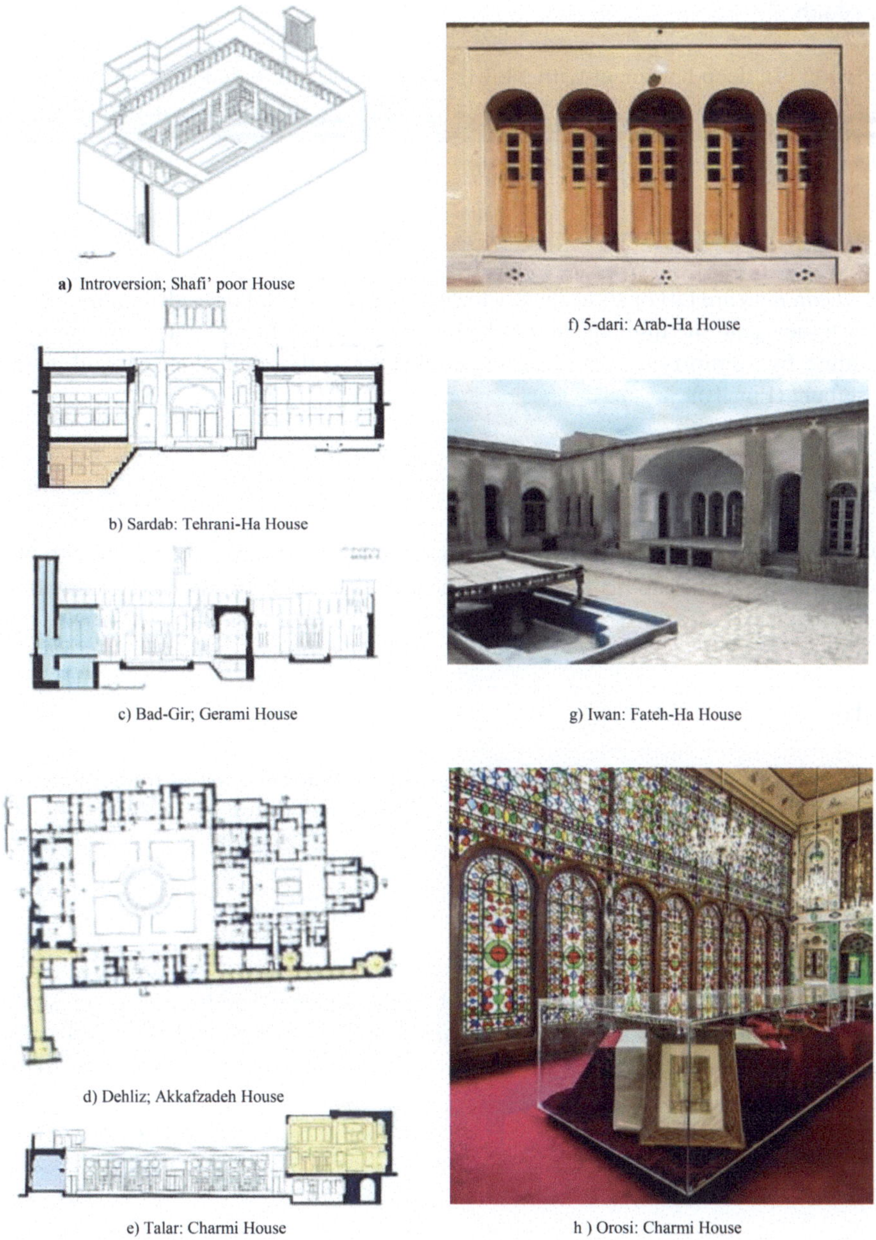

Fig. 1 Sustainable solutions in Iranian traditional houses (Haji-Qhasemi 1999; Hajighasemi 1998). (**a**) Introversion: Shafi' Poor House. (**b**) Sardaab: Tehrani-Ha House. (**c**) Bad-Gir: Gerami House. (**d**) Dehliz: Akkafzadeh House. (**e**) Talar: Charmi House. (**f**) 5-dari: Arab-Ha House. (**g**) Iwan: Fateh-Ha House. (**h**) Orosi: Charmi House

Sardaab

Sardaab is a deep basement with a lower temperature compared to the ground level due to the earth thermal mass; this space is appropriate for daily activities in hot summer days (Fig. 1b).

Bad-Gir (Wind Catcher)

Wind catchers are tall or short towers located above the southern side (summer-time zone); they grab the high-speed winds flowing at higher levels, reduce the air temperature by humidifying, and lead the cooled air to the main floor as well as the basement (Fig. 1c).

Dehliz

Dehliz is a complex corridor which accesses the hot temperature of the street to the inside humid moderate air and inhibits heat exchanges between the two environments (Fig. 1d).

Talar

Talar is a summer-time room with ceilings higher than other sides to facilitate the hot airflow upward and provide a more pleasant condition at the living height (Fig. 1e).

3-dari/5-dari

3-dari/5-dari are principal rooms placed in the shaded (southern side) or sunlit (northern side) facades allocated to the main daily activities with service rooms/secondary spaces in lateral sides. Three or five connected doors are faced toward the courtyard (Fig. 1f).

Iwan

Iwan is a semi-open space at the southern side of the courtyard and adjacent to the main room, suitable to be used for family activities in warm seasons. It protects the underneath area from the intense sun radiation meanwhile exposed to the humid breeze flowing from the outside (Fig. 1g).

Orosi

Orosi are wide and high wooden windows, mostly at the length of the room. In mild seasons, panes are rolled upward and rooms are directly connected to the courtyard and as the sun protection is desired for interior spaces, they are fully closed. Colored glass pieces bring adequate light inside even though panes are down (Fig. 1h).

Selection of Houses

Two historical cities—Yazd and Isfahan—which are well known for their rich architectural heritage were selected as the context of the study. Ten famous houses of each city are the target of the study. Selection of houses covers small-, medium-, and large-sized courtyards. In houses with more than one courtyard, the main and larger one (often outer) was the subject of analysis.

Yazd

Yazd province has an area of 131,575 km^2 and is situated in the internal part of Iranian plateau bounded by latitudes of 29–34N and longitudes of 52–56E. The lowest temperature is −20 °C at mountainous regions and the highest temperature may reach 47 °C during the day in summers (Dehghan 2011). Its annual average temperature is 19.1 °C and total annual precipitation average is 60 mm (climatemps 2019) (Fig. 2).

Isfahan

Isfahan province with the total area of 106,786 m^2 is at the east side of Yazd province within latitudes of 30–34N and 49–55E. The annual precipitation is about 120 mm, most of which is during winters. The temperature goes down to −10.6 °C and reaches the highest values of 40.6 °C (National Meteorological Organization 2019).

Analysis of Natural Factors

As an ecosystem, the Iranian traditional courtyard consists of natural bodies (Fig. 1). As the main aim of this research, characteristics of water and greenery which are natural bodies in the central courtyards of the selected houses were studied. Courtyard dimensions vary in different contexts and mostly depend on the latitude, climatic parameters, as well as available land. The proper ratios of the allocated areas to water and greenery to the total courtyard area were determined (Fig. 3, Tables 2 and 3).

Fig. 2 Location of studied cities and province division

Fig. 3 Charmi House, Isfahan (Sameti 2019)

Table 2 Selected houses in the city of Yazd (Haji-Qhasemi 1999)

No.	Name	Plan	Image
Y_1	Akhavan Sigari House	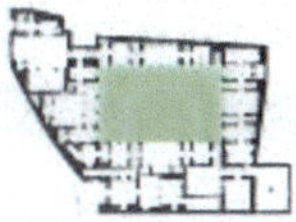	
Y_2	Tehrani-Ha House	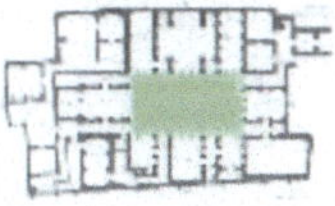	
Y_3	Rasoolian House	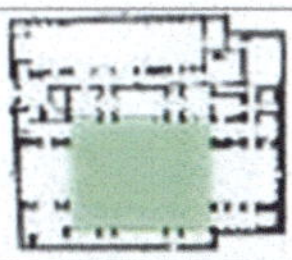	
Y_4	Rismanian House	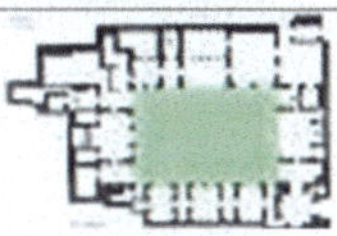	
Y_5	Semsar House	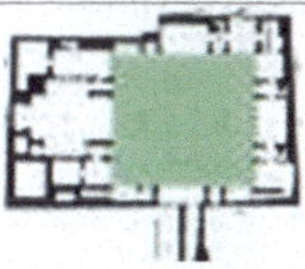	
Y_6	Arab Kermani House	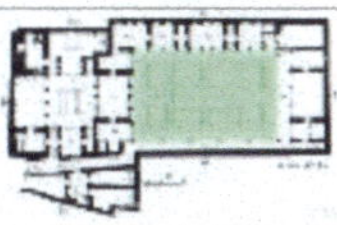	
Y_7	Lari-Ha House	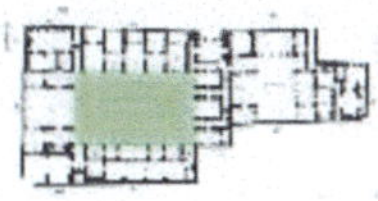	
Y_8	Mortaz House	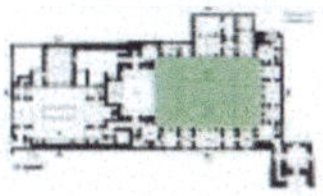	
Y_9	Meshkian House	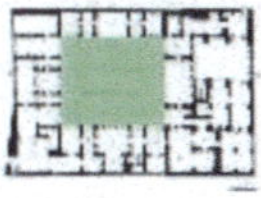	
Y_10	Malek House	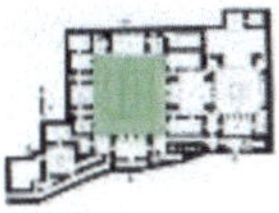	

Table 3 Selected houses in the city of Isfahan (Hajighasemi 1998)

No.	Name	Plan	Image
I_1	Dehdashti House	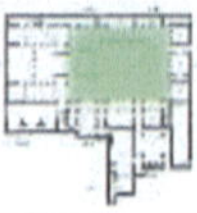	
I_2	Dr. A'lam House	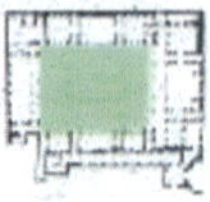	
I_3	Charmi House	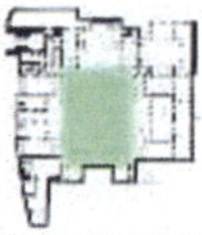	
I_4	Haj Mosavver-ol-Molki House		
I_5	Haj Rasuli-Ha House	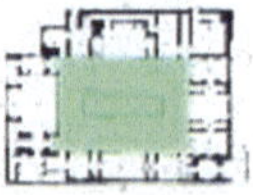	
I_6	Zovelian House	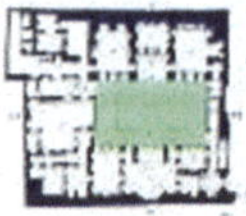	
I_7	David House	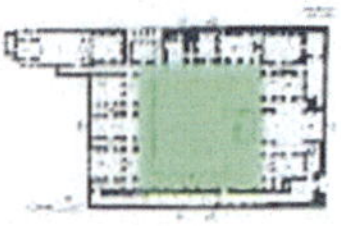	
I_8	Labbaf House	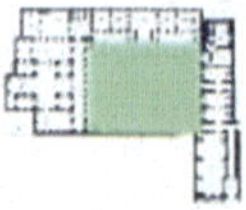	
I_9	Qodsiyeh House	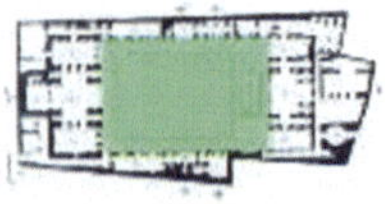	
I_10	Sheikh-ol-Eslam House	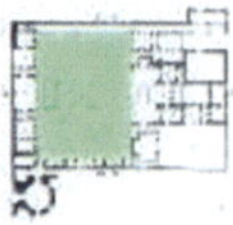	

Water and Greenery

Moisture, as the vapor in the air mixture, is enhanced by the vegetation, trees, and shallow ponds arranged in the central courtyards. High amount of thermal capacity and specific heat of water compared to other materials are key factors in moderating diurnal temperature in the microclimate of buildings. Native and low-water-use plants are cultivated in this region to resist the high level of dryness. They play an important role in providing shaded areas in the open space which consequently increases natural ventilation. Also, they aid in the storage of sun heat during the day and release it at night in cold seasons (Soflaei et al. 2016).

Heidari (2010) argued that humidity in the studied courtyards could be improved through landscaping, and a pond could significantly affect thermal comfort. The depth-to-width ratio of a courtyard is an important factor of airflow pattern. Zare et al. (2012) studied the relation between the nature and central courtyard in particular in traditional houses of Kashan, Iran; they showed that the water area increases as the courtyard enlarges, however in a considerably milder gradient. They also showed that the water area is positively related to the green area, particularly in larger courtyards. Tofan (2006) explored the role of water in Iranian courtyard houses. He concluded that the pools are nearly rectangular and the depth of water is generally low in order to maximize solar absorption and evaporation. He states that they are often located at the center of the courtyard and along one of its main axes.

In this regard, in order to determine the role of these two natural parameters in Iranian traditional courtyard houses, the qualitative and quantitative features of their design are analyzed in the following.

Area (Table 4)

The lowest and greatest water and greenery areas go for Y_4 (24%) and Y_6 (41%) in Yazd, respectively. In Isfahan, the minimum and maximum water and greenery ratios are assigned to I_8 (14%) and to I_9 (56%), respectively. As observed in Figs. 4 and 5, the variations in the assigned area to water and greenery in Isfahan (tolerance of 42%) are greater than Yazd with 17% of change.

In Rismanian house (Y_4), the total of (Aw + AG) is rather small compared to the courtyard area and some alterations during decades are probable to be applied in the courtyard design. In Meshkyan house (Y_9), due to the large size of the pond area, a limited space is left to the daily activities outdoor. It seems that the semi--open spaces in this house (Iwan) are proper substitutes for this shortage.

As illustrated in Fig. 6, it is interestingly observed that the average area assigned to water and greenery in both cities is approximately similar; however the water area is quite smaller in Isfahan. The courtyard houses in Yazd have lower allotted area for plants because of the less precipitation and water shortage while in Isfahan—which is close to the Zayandeh Rud River that irrigates plants in the entire city—planted areas are quite larger (about 6% greater than Yazd).

Table 4 Area analysis of water and greenery in courtyard houses

No.	Ac	Aw	AG	Ap	Aw/Ac (%)	AG/Ac (%)	Ap/Ac (%)
Y_1	345	48	65	232	14	19	67
Y_2	162	12	42	108	7.5	26	66.5
Y_3	310	44	65	201	14	21	65
Y_4	330	40	42	248	12	12	76
Y_5	272	37	35	200	13.5	13.5	73
Y_6	416	35	136	245	8.5	32.5	59
Y_7	485	63	83	339	13	17	70
Y_8	455	63	82	310	14	18	68
Y_9	323	48	69	206	15	21	64
Y_10	195	24	35	136	12	18	70
Average	329.3	41.4	65.4	222.5	12	20	68
I_1	90	5.5	17	67.5	6	19	75
I_2	374	40	70	264	11	19	70
I_3	462	58	125	279	12.5	27	60.5
I_4	168	20	38	110	12	22.5	65.5
I_5	336	45	81	210	13.5	24	62.5
I_6	123	4	21	98	3	17	80
I_7	552	26	245	281	5	44	51
I_8	616	20	70	526	3	11	86
I_9	455	33	225	197	7	49	44
I_10	661	26	195	440	4	29.5	66.5
Average	383.7	27.75	108.7	247.25	7.7	26.2	66.1

Ac courtyard area, *Aw* water area, *AG* greenery area, *AP* pavement area

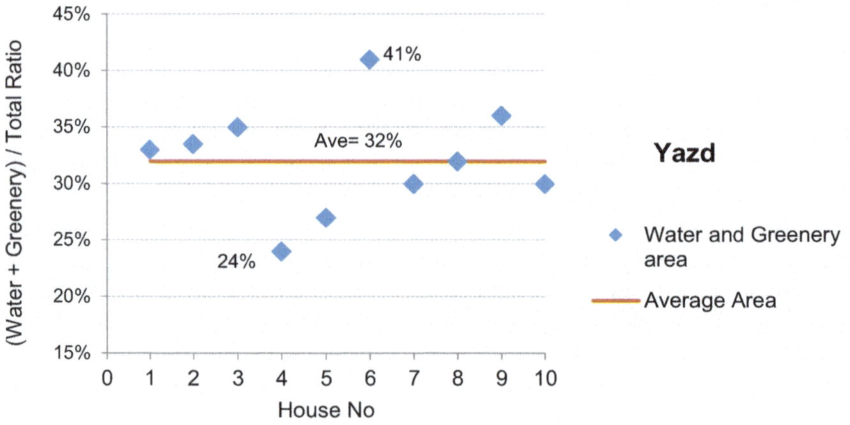

Fig. 4 Water and greenery area analysis in city 1 (Yazd)

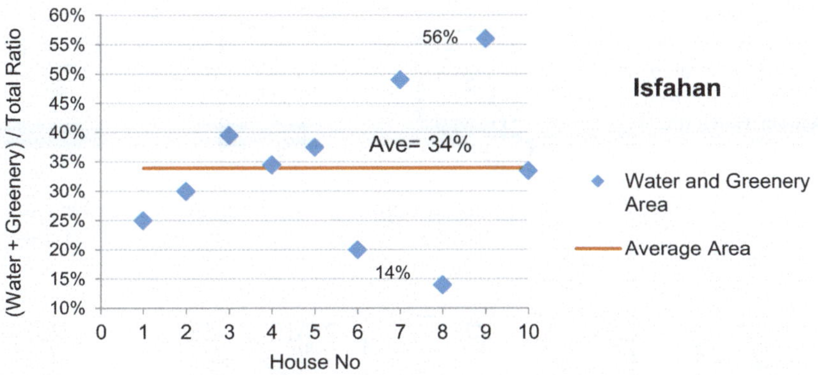

Fig. 5 Water and greenery area analysis in city 2 (Isfahan)

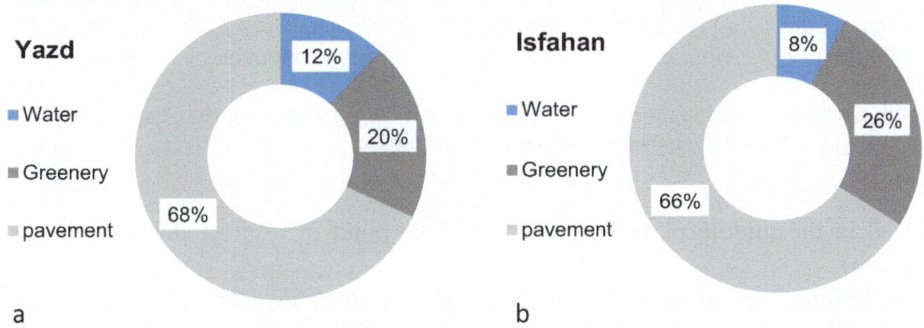

Fig. 6 Average areas; (**a**) Yazd, (**b**) Isfahan

Configuration

Most of the houses in this analysis are of the single-courtyard type. However, there are also a number of multi-courtyard houses among the analyzed houses, particularly in Yazd.

In the first city, the most frequent configuration is recognized in eight houses out of ten; the linear pond is located in the central longitudinal axis (north–south direction) and long vegetation zones are aligned on its two sides (Fig. 7a). In Semsar House (Y_5), the long pond is rotated toward east–west direction as an emphasis on the west and east internal facades. Additionally, two small-sized ponds are on the sides in front of the north and south facades (Fig. 7b). As an exception, the pond is located in front of the main Iwan, instead of the central zone. The shapeless green volume of trees and shrubs in the middle area is contradictory to the symmetric geometry and order of the facades (Fig. 7c).

The typical layout of Yazd courtyards is also observed in Isfahan; for example in Dr. Alam House a long pond in the middle and four equal-sized flower beds are aligned around it (Fig. 8a). In the most frequent design, a large-sized flower bed is

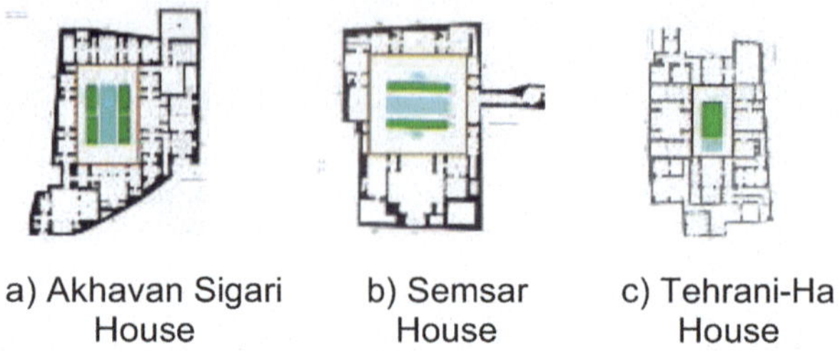

Fig. 7 Courtyard configuration in city 1 (Yazd). (**a**) Akhavan Sigari House. (**b**) Semsar House. (**c**) Tehrani-Ha House

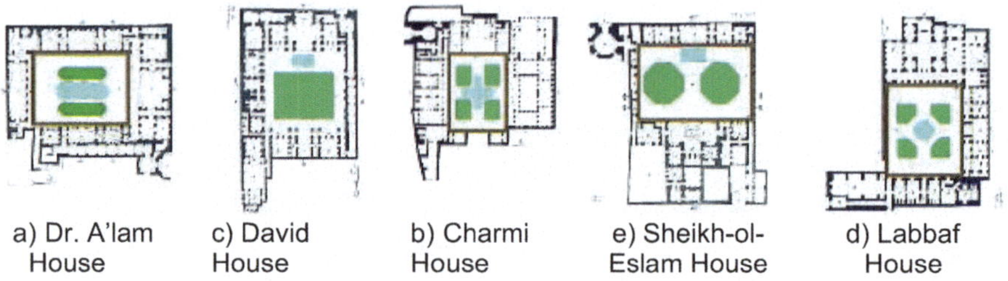

Fig. 8 Courtyard configuration in city 2 (Isfahan). (**a**) Dr. A'lam House, (**b**) Charmi House, (**c**) David House, (**d**) Labbaf House, (**e**) Sheikh-ol-Eslam House

placed in the middle of the courtyard and the pond in front of the main facade (Fig. 8b).

In addition, a number of less frequent configurations are seen in Isfahan houses. Charmi House has a cross-shaped pond and four flower beds around it in a central layout (Fig. 8c). Sheikholeslam house (I_10) has a large courtyard used for religious ceremonies. Its landscaping is different from other houses and the flower beds are designed in polygonal shapes (Fig. 8d). Labbaf House has a circular pond in the center and flower beds in its borders which is scarce in the courtyard layouts of this era (Fig. 8e).

Sun

Thermal condition provided by courtyards significantly depends on the amount of gained solar radiation. Al-Masri and Abu-Hijleh in a research in 2012 (Al-Masri and Abu-Hijleh 2012) deduced that the orientation, size of footprint, and height of walls surrounding the courtyard highly affect the thermal conditions inside the courtyard and the boundary spaces.

Muhaisen (Muhaisen and Gadi 2006) also investigated the effect of rectangular courtyard proportions on the shading and exposure conditions in various climates. The optimum courtyard height to obtain a reasonable performance in summer and winter was found to be two-story in hot-dry climate.

Courtyards in Yazd and Isfahan do not have exactly similar properties because of their different latitudes and local environmental conditions. Although the absolute values for width, length, and height dimensions are not generalizable to the contemporary courtyards, the different courtyard shape factors (W/L, H/L, and H/W ratios) are investigated to be applied in new design geometries. According to Table 5, planar proportions are quite equal in both cities—0.73 in Yazd versus 0.74 in Isfahan—while the height-to-length ratio is different and surrounding walls are generally taller in Yazd. Also, the height of surrounding facades is greater in west–east direction compared to the north–south one in both cities which is due to the higher importance of sun protection in west and east sides.

The area assigned to the courtyard varies between 14 and 35% in Yazd and 18 and 52% in Isfahan. The average values are specified in Fig. 9 and point out that the average areas are 25% and 31%. No significant correlation between courtyard size and water/greenery area was found.

Table 5 Courtyard proportion analysis

No.	L (m)	W (m)	H1 (m)	H2 (m)	ACy/AT (%)	W/L	H1/L	H2/W
Y_1	23	15	6.8	5.8	24	0.65	0.30	0.38
Y_2	17	9.5	6.3	5.6	14	0.59	0.37	0.59
Y_3	20	15.5	8.2	6	23	0.77	0.41	0.73
Y_4	22	15	8	5.6	25	0.68	0.36	0.37
Y_5	16.5	16.5	6.5	6.5	32	1	0.39	0.39
Y_6	26	16	5.3	5.3	26	0.61	0.20	0.33
Y_7	28.5	17	8.5	6	28	0.60	0.30	0.35
Y_8	26	17.5	7.5	5.2	35	0.67	0.29	0.30
Y_9	19	17	5.5	5.5	22	0.89	0.29	0.32
Y_10	15	13	7	6	24	0.87	0.47	0.46
Average	21.3	15.2	6.96	5.75	25	0.73	0.34	0.42
I_1	10.5	8.5	2.5	2.5	33	0.80	0.24	0.29
I_2	22	17	5.5	5.5	30	0.77	0.32	0.25
I_3	25	18.5	5.5	5.5	24	0.74	0.22	0.30
I_4	16	10.5	5.2	5.2	20	0.65	0.32	0.49
I_5	21	16	7.3	7.3	34	0.76	0.35	0.46
I_6	14.5	8.5	6.3	6.3	18	0.59	0.43	0.74
I_7	23.5	23.5	6.3	5.3	33	1	0.27	0.22
I_8	28	22	5.8	5.8	52	0.78	0.21	0.26
I_9	26	17.5	7	5.2	33	0.67	0.27	0.30
I_10	31.5	21	4.8	7.6	36	0.67	0.15	0.36
Average	21.8	16.3	5.62	5.62	31	0.74	0.28	0.37

Acy courtyard area, *AT* total area, *L* length, *W* width, *H* height

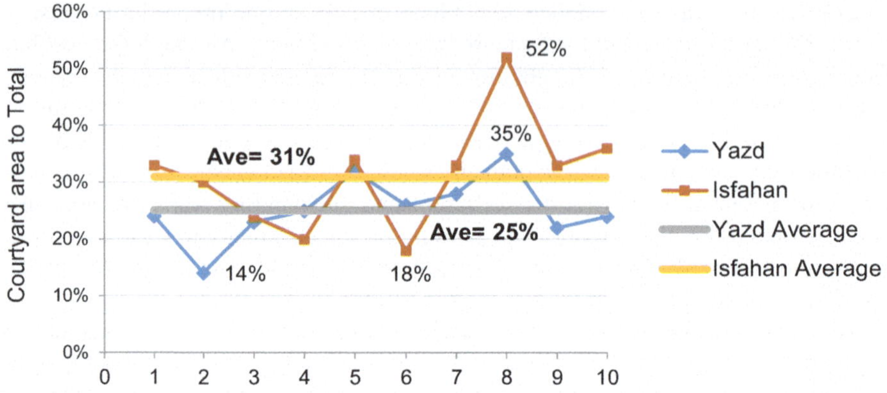

Fig. 9 Courtyard area proportion

Conclusion

Traditional courtyard houses of the studied region have many architectural features usable as passive techniques in new buildings. This chapter has provided an overview on the characteristics of natural resources utilized in the central courtyards. A summary of main features of water and greenery discussed in this study is as follows:In term of areas and proportions:

- Courtyard to total area: 25–31%
- Water to courtyard area: 8–12%
- Greenery to courtyard area: 20–26%
- Pavement to total area: 66–68%
- Width-to-length ratio: 0.73–0.74
- Height-to-length ratio (south–north): 0.28–0.34
- Height-to-width ratio (west–east): 0.34–0.42

In addition, two major configurations were identified:

- Longitudinal pond along the main axis and in the middle of the courtyard surrounded by greenery zones
- Large-sized flower bed in the center of the courtyard and a pond adjacent to it in front of the main facade

Despite the abovementioned predominant designs, a number of exceptions are observed. Some courtyards are arranged differently mostly due to the physical alterations through time or because of specific social or religious functions.

The achieved principles could also be generalized for designing future climate--responsive courtyard buildings in similar climatic conditions in other locations.

References

Agha, R. (2015). Traditional environmental performance: The impact of active systems upon the courtyard house type in Iraq. *Journal of Sustainable Development, 8*(8), 28–41.

Aldawoud, A. (2008). Thermal performance of courtyard buildings. *Energy and Buildings, 40*(5), 906–910.

Al-Hemiddi, N. A., & Megren, A. (2001). The effect of a ventilated interior courtyard on the thermal performance of a house in a hot-arid region. *Renewable Energy, 24*(3–4), 581–595.

Al-Masri, N., & Abu-Hijleh, B. (2012). Courtyard housing in midrise buildings: An environmental assessment in hot-arid climate. *Renewable and Sustainable Energy Reviews, 16*(4), 1892–1898.

Almhafdy, A., Ibrahim, N., Ahmad, S. S., & Yahya, J. (2013). Analysis of the courtyard functions and its design variants. *Procedia—Social and Behavioral Sciences, 105*, 171–182.

Almhafdy, A., Ibrahim, N., Sh Ahmad, S., & Yahya, J. (2013). Courtyard design variants and microclimate performance. *Procedia—Social and Behavioral Sciences, 101*, 170–180.

climatemps. (2019). climatemps [Online]. Retrieved October 5, 2019, from http://www.yazd.climatemps.com/

Dehghan, A. A. (2011). Status and potentials of renewable energies in Yazd Province-Iran. *Renewable and Sustainable Energy Reviews, 15*, 1491–1496.

Edwards, B., Sibley, M., Hakmi, M., & Land, P. (2005). *Courtyard housing: past, present and future*. New York: Taylor and Francis.

Ghobadian, V. (2006). *Survey on the climatic design of the Iranian traditional buildings (in Persian)*. Tehran: Tehran University.

Hajighasemi, K. (1998). *Ganjnaameh houses of Isfahan*. Tehran: Shahid Beheshti University.

Haji-Qhasemi, K. (1999). *Ganjnaameh houses of Yazd*. Tehran: Shahid Beheshti University.

Heidari, S. (2010). A deep courtyard as the best building form for desert climate, an introduction to effects of air movement (Case study: Yazd). *DESERT, 15*, 19–26.

Memarian, G. & Brown, F. (2000). The shared characteristics of Iranian and Arab courtyard houses. In *First international Arab courtyard houses* (pp. 33–51). Homs University Syria and Huddersfield University.

Memarian, G., & Sadoughi, A. (2011). Application of access graphs and home culture: Examining factors relative to climate and privacy in Iranian houses. *Scientific Research and Essays, 6*(30), 6350–6363.

Muhaisen, A. S., & Gadi, M. B. (2006). Effect of courtyard proportions on solar heat gain and energy requirement in the temperate climate of Rome. *Building and Environment, 41*, 245–253.

National Meteorological Organization [Online]. (2019). Retrieved November 30, 2019, from http://www.irimo.ir/far/services/climate/796

Pirnia, M. K. (2005). *The style of Iranian architecture*. Tehran: Shoroosh Danesh Publications.

Safarzadeh, H., & Bahadori, M. (2005). Passive cooling effects of courtyards. *Building and Environment, 40*(1), 89–104.

Sameti, M. (2019). *Art hut magazine* [Online]. Retrieved October 1, 2019, from https://www.arthut.ir/house/1202- isfahan-charmi-house.html

Shaheen, R., & Bahjat, A. S. L. (2011). He social—environmental and aesthetic integration in the designing of the multifamily housing developments: Evaluating the Iraqi experience. *The Journal of Engineering, 17*(4), 156–176.

Soflaee, F. & Shokouhian, M. (2007). Environmental effect of courtyard in sustainable architecture of Iran (hot-arid regions, mesoclimate BWks) (Case study: Courtyard houses in Isfahan & Kerman). In *Proceedings of the 2nd PALENC conference and 28th AIVC conference on Building low energy cooling and advanced ventilation technologies in the 21st century*, Crete Island, Greece.

Soflaei, F., Shokouhian, M., & Mofidi, S. M. (2015). Investigation of Iranian traditional courtyard as passive cooling strategy (a field study on BS climate). *International Journal of Sustainable Built Environment, 5*(1), 99–113.

Soflaei, F., Shokouhian, M., & Mofidi Shemirani, S. M. (2016). Investigation of Iranian traditional courtyard as passive cooling strategy (a field study on BS climate). *International Journal of Sustainable Built Environment, 5*, 99–113.

Toe, D., & Kubota, T. (2015). Comparative assessment of vernacular passive cooling techniques for improving indoor thermal comfort of modern terraced houses in hot- humid climate of Malaysia. *Solar Energy, 114*, 229–258.

Tofan, S. (2006). Recognition of water's role in Iran' traditional courtyard houses. *Garden View, 6*, 75–77.

Zare, L., Naghizadeh, M., & Hariri, S. (2012). The relation of nature and central courtyard (with the emphasis on the Iranian residential architecture—Kashan). *Identity of City, 6*(12), 49–60.

Adaptive Design of Green Facades and Vertical Farm: Examples of Technological Integration of Microalgae for Energy Production in Resilient Architecture

Antonella Trombadore, Beatrice Paludi, and Michele D'Ostuni

Algae as New Energy and Architectural Frontier

In recent years there has been a continuous search to succeed in bringing nature into the cities, the productivity of the countryside in the anthropized environment. After the first experiments with vertical, productive or just decorative greenery, we came to look for increasingly integrated solutions to produce food and energy.

About 30 years ago NASA[1] began to look for a way to break down carbon dioxide and produce oxygen in a closed environment, but also a way to recycle water and organic fluids. Today in Spain, Israel, Italy and the United States there are many producers of microalgae, which mainly use two cultivation systems: open ponds or photobioreactor. The new frontier is integrating the microalgae production into architecture, creating green building, not only to reduce carbon dioxide, but also to produce biomass.

How Algae Work

Microalgae are microscopic plants that live in water. Microalgae have the same thermal needs as humans; if kept at a temperature between 16 and 25 °C, they will grow and multiply. By placing algae crops in the glass of building facades, we could

[1] NASA Envisions "Clean Energy" From Algae Grown in Waste Water, https://www.nasa.gov/topics/earth/features/clean_energy_042209.html

A. Trombadore (✉) · B. Paludi · M. D'Ostuni
DiDA Department of Architecture, University of Florence, Florence, Italy
e-mail: antonella.trombadore@unifi.it

A. Sayigh, A. Trombadore (eds.), *The Importance of Greenery in Sustainable Buildings*, Innovative Renewable Energy,
https://doi.org/10.1007/978-3-030-68556-0_10

exploit the building's thermal atmosphere which is already designed for human comfort, to keep the microalgae warm in the walls. This coexistence generates a double advantage, as it saves 50% of thermal energy for the inhabited environment and 80% of thermal energy required for the cultivation of microalgae (Blinová et al. 2015).

One of the algae production methods is in photobioreactors, usually in plastic or borosilicate glass, exposed to sunlight. Every single photobioreactor allows to cultivate a different species, unlike the open pond in which only one species is cultivated. Nutrients and carbon dioxide are supplied by controlled artificial systems, which allow an optimal yield.

The algae could be grown on roofs, in deserts and even in waste water and, compared to plants in general that are already used to produce biofuels, are potentially more productive because algae use the entire biomass.

The collection process is divided into three phases: flocculation, filtration and centrifugation. These three processes are aimed at separating the biomass from the water, but the processed algae will still contain 97–99% of water; for this they undergo a drying process. Subsequently the oil is extracted by pressing or chemical extraction. Microalgae are used in human nutrition, animal feed and aquaculture, but also in cosmetics, fertilizers, bioplastics and biofuel.

Microalgae can divide up to once a day, thus doubling their biomass. The biomass of microalgae is an energy carrier: 1 g of dry biomass contains about 23 kJ of energy. Microalgae are one of the most promising biofuel crops of today, yielding over 30 times more energy per acre than any other fuel crop (Chinnasamy et al. 2012).

Algae need only sunlight, water and nutrients, to grow up, but they have a very high production potential (Fig. 1).

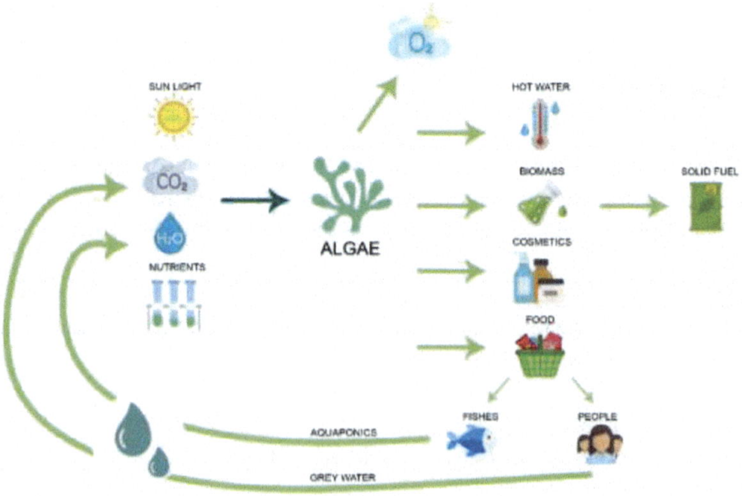

Fig. 1 Life cycle of algae

Algae Architecture: Case Studies

Vertical greenery is no longer satisfied with being relegated to building redevelopment. The sector seeks to make its way into modern architecture with innovative design solutions, moving from traditional garden walls to new symbiotic systems. In the last years many pilot projects and prototypes have been developed, with the aim to optimize the architectural integration of algae in dynamic facades as living skin, as well as to foster the aesthetical value and to promote the message of sustainability in cities.

As a research project we based our design on the literature and the analysis of international case studies: from the pavilion "Algae Folly" realized as a prototype from the Expo in Milano in 2015 to one of the largest example, offered by the bio-facades of French Dream Towers, designed by the Parisian studio XTU architects for the Chinese city of Hangzhou, as a complex of glass towers with curved and light lines, equipped with a second vegetable skin of microalgae.

Algae Folly: ecoLogicStudio

Urban Algae Folly,[2] a project by ecoLogicStudio: Marco Poletto and Claudia Pasquero (London, UK), is an interactive pavilion for the urban cultivation of microalgae. Soft tubes branch along the metal-bearing elements that rise to form the triangular covering modules, in whose semi-transparent surfaces a greenish fluid flows.

The Urban Algae Folly production system is a closed system, divided into three circuits, each consisting of an underground tank for the cultivation of microalgae, a pump that pushes the fluid into the EFTE modules located at the highest altitude and the triangular photobioreactors housed on the metal structure within which the fluid flows slowly to allow photosynthesis and then returns to the tank, at the base of the structure, through underground ducts.

The three circuits are independent and therefore can be used for different crops. In each circuit the fluid is in continuous motion, taking advantage of the force of gravity with which it reaches the tanks, to then return to the highest level of the structure with the help of the pumps.

Cultivation has a sensor system that allows the monitoring of the fluid in real time, controlling its PH and the temperature of the water and the air.

Viewers can benefit from this shading structure and interact with it: at any time, the actual potential for transparency, colour and shading of the cultivation will be the product of this complex set of relationships between climate, microalgae, visitors and system of digital controls.

In the small deposit underneath the cultivation, the spirulina collection point, there is a microfibre fabric that is used to filter the culture liquid and to retain the

[2] Algae Folly: ecoLogicStudio (2015), Expo Milan, Italy, https://www.photosynthetica.co.uk/

Fig. 2 Algae Folly at Milan Expo 2015

fresh algae before reaching the drain. Here the public is given the opportunity to touch the extracted product and use it directly.

This simple structure produces the effect of 25 tall trees. The spirulina produced in 1 day is equivalent to 2 kg of meat (Fig. 2).

BIQ House: Arup

The project BIQ House[3] was presented on the occasion of the "International Building Exhibition" located in Hamburg and was built in 2013.

It is a unique building; its 129 photobioreactors (2.5 m × 0.7 m in size) with algae in the facade, facing south-east and south-west, generate biomass (and consequently biofuel) from the microalgae contained in it and also act as a solar shield and thermal solar collectors (Fong et al. 2013).

The algae transform the CO2 coming from the combustion gases of the heat generator into oxygen. When enough quantity of algae accumulates, they are collected and sent to the technical room for processing into biofuel (biogas). The photobioreactors are also used as solar collectors recovering the heat accumulated through a heat exchanger.

The energy produced by the biomass system and by heat exchange is around 30 kWh/m2 for each system (about 60 kWh/m2 total). The property consists of

[3] BIQ—Das Algenhaus (2011–2013), Splitterwerk—Hamburg, Germany, http://www.biq-wilhelmsburg.de

about 15 units, with cuts ranging from 50 to 120 m2, which are intended to present an example for the apartments of the future.

According to the information found on SSC GmbH website, annual net energy from a 200 m2 bio-facade is 4540 kWh from methane and around 6000 kWh from heat. In comparison, an average household in Germany consumes 3100 kWh per year, so the system cannot be enough for all the energy needs of BIQ House which has the exact bio-facade surface. The system itself produces 30% more energy from methane, but it is used as auxiliary power or lost during the conversion.[4]

Energy generated can be used to modulate temperature or supply hot water. In this case, the system services 1/3 of the building's heating needs. The biomass works as a dynamic shading and acoustic buffering system that responds naturally to external changes. The more the sunlight the system gets, the more the biomass that grows and blocks off excess natural light.

"SOLARLEAF" is the name of the photobioreactor system, installed for the first time in BIQ House. The panels are composed of four layers of glass; the two innermost levels are those that contain algae; they have a capacity of 24 L. The external panel is a white glass with anti-reflective treatment. The algae are continuously fed with nourishing liquids and carbon dioxide through a separate water circuit that crosses the facade. The grown and mature algae are then separated from the rest of the algae still growing and transferred to the technical compartment of the BIQ; they will then be fermented in an external biogas plant so that they can be reused to generate biogas. The BIQ has a global energy concept: it draws all the energy needed to generate electricity and heat from renewable sources (Figs. 3 and 4).

Fig. 3 Facade of the building

[4] Bioreactive facade with algae-filled panels, Damir Beciri, http://www.robaid.com/tech/bioreactive-facade-with-algae-filled-panels.htm

Fig. 4 Facade detail

PhotoSynthetica, ecoLogicStudio

PhotoSynthetica,[5] conceived by ecoLogicStudio, a London-based architecture and urban design studio, in collaboration with Urban Morphogenesis Lab—UCL and Synthetic Landscapes Lab—University of Innsbruck, was presented in Dublin during the Climate Innovation Summit 2018. This skin captures CO2 from the atmosphere: about a kilo of CO2 per day, equivalent to that of 20 large trees. PhotoSynthetica is a photosynthetic building cladding system which uses solar energy to remove CO2 and pollutants from the atmosphere and produce a valuable food resource in the form of algae.

Designed to be integrated into existing and new buildings, it is composed of 16 modules (2 × 7 m), each of which functions as a photobioreactor—a digitally designed and tailored bioplastic container—using sunlight to nourish living micro-algal cultures and artificial light at night to create a dramatic effect. The air of the city is introduced directly into the lower part of the facade and the air bubbles naturally rise through the liquid inside the bioplastic photobioreactors. CO2 molecules and air pollutants are captured and stored by algae and transformed into reusable biomass. The photosynthesized oxygen is then released from the top of each module. Their serpentine scheme serves to hold the carbon for as long as possible, so that the algae can process it. The algae are bioluminescent, emit a faint glow at night and are very scenic.

[5] PhotoSynthetica: ecoLogicStudio (2018) Printworks Building, Dublin, Ireland, http://www.ecologicstudio.com/

The architects exhibited a prototype of the system during the Climate Innovation Summit 2018 in Dublin in early November. The installation was extended to first and second floors of the main facade of the Printworks Building at Dublin Castle. As in other ecoLogicStudio projects, the curtain is a form of biomimetic, a design that copies structures and processes from nature (Fig. 5).

WaterLilly: Cesare Griffa

WaterLilly,[6] designed by Cesare Griffa, is a system of intelligent architectural components designed to act as photobioreactors for cultivating microalgae inside buildings. It is a project started in 2012 in collaboration with a team of microbiologists from the University of Florence.

Integrating microalgae crops into architecture is a way to create virtuous symbiotic behaviours, such as reducing CO2 emissions, purifying air and grey water, improving passive behaviours through the algal thermal mass and producing food from use by local communities.

WaterLilly 2.0 is a system for the cultivation of microalgae on architectural facades. The photosynthetic activity of microalgae is much more intense than that of more complex plant organisms, with a greater ability to fix CO2 and a greater production of O2. The nutrients needed for fertilization must be rich in nitrates and phosphates, normally found in domestic waste water. Thus, growing in an urban environment, microalgae purify air and water.

The microalgae grow to saturate the aqueous solution. At that point, they must be collected, and the biomass obtained can be used to produce proteins for the food industry, nutraceutical industry, cosmetic and pharmaceutical molecules, bioplastics and biofuels such as ethanol and biodiesel. WaterLilly Ma & Pa (2012) are a pair of twin prototypes designed to be used in the near future to populate building surfaces.

WaterLilly Ma & Pa differ from other systems because all the components necessary for algae growth have been incorporated into a single element, without the need for a service module. Other differences consist of the presence of RGB LED strips and a PIR motion sensor that makes the Lillies aware of the movement of people around them, so metabolic changes occur only through the modulation of LED light (Fig. 6).

Eco Pod: Höweler + Yoon Architecture and Squared Design Lab

The work team composed of the two American studios Höweler + Yoon Architecture and Squared Design Lab has developed a project of temporary vertical structures, to beautify the city of Boston, but at the same time purify the air and produce energy.

[6] Waterlily: Cesare Griffa (2012), https://cesaregriffa.com/waterlilly

Fig. 5 (**a–c**) The prototype installation of PhotoSynthetica created by ecoLogicStudio, covering the first and second floors of the historic building in Dublin. General view and details

Eco Pod[7] is installed on unused buildings, waiting to be redeveloped, thus making buildings unused and often in a state of decay. In this project the photobioreactors are prefabricated capsules, which are installed on the existing structure.

Taking advantage of the stalled Filene's construction site at Downtown Crossing, Eco Pod is a proposal to immediately stimulate the economy, and the ecology, of downtown Boston. The capsules are modular, so they have been designed to be easily moved, with mechanical arms, so that they are always in the position that most favours the growth, and therefore the yield, of algae. The modular units allow the structure to be transformed to meet programmatic and economic needs.

The individual elements will be sources of biofuel and micro-incubators for research and development programmes. Here scientists can test species of algae and fuel extraction methods, including new techniques for using low-energy LED lighting to regulate algae growth cycles. The centrality of the position is very important, as it brings together many people with a little-known theme. The very structure of

[7] Eco Pod - Höweler + Yoon Architecture and Squared Design Lab (2009–2012), http://www.squareddesignlab.com/projects/eco-pod

Fig. 6 Prototype of assembled elements

Eco Pod being very flexible and modifiable creates public parks with unique plant species, bringing an ordinary citizen to the production of biofuel even closer.

As a productive botanical garden, it also acts as a pilot project, a public information centre and a catalyst for ecological awareness. The designers intend to use the structure, called Eco Pods, to inform the public about the potential of microalgae, a biofuel that can be grown vertically (Figs. 7 and 8).

In Vivo: XTU Architects and Symbio2

The In Vivo[8] project involves the insertion of 932 m^2 of bio-facades in the heart of Paris and photobioreactor panels in the facade of a tower, creating an integrated and efficient system, to reduce consumption and also be able to reuse water, creating a continuous cycle.

Two patents have been filed; In Vivo aims to become a showcase of French biotechnology, hoping to attract foreign investors.

The building owner will be able to customize the panel; if a transparent panel is chosen, it will be possible to observe the perpetual movement of water and microalgae. The movement generated by the airlift system allows the algae to always be in the sun, and small air bubbles are created which are very relaxing in their movement. As for maintenance In Vivo provides a collaboration with GPEA, which has been working on self-cleaning systems for several years. Thanks to the movement of the water it is possible to keep the glass surfaces clean.

[8] In Vivo: XTU Architecture, https://www.xtuarchitects.com/work-1#/invivo-xtu/

Fig. 7 Eco Pod render

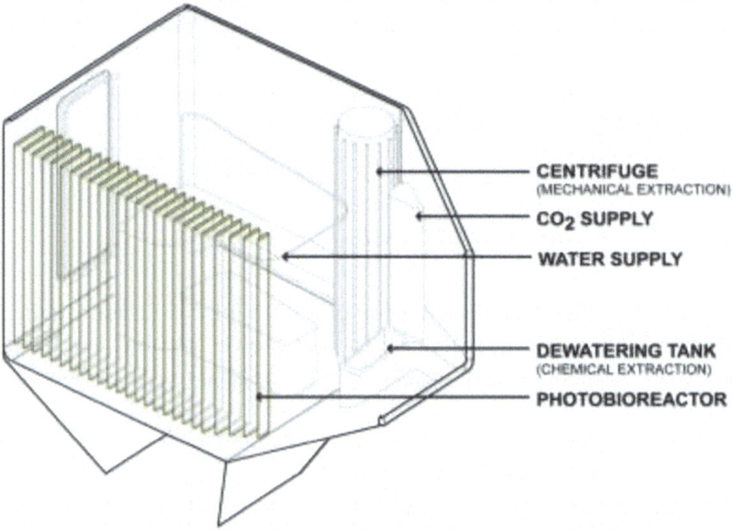

Fig. 8 Eco Pod detail, a single cell

A further gain will be made from the production of algae, entrusted to a "farmer" who, by paying a fee, will be able to manage the production and sell it later.

Besides being a productive building, it is also a place for research and experimentation. At the foot of the building, the association La Paillasse, which works on life and research, will be established. The building will also feature a European research programme with several universities, including Paris VII, but also Nantes and Warsaw, which will study the future markets for microalgae in pharmacy and medicine.

The idea is to bring out a new economic sector. The world of microalgae is a developing world; for the moment we know well only about 20 microalgae.

It is a new mission for the city not only to consume what the countryside produces, but also to be productive thanks to urban algae farming (Fig. 9).

French Dream Towers: XTU Architecture

The French studio XTU Architects[9] has developed a concept of four glass towers in Hangzhou, China, with facades covered with panels impregnated with microalgae. The algae facade will contribute to thermal and acoustic insulation, as well as reducing the building's carbon footprint by absorbing carbon dioxide and releasing oxygen.

Glass towers can be problematic from an energy point of view, as they create a chimney effect that brings out heat, but at the same time the glass surfaces generate greenhouse effect, raising the internal temperature; for this reason the insertion of algae can be a good compromise, generate shading and create thermal energy. In addition to the energy function, algae will be collected and used for cosmetic or medicinal purposes.

Each of the towers would focus on themes chosen to demonstrate "French competence". One is for the kitchen, with fused gourmet restaurants and a panoramic

Fig. 9 In Vivo render

[9] French Dream Towers in Hangzhou: XTU Architects, https://www.xtuarchitects.com/work-1#/french-dream-towers-hangzhou-china-1/

bar. The hospitality tower would include a hotel, a spa and a beauty salon. The tower dedicated to art would have its gallery and the artists' residences; the fourth tower will be dedicated to offices, start-up accelerator and co-working spaces.

In addition to the algae, other green elements of the French proposal Dream Towers include panels covered with vegetation and greenhouses in the tops of the tallest glass towers, full of trees that would filter the air and offer occupants a green place to relax.

"Hangzhou is a city where water is very present, we want to continue, update and highlight this tradition", the architects explained; to this we owe the organic shape of the towers and the rippled podium that connects them all to their base, so that the rain water slides along the sides, through the opening in the centre and in the basins below (Figs. 10 and 11).

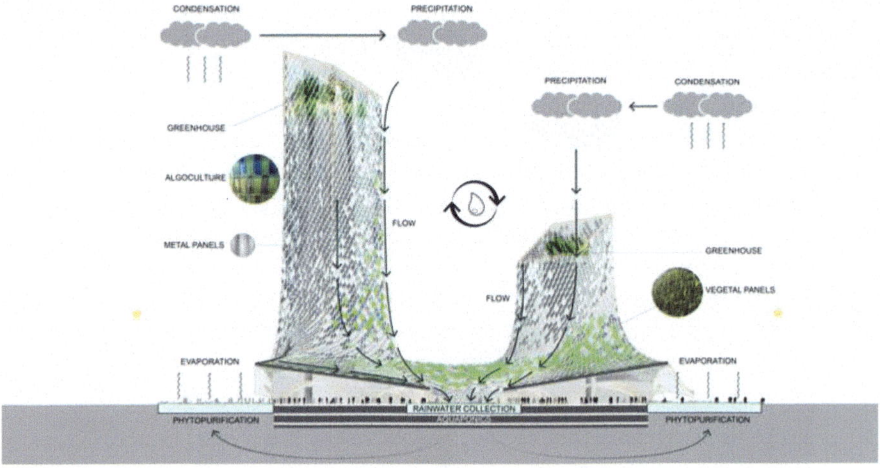

Fig. 10 Water cycle scheme in French Dream Towers

Fig. 11 French Dream Towers render

The Living Green Tower: The Concept Design for Food Technologies and Energy Concepts

This concept design, elaborated by thesis students and researchers of DIDA Department of the University of Florence, was presented at the competition "Could you design an ultimate urban greenhouse?" At the University of Wageningen, the challenge consisted of a project to regenerate one of the existing towers of the former prison complex of Bijlmerbajes in Amsterdam. With a multidisciplinary team we have created a totally integrated project between architecture and production, like a manifesto of urban agriculture.

In addition to rethinking the inside of the tower with an integrated aquaponics, hydroponics, aeroponics and terrain, we have gone so far as to look for a way to visibly declare the nature of the green tower, and the idea of a cladding in ETFE cushions was born, inside which algae are grown (Le Cuyer 2008). This second skin has left the supporting structure of the prison unaltered, which remains a memory of the old function, but is transformed and opened to the outside.

Concept: Origami Green Skin

The algae on the facade of the tower behave just like any plant, being able to transform carbon dioxide into oxygen, completing photosynthesis, growing and multiplying. This closeness to nature has led us to reflect on the positioning of the algae. In need of sunlight, the optimal position is on the south side; in our case we tried to take advantage of the angle occupying the facade from south-east to south-west. Some species of algae can grow even with less light or at colder temperatures, so a distinction has been made: the central corner, which ideally continues to become a roof covering, is distinguished from the facings that occupy the facades to the east and to the west. Only the structure that occupies the south corner is mobile; to the east and west the covering remains fixed. The movement of the south corner is justified by the desire to make maximum use of sunlight; in fact the coating bends, curves and adapts throughout the day, to always expose the algae in the sun (Fig. 12).

Given the naturalization of the architectural element, born from the concept we tried to geometrize something that had a correspondence in nature. The element closest to our concept is identified in the ivy. In fact, the ivy leaves the structure of the buildings unchanged, but rises and climbs on them. In our case the single ivy leaf is associated with the elementary figure of the triangle.

To be able to create the movement of the entire element, we worked to find an origami scheme (Fig. 13) that can be traced back to triangular shapes. With the lateral compression, the chosen scheme allows to obtain a vertical compression; with just one movement another consequent one is generated. The horizontal bands that cover the south corner can curve and by curving, the individual triangles tilt up to reach almost 90°. In addition to allowing better growth of algae, this movement

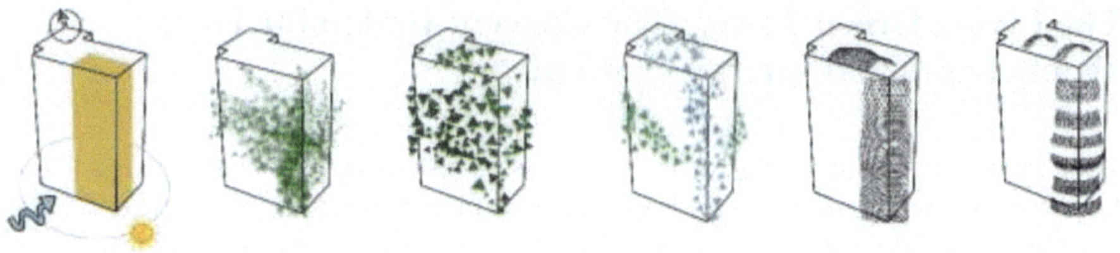

Fig. 12 Concept diagram. Choose the sunniest part; here the nature “eats up” the building and after that the ivy becomes a triangular form. Differentiate the geometric ivy; the blue one is mobile, and the green one is fixed. The roof cover and the angular skin are mobile: the two different options, open and closed surfaces

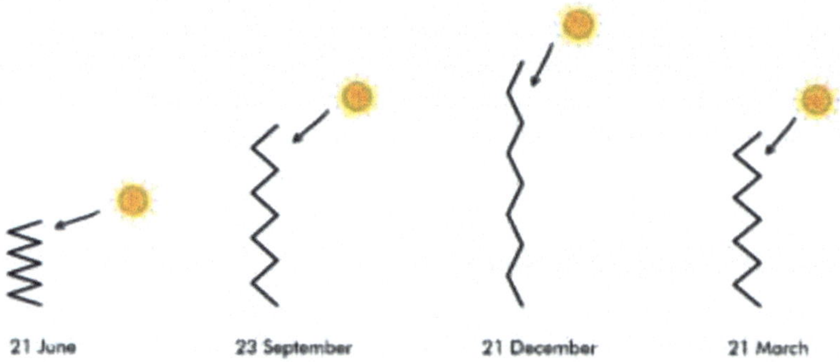

Fig. 13 Scheme of how the inclination of the curtain changes during the year to maximize solar radiation

makes it possible to use the covering as a curtain, to open or close as needed. If the coating is at its maximum curvature, it will also be very flattened and therefore will allow a more open view to the outside; if it is totally open and straight, it will be an excellent solution to shield the sun’s rays.

Structure of the Mobile Envelope

The facade system brings together three goals: origami structure, movement and production, creating a living greenhouse (Fig. 14).

When we started our research, we were looking for a material that could adapt to our needs; it had to be sustainable, modular, light, mobile, permeable to light and UV rays, insulating and resistant. Most of the previously studied cases involve the use of glass panels but considering the weight of the glass itself and the weight of the algal liquid inside, we had to discard this option, which would have created a structure too heavy and impossible to move. We have identified the ideal solution in plastic materials, to be precise in the ETFE (ethylene tetrafluoroethylene, a fluorine-based plastic), a material that meets all our requirements. The only problem with ETFE is that once a triangular element is filled, the pressure of the liquid would be too strong, which is why we created textures, like a path created by welding the two layers of the cushion. The liquid therefore does not fill the entire pillow but occupies only a percentage; this allows to reduce the pressure and to create a natural flow, which transmits the liquid from one triangle to another.

The advantages of using an ETFE facade for algae production are[10]:

- *High Light Transmission and UV Transmission:* ETFE shows high translucency by transmitting up to 94–97% of visible light (380–780 nm) and 83–88% of UV range (300–380 nm) and absorbs a big part of the transmitted infrared light, a quality which can be used in order to improve the energy consumption of buildings. ETFE allows in comparison to glass a higher light exploitation, better growing conditions and an abdication of pesticides, because of the UV's bactericidal effect.
- *High Resistance:* ETFE was designed to have high corrosion resistance and strength over a wide temperature range. It has a relatively high melting temperature (hardly inflammable), and excellent chemical, electrical and high-energy radiation resistance properties.
- *Sustainability:* ETFE is ecologically friendly: Its production consumes little energy in comparison to other cladding materials and after its long life it is easily recyclable.
- *Weight:* ETFE film has approximately only 1% of the weight of glass and thereby allows higher spans with filigree substructures. Being light and small storage needy, it is easy to transport the material to site, again lowering its carbon footprint.
- *Insulation:* A standard four-layered ETFE cushion presents a U-value of around 1.5 W/(m2K) that gets further improved by the trapped air inside and thus lowering heating costs.
- *Low Maintenance:* The surface of the ETFE pillow is anti-adhesive and thereby self-cleaning, because dust and mud will not stick onto the surface. Low clean-

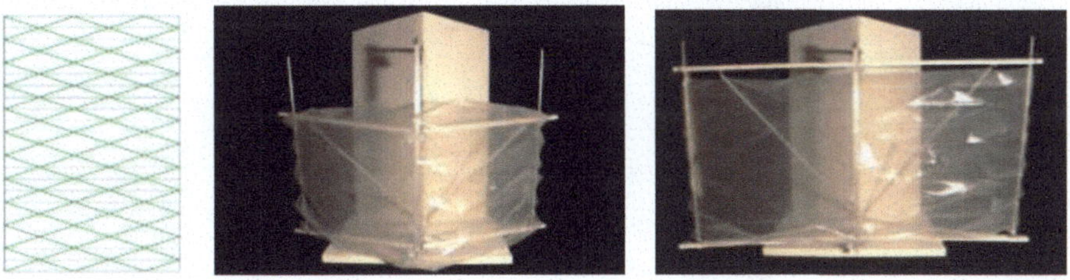

Fig. 14 (**a–c**) Origami scheme: in correspondence of the green lines we will have some hills, and the green lines will give the valleys. Structure study model. First configuration, open and plain. Second configuration, curve and compressed

[10] Le Cuyer, A. "ETFE - Technology and Design", Birkhauser Verlag AG (January 2008)

ing/maintenance costs are the result. ETFE has a long service life (at least 25–35 years).

The ETFE cushions can also have many customizations; in addition to algae, solar cells and LEDs can be inserted; this flexibility increases the possibility of use. The kinetic facade (Fortmeyer and Linn 2014) gives the building a double benefit: algae can find the best growing conditions following the sun path; the skin acts as a solar shield and as thermal insulation (Figs. 15, 16, and 17).

Each horizontal strip in which the ETFE skin is divided has a separate structure. The element that holds the various components and allows to stick to the existing facade is a simple square-section tubular placed vertically. At the upper and lower ends there are hinges. The upper hinge is free to slide along the vertical rod. At the two ends of the hinges four metal arms are attached, with slots in the terminal part. Two vertical slats are housed at the slots, which leave more freedom of movement. Each covering strip connects to the structure with rings, which run along the vertical slats. Telescopic elements have been inserted in a diagonal position to better support the metal arms, which are considerably long.

To reproduce the movement of the hands, which force on the external sides generating the consequent crushing, the surface is attached by means of rings, which in turn slide on vertical rods. The vertical rods must be able to move, because when the surface closes, it generates two movements: it bends to close and moves towards the centre, holding onto horizontal slots. The movement starts from telescopic arms; you can retract the arms closer to the building and extend them away. During this movement you create a semicircle, but for the characteristic of the "origami"

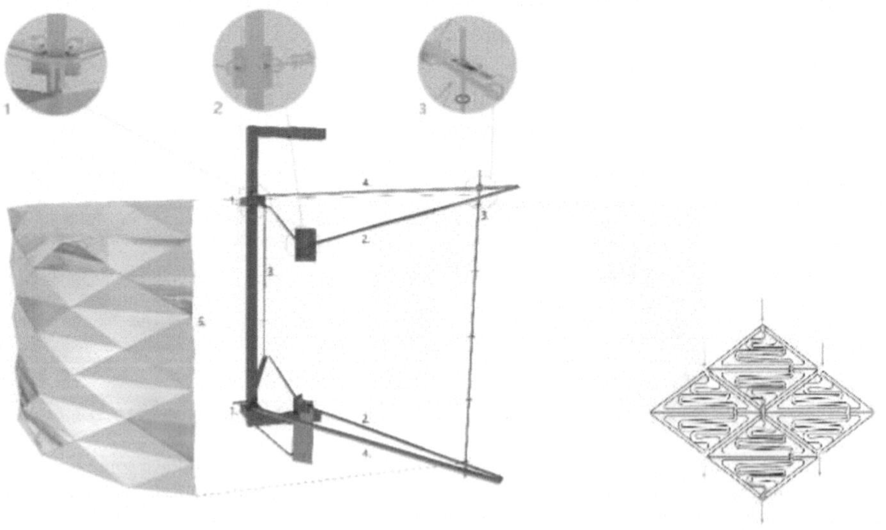

Fig. 15 Facade system: (1) hinge, (2) telescopic element, (3) sliding rod, (4) mechanical arm, (5) origami skin, (6) algae dynamic flux

Fig. 16 Facade elevation

scheme, when you bring the ends together, the skin is crushed, and turns from sheet to strip or the opposite (Trombadore et al. 2019).

Bioreactor: Facade and Energy Production

Usually we do not think of algae in comparison with terrestrial plants, but they perform the same function; indeed they are much more efficient, since they grow faster and produce more oxygen. Algae can be produced in soils or waters not suitable for other types of plants, for example they grow in salt water; in addition to growing very quickly, they need little space, compared to the land necessary for traditional agriculture. Inserting algae into architectural elements has many advantages in

Fig. 17 Building section

addition to simple production, aimed at selling for food, cosmetic or pharmaceutical purposes.

In our case we have tried to think of a totally integrated system, in which algae can also be part of the water reuse cycle. They are in fact able to purify grey water, which can then be used again for toilet flushes or plant irrigation. The nutrients contained in the recycled waters help to increase the nutritional value of algae, for the benefit of their future use. In the Green Tower an aquaponic system has also been planned that will complete the water reuse cycle.

The water in which the algae live and grow in the triangular elements is continuously heated by the sun; thanks to this natural process, there will be a liquid that, once filtered, can also be used for heating the inside of the tower. Another important advantage of algae is that the whole "plant" can be used for the production of biomass, without any kind of waste. The biomass can be transformed into biogas and thereby electricity, which is especially interesting given the high yield of biofuel produced by algae. The algae can produce more than 2000 L of fuel per acre per year, compared with palm 650 L per acre per year, sugar cane 450 L per acre per year, corn 250 L per acre per year and soy 50 L per acre per year, which is a very high yield.

The algae inside the cushions complete their growth and multiplication process; this process can be fruity to shield the sun's rays. As the algae grow, they will create a natural shield, allowing them to maintain a cooler temperature inside the environment, working together to create climate comfort (Fig. 18).

Unlike other renewable energies, such as photovoltaic panels, biomass can be used in a more versatile way and there are no energy losses during processes. The system can be used all year round. The efficiency of converting light into biomass is currently 10% and light in heat is 38%. For comparison, photovoltaic systems have an efficiency of 12–15% and solar thermal systems of 60–65%.

The bioreactive facade aims to create synergies by connecting different systems for construction services, energy and heat distribution, different water systems and

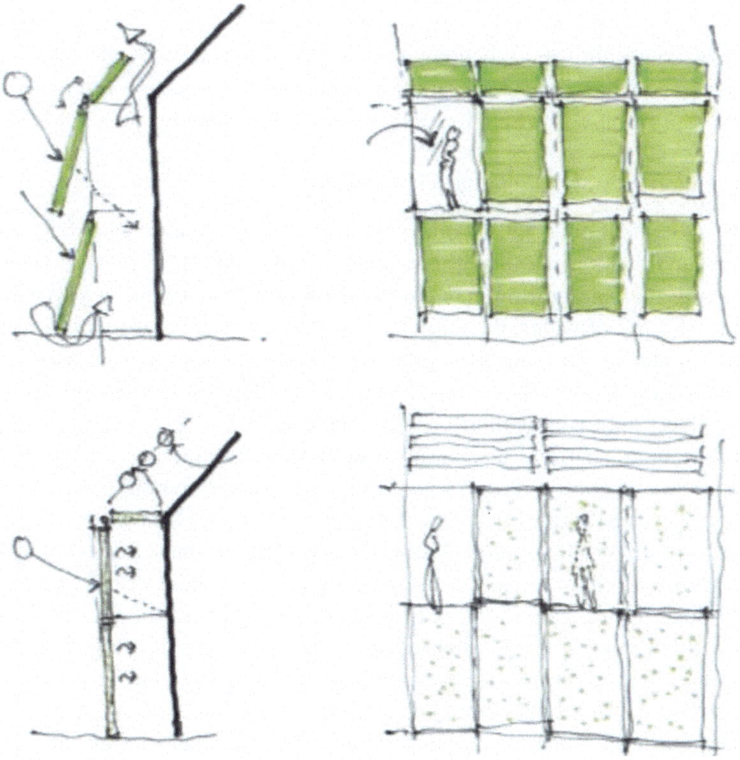

Fig. 18 Shading power and ventilation. Sketches for SOLARLEAF—bioreactor facade by Arup, an installation for the International Building Exhibition, 2013

combustion processes. There are many advantages to using algae, but there are still some limits that must be overcome: algae produce little biomass which is why it is currently not convenient to use them for biodiesel production. Convenience and sustainability are achieved in cases where the environmental and social dimensions intersect. Research is currently underway to increase algae productivity.

Tomas Morosinotto (Larosa et al. 2018) (Associate Professor in Biochemistry, Department of Biology, University of Padova) studies the question and the results achieved so far seem promising: the researchers were in fact able to produce three genetically modified variants of a particular type of unicellular green alga belonging to the genus Nannochloropsis that have a productivity greater than 25% compared to the original stock. The future hope is to be able to experience the productivity of these species first in a greenhouse and later in a more natural environment. Currently one hectare of algae cultivation does not produce more than 20–30 tons of biofuel; this leads to having a price of the final product too high to be competitive on the market.

Conclusion

The idea of combining the advantages of a kinetic facade with the cultivation of algae offers a new horizon for research; our theoretical project would need the realization of a prototype in order to really quantify the benefits it would bring to a structure.

Some important points remain to be verified:

- ETFE is the ideal material for our objectives, but as already described, an ETFE cushion could not withstand the pressure of the liquid, if it were totally full. Estimating that each pillow has a dimension of 179 × 45 cm, we would have a volume of 4 m3 of water, the equivalent of 4 tons, a remarkably high weight for a plastic material. To reduce the pressure, we thought of creating paths, through which the liquid flows, but it remains to define the best technology to weld the plastic layers and what is their best distribution.
- Each horizontal strip has a separate structure and as much as this facilitates the movement, a discontinuity is created that we must understand how it is better to manage. The algal liquid descends by gravity from one triangle to the other but reaches a point of interruption. Currently two solutions have been highlighted; in the first case it would be necessary to block the movement of the facade, during the production of algae, in the configuration in which all the triangles are adjacent, without interruptions. This solution would be very limiting and would not allow the algae to exploit the variation of the triangles during the day. In the second case we should consider each strip as a separate system, so as not to create temporary discontinuities, but to always manage them separately. This system would have the advantage of being able to produce different types of algae at the same time, since the circuits will be separated. The difficulty would be

encountered in managing the sorting pipes and the collection tanks, which could be indicatively every three or four floors, going to subtract surfaces inside the tower.

Careful research should be carried out for the selection of algae species, which require different climatic conditions. In Amsterdam for example you could grow “Spirulina Platensis” or “Spirulina Pacifica” from May to October; during the winter in the period from November to April we expect the cultivation of the ice algae, more suited to the winter climate.

Once this research has been developed it is hoped to be able to create a modular system, adaptable to any existing or new building, going to expand the panorama of alternative energy and making algae production an innovative integrated system in inhabited areas.

Link for Notes of Case Studies

1. NASA envisions “clean energy” from algae grown in waste water. Retrieved from https://www.nasa.gov/topics/earth/features/clean_energy_042209.html
2. Algae Folly—Ecologicstudio. (2015). Expo Milan, Italy. Retrieved from https://www.photosynthetica.co.uk/
3. BIQ Das Algenhaus—Arup. (2011–2013). Splitterwerk—Hamburg, Germany. Retrieved from http://www.biq-wilhelmsburg.de
4. Bioreactive façade with algae filled panels, Damir Beciri. Retrieved from http://www.robaid.com/tech/bioreactive-facade-with-algae-filled-panels.htm
5. PhotoSynthetica—EcoLogicStudio. (2018). Printworks Building, Dublin, Ireland. Retrieved from http://www.ecologicstudio.com/
6. Waterlily—Cesare Griffa. (2012). Retrieved from https://cesaregriffa.com/waterlilly
7. Eco Pod—Höweler + Yoon Architecture and Squared Design Lab. (2009–2012). Retrieved from http://www.squareddesignlab.com/projects/eco-pod
8. In Vivo—XTU Architecture. Retrieved from https://www.xtuarchitects.com/work-1#/invivo-xtu/
9. French dream towers in Hangzhou—XTU Architects. Retrieved from https://www.xtuarchitects.com/work-1#/french-dream-towers-hangzhou-china-1/

References

Blinová, L., Bartošová, A., & Gerulová, K. (2015). *Cultivation of micro-algae (Chlorella vulgaris) for biodiesel production*. Faculty of Materials Science and Technology in Trnava, Slovak University of Technology in Bratislava, vol. 23, no. 36. [1]

Chinnasamy, S., Hanumantha, R., Bhaskar, S., Rengasamy, R., & Singh, M. (2012). Algae: A novel biomass feedstock for biofuels. In R. Arora (Ed.), *Microbial biotechnology* (pp. 224–239). Oxfordshire: CABI. [2]

Fong, Q., Snijders, A., Bilow, M.. (2013). *Algae architecture in Architectural Engineering Lab 10, TU Delft BIQ—Das Algenhaus (2011–2013).* Splitterwerk, Hamburg. Retrieved from http://www.biq-wilhelmsburg.de [3]

Fortmeyer, R., & Linn, C. (2014). *Kinetic architecture: Designs for active envelopes.* Alfreton: Images Publishing. [5]

Larosa, V., Meneghesso, A., La Rocca, N., Steinbeck, J., Hippler, M., Szabo, I., & Morosinotto, T. (2018). Mitochondria affects photosynthetic electron transport and photo-sensitivity in a green alga. *Plant Physiol, 176*(3), 2305–2314. [7]

Le Cuyer, A. (2008). *ETFE—technology and design.* Birkhauser Verlag AG. [4]

Trombadore, A., Paludi, B., & Dostuni, M. (2019). The energy of the green: Green facades and vertical farm as dynamic envelope for resilient building. *J Phys Conf Ser, 1343*, 012172. [6]

Significance of Courtyard House Design in the Arab World

Falah AlKubaisy

Introduction

Bahrain is gifted with a large diversity of economic activities by virtue of its geographical location and historical value. This historical fortunate position has been very influential on the country's community activities. In the last three decades, the country has witnessed a very active process to revive and safeguard heritage, as well as enhance the quality of life for the inhabitants, by building parks and gardens within its neighborhoods as much as possible.

Establishing a National Planning Development Strategy (NPDS) in the Kingdom of Bahrain has created a unique opportunity to ensure that sustainability principles are applied for urban development over the next 25 years which will provide a superior quality of life for generations to come. This chapter aims at providing a framework for the development that can integrate the many elements of the NPDS, so that these policies and strategies are consistent with the principles of sustainability and can be implemented in an integrated and coherent manner to help Bahrain become a nation and a society of mixed communities that are sustainable in the long term. Sustainable development has been recommended to "Meet the needs of the present without compromising the ability of future generations as well as meet their own needs" (AlKubaisy n.d.).

A green building regulations and specifications document has been tailored for Bahrain and is needed to be implemented soon. They are complementary to the NPDS and they benefit the below purposes:

1. To improve the performance of buildings in the country by reducing the consumption of energy, water, and materials; improving public health, safety, and general welfare; and enhancing the planning, design, construction, and operation

F. AlKubaisy (✉)
Muharraq City, Kingdom of Bahrain

A. Sayigh, A. Trombadore (eds.), *The Importance of Greenery in Sustainable Buildings*, Innovative Renewable Energy,
https://doi.org/10.1007/978-3-030-68556-0_11

of buildings to create an excellent city that provides the essence of success and the comfort of living.
2. The regulations intend to support NPDS, create a more sustainable urban environment, and extend the ability of the Kingdom's infrastructure to meet the needs of future development.
3. Green building is the practice of creating structures and using processes that increase the efficiency of resource use—energy, water, and materials—while reducing building impacts on human health and the environment during the building's life cycle and through better setting, design, construction, operation, maintenance, and removal.

Sustainable development must be an overriding goal of government policy, whereas sustainable construction is one of the principle contributors to sustainable development. Economic growth and social progress are products of a setting/society created from the effective use of resources. Bahrain needs to address local and regional sustainable development while taking into account economic, social, and environmental issues as these are seen as the pillars of a holistic and integrated approach to sustainable development.

The inhabitants of Bahrain are distributed over towns, villages, and neighborhoods close to the traditional cities. Due to the urbanization expansions, surrounding villages have gradually become dwelling suburbs of the city. The population of Bahrain included 12,347 million inhabitants in 2010, with an area of over 770 square kilometers. The population is projected to be 2.2 million by 2030. The number of built green areas that had been planted by the government before 2010 was equal to 6.94 m per inhabitant, with a massive effort by the Ministry of Municipalities Affairs and Urban Planning (MOMUP) which has increased during the last 4 years; Bahrain currently has a 7.76 m^2 per capita ratio, bearing in mind that the population has increased by 3.2% in Bahrain, which was estimated to have 1.4 million in 2014.

A fundamental criterion of sustainability and visibility in Bahrain's community policies is the preservation, creation, and management of a sustainable society. Sustainability within this context is demonstrated by the past, present, and future conditions of landscapes as well as heritage that are developed to encourage a highly livable community. Despite Bahrain's cities being with a compact nature of urban structure, especially in the traditional towns, the community has implemented a flourishing green policy and strategy for every land available to the public.

These policies have contributed to the community's sustainability and are the tools that have managed to produce a sustainable countryside for community policies, especially those close to the waterfront. Furthermore, in order to treasure the area's history, roots, and traditional practices, the municipal council has adhered to the public's wishes and renovated the fishery ports in Bahrain. Traditional places that live in memories are precious to the community, and thus have been protected. Upgrading, regularization, relocation, and other activities related to housing and informal settlements are all determined to a large extent by the leadership role of local authorities and strategies followed at the local level, which should be comprehensive,[1] Fig. 1.

This chapter highlights the best practices ensuring a sustainable environmental community in Bahrain and diagnoses as well as analyzes the key issues of the quality of life by protecting and making the enhancement of the natural and built landscapes as well as environmental sustainability of buildings.Additionally, it covers their impact on the city's development and prosperity. The conclusion draws that design and arrangement of new houses along with buildings contain a spacious courtyard that is important in reducing heat.

Sustainability and Infrastructure Strategies of NPDS in Bahrain?

Infrastructure is the basic physical and organizational structure needed for the operation of a society, enterprise, or services and facilities necessary for an economy to function. It can be generally defined as the set of interconnected structural elements that provide a framework supporting an entire structure of development. It is an important term for judging a country's or region's development (from Wikipedia, the free encyclopedia).

Fig. 1 A traditional house with a courtyard and a wind tower to relieve the harsh climate of the inner spaces of houses. Muharraq, Kingdom of Bahrain (source: photograph by the author, 2016)

[1] Ibid., page 157.

There is an inherent restraint that must be applied to development, commonly known as the "Triple Bottom Line" or "Three-Legged Stool" that applies economic, social, and environmental components to sustainable development. In other words, development must be measured in economic, social, and environmental terms if the outcome is to be regarded as sustainable, Fig. 2.

Establishing the NPDS of Bahrain has a unique opportunity to ensure that sustainability principles are applied to development over the next 25 years, which will provide a better quality of life for generations to come. However, the aim of the Sustainable Development Framework is to provide focused sustainable development guidelines and targets that relate closely to the existing and future situation in Bahrain. A principal objective can be seen from this chapter which is to provide a framework that integrates the many elements of the NPDS so that policies and strategies are consistent and can be implemented in a coherent sustainable method.

Environmental Best Practices

In recent years, public agencies and NGOs have been exploring and adopting best practices when delivering environment, health, and human services. In these settings, and in this context, the use of the term "best practices" is often used interchangeably; several best practices and environmental initiatives, small or large, have been carried out in all the cities of Bahrain. The first of these is an initiative related to the Arad Bay project—called Signature Bahrain. It has started a movement to support the environmental organizations and individuals who are making a

Fig. 2 Compact buildings in a traditional town increase the shade and wind movement on roads

difference to Bahrain's environment. Other initiatives like Bahrain Green Awards have been created as a platform to enable green initiatives to be showcased to inspire and stimulate others for the benefit of the environment locally, regionally, and globally. Bahrain Green Awards is a contribution to the cause; however, it requires collective efforts and active participation of Bahrain's business community to sustain and make it a national success.

Health and Lifestyle

Quality of life in Bahrain means achieving high level of human development. Assessing the quality of life in Muharraq will be based on the review of what has been accomplished in meeting the UN Millennium Development Goals. More importantly, economic and social data as well as survey findings have been extensively set, showing that there are currently no issues of insufficient income that exist or what is referred to as extreme poverty. This has been achieved by government efforts and poverty eradication programs which were initiated in the previous decade to provide in every field and for every segment in the society. The free-for-all healthcare and education services through the state's social welfare services include productive families or "Youth Employment" programmers—identified as Tamkeen (an empowerment agency).

Furthermore, almost two-thirds of adults in Bahrain are either overweight or obese, which can lead to cardiovascular diseases, cancers, and diabetes. The combined burden of these diseases is challenging to the country. The Arad Bay project provides an opportunity and a place for children across Muharraq to exercise and enjoy playing outdoors in the fresh air, while adults exercise. The Ministry of Health in collaboration with the Ministry of Education created a joint committee in order to monitor health promotion at schools. Other programs included:

1. Healthy Living Program: This program is directed to women to supply them with best practices and tools that will help them maintain a healthy lifestyle—emotionally, mentally, and physically. Working in an alignment with Michelle Obama's anti-obesity campaign "Let's Move" participants will learn the nutrition and benefits of an active lifestyle.
2. "Lose Weight and Begin a Fresh Start" is a campaign established by the community to offer advice (via smartphone devices) on a daily and weekly basis in order to slim down healthily as well as attain better concentration and energy along with faster metabolism.

The community is encouraged to take care of gardens and walkways to ensure positive feedback on cleanliness and order by the larger public. There is also a strong vibrant citizen sector in Bahrain—individuals, families, and communities who devote their time and energy to public causes. The objectives of the Bahraini communities have been achieved by turning the Bay into a natural park for marine

Fig. 3 One of the best-practiced projects that were developed was the Protectorate Areas. This project was the Bronze Prize winner in the LivCom Community Competition in China, 2013

plants, fish, prawns, and migratory birds as well as a site for gathering and fitness exercises, Figs. 3 and 4.

Green Areas

The per capita share of gardens and parks has drastically shrunk. However, as a result of recycled and treated water being provided since a few years back, Bahrain's municipality was able to expand green areas and roadside planting. Main streets have been planted on either side and in the middle curbs, resulting in small areas of greeneries, gardens, and parks in urban areas. Furthermore, the country also currently has a 0.5 m^2 per capita ratio, in gardens and parks (ARCHNET n.d.).

Fig. 4 The project provides 24-h daily walking and sports

New projects in Bahrain have been undertaken as part of the municipality urban development strategic plan supported by the two strategic initiatives which are set by the municipality. These will increase the number of green spaces and general coastal beaches. Thus, the per capita share of inhabitants for gardens and parks will reach 1.25 m^2 in 2015, an acceptable rate for a country that lacks freshwater and has very limited rainfall. Despite land and water scarcity, intensive efforts have been made by municipal authorities to provide green areas through increasing the number of parks and gardens (5).

Sustainability Context

A fundamental criterion of sustainability and visibility in Bahrain's community policies is the preservation, creation, and management of a sustainable society. Sustainability within this context is demonstrated by the past, present, and future conditions of landscapes and heritage that are developed to encourage highly livable communities. Despite Bahraini towns being a town with a compact nature of urban structure, especially in the traditional towns, the community has implemented a flourishing green policy and strategy for every land available to the public. These policies have contributed to the community's sustainability, and are the tools that have managed to produce a sustainable countryside for community policies, especially those close to the waterfront or desert. Furthermore, in order to treasure the area's history, roots, and traditional practice, the municipal council has adhered to the public's wishes and renovated the fishery ports in Bahrain. Traditional places that live in memories are precious to the community, and thus have been protected (Fig. 5).

Fig. 5 Such landscape in harsh climate is available thanks to recycled treated water. Conservation, rehabilitation, traditional buildings, and buildings converted to decent activities are parts of sustainable policy in Bahrain

Design and Layout of Traditional Houses in Bahrain's Old Towns

This section shows how elements like social-economic conditions, regional climate, and building materials have influenced the design and layout of the traditional houses in Bahrain old towns. The most important difficulty in the decentralization process is the limitations of authority transfer. Legal and administrative frameworks should promote autonomy over the acquisition and expenditure of public revenues. On the other hand, lack of participatory planning processes, limitations in the capacity of civil society organizations, and lack of modalities to involve the most vulnerable groups in decision-making appear as factors at the local level hindering the effectiveness of decentralization. Local level actions within an enabling environment are fundamental to shelter development, particularly for low-income groups.

Environmentally, the courtyard has gone a long way to mitigate, if not to overcome, the conditions of the climate of Bahrain, which falls into the category of hot and dry. Summers are very hot; winters are mild with short transitional periods of autumn and spring. There is a high percentage of sunshine and a high summer diurnal range (i.e., the difference between day and night temperatures) accompanied by very low relative humidity. There is very little rainfall over much of the country and the prevailing wind is mainly from the north-west. Dust storms occur on 60 days on a yearly average and take place in every month with the exception of December and January (ARCHNET n.d.).

The spacious courtyard is the focal point of the house. It acts as an extension to the surrounding covered terraces and the rooms beyond, giving a sequence of open space (courtyard) covered or in-between space (covered terrace) and enclosed space (rooms). On the account of its central position, the courtyard acts as a general space where nearly every movement between various elements of the house begins, ends, or passes through. Hence, the courtyard helps to reduce circulation space within the house. Incidentally, the courtyard is an ideally safe area for babies and children to play, as they can be easily watched by their mothers. In addition to the bent entrance, reception room, and courtyard, the ground floor includes covered terraces on two, three, or even four sides of the courtyard, a family or living room/dining room, kitchen, staircase, and a toilet. The widest covered terrace lies between the family room and the courtyard. It is in this terrace that the families spend most of their days and have all their meals during the summer season. In a sense, this terrace is an open living room and contains only essential and easily portable furniture. The favored Arab custom of sitting and eating on mats that are spread on the floor is still practiced (ARCHNET n.d.).

Few houses in Bahrain have been influenced by the Baghdadian style of a house with regard to ventilation. Whereas in Baghdad natural ventilation takes place by means of air scoops, known as Badgeer, other systems of ventilation have been influenced by the Arabian Gulf climate, and the wind tower, which is rarely found in Bahrain.

The heat that is lost during the night to the clear sky by radiation allows the courtyard to remain cool most of the day. The covered terraces, usually on two or three (occasionally four) sides of the courtyard, and the identical first floor covered gallery that is immediately above, help to reduce the quantity of heat gain during the day by obstructing the direct solar radiation (see Fig. 6).

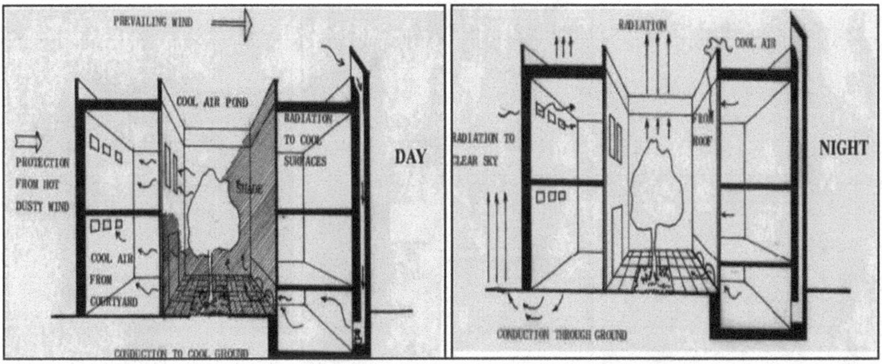

Fig. 6 Diagram of the thermal system for courtyard buildings. Drawing by the author (source: Mosul: the Architectural Conservation in Mosul Old Town—Iraq, 2010, page 35; the author noticed that the traditional Bahraini house layout was affected by Baghdadian house layouts; however, few houses in Bahrain towns contain Badgeer)

Since the height of the courtyard is greater than any of its plan dimensions, the area exposed to this radiation is reduced to a minimum, leaving adequate space in the shade, even at midday when the Sun is near the zenith. By means of a fountain, plants, or both, the very low relative humidity of the air is raised to a comfortable level. The courtyard is much quieter than the alleyways because of its position within the house. The enclosing rooms, which are built with thick walls, act as a buffer against noise (ARCHNET n.d.).

Greenery Courtyard Houses Demanded for Heritage Tourism

While tourism will be one of the fastest growing segments of the world's economy in the twenty-first century, not every city can or should look at tourism as a major portion of its economic base. There are cultural, economic, logistical, and sometimes even religious reasons why tourism is not appropriate for every community. Further, it would be a mistake to only connect historic buildings with tourism—there are many more ways that historic buildings can be used as a local resource. In the USA, for example, 95% of all the historic resources in productive use have nothing whatsoever to do with tourism (Alsammarae 2019).

At the same time, when tourism is identified locally as a component of an overall economic development strategy, the identification, protection, and enhancement of the city's historic resources will be vital for any successful and sustainable tourism effort. Worldwide, heritage visitors stay longer, spend more per day, and have a

Fig. 7 Good examples of preserving traditional buildings by private sector initiatives (source: photos taken by the author, 2006)

significantly greater economic impact per trip. An even more important conclusion emerges: when heritage tourism is done right, the biggest beneficiaries are not the visitors but the local residents who experience a renewed appreciation and pride in their local community and its history, Fig. 7.

Inspired Historical Roots and Adapted for the Future

Any building must be committed to the ideas and content depending on the identity and heritage, be responsive to the surrounding environment, and meet the needs of the community. This is what architectural practices should have been these days, that past is important to be studied and takes the concepts of design for the present—looking forward to the future—"There is No Architecture Without Thinking." Although perceiving around the world that modern technological tools are changing and evolving with time, the thought of urban architecture must be solid.

Clients' requirements cannot be achieved alone. A designer needs to convey those ideas embodied in the concept of an architect. This is because the client does not understand the tools of architecture, whereas the designer must have architectural thoughts and tools.

What we see now is the phenomena of tools given to architects without any thought—as a pioneer architect once quoted: "An architect should wear the shoes of his client." There are important educational parts to raise awareness among architects first, and the community later.

Contemporary Architecture in Arab States

These nonserious pieces of architecture are being designed by Arab architects nowadays and should not be copied from the West. These buildings are nothing but work of glass structures not related to this environment or society. It is clear in the final outcome that the goal is characterized by the superficiality of emotion and intellectual failure in the lack of knowledge of the cultural transformations of vital communities which must be inspired by the past in symbols or traditional characters and employed in new projects, not to reproduce elements for the purpose of decorating buildings in a fake heritage. Here is a good example of adapting the courtyard concept of design for a house in Baghdad, Kuwait, and Bahrain where all of the spaces in the house are open to four courtyards with solid walls that are exposed to the outdoor. Particular plants that grow in the desert and harsh climates have been planted in these courtyards (see Figs. 8, 9, 10, 11, and 12).

German art historian Udo Kultermann has noted in his book: "In the 1960s, oil was discovered in the Middle East, and the economic and cultural impact of that discovery has caused a building boom unparalleled elsewhere in the world at any time in history" (Alsammarae 2019). Fortunately, these days, a seminal work on

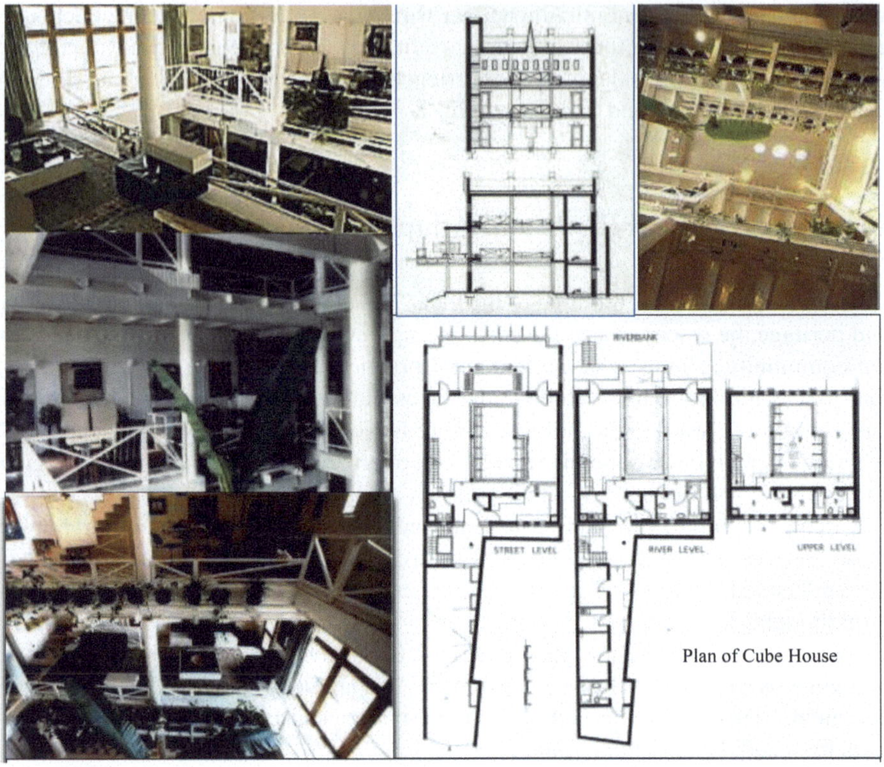

Fig. 8 A cube house in Baghdad, Iraq (above). A hospital and a private house (bottom). (Source: Design by Maath Alousi from the Architect Collection)

Middle Eastern architecture is focusing not on the plethora of Western companies and architects doing work in the Middle East but on local architects who are mostly unknown in the West, despite years of brilliant accomplishments. Such names include Hassan Fathy, Mohamed Saleh Makiya, Rasem Badran, Rifat Chadirji, Maath Alousi, AbdelWahed AlWakil, and Kamal Al-Kafrawi along with many others. Many excellent buildings are designed by Western architects who have shown a sensitivity to the local. The focus on local traditions and the cultures that spawned Arab architect as well as disdaining for the international architecture has no connection to the past or to the people it is designed for.

Arab architects who are looking for inspirations from their predecessors as well as harmonious with the merging tradition with contemporary requirements can be referred to the Architect's Cube House designed by Maath Alousi in Baghdad. The Cube House is a project based on the principle of a cube while surrounding a central internal courtyard, as an example (4). A private residence is composed of three independent houses on a plot wherein the nature of the layout is designed for family interaction. Each of the houses is organized around a central courtyard which is positioned on the axis and is identical in each house. This contemporary space

Fig. 9 (Above)—Examples of six row houses close to Tigris River in Baghdad, which use the courtyard design in two floors covered by a dome, and windows from all sides to provide cross ventilation and wind breeze by using traditional materials only. Designed by the author in Baghdad, 1992. (Down)—Souq Hamada neighborhood design by IDRISI Center for Engineering Consultant

arrangement is sensitive to tradition (4). Other public building projects in Baghdad had the courtyard principle which is adopted by the same abovementioned architect (see Fig. 8).

Conclusion

From thousands of years till the 1950s the courtyard-type houses dominated the building design in the Middle East. This phenomenon was driven by two significant factors: firstly, the internal privacy of the family activities, and secondly, to reduce the harsh hot weather and dust effect which occurred most of the year in the region.

This chapter showed how modern architecture has deviated from the Arab traditional architecture which resulted in dwellings not even fit for animals to live in

Fig. 10 Architect Ahmed Al Jowader built his house in the Hidd area (Bahrain) to contain a wind tower and a courtyard adapting traditional materials to modern use as well as save a massive amount of energy

Fig. 11 Good example of using courtyard type of houses in harsh climate like in Kuwait by DeZone: https://bit.ly/2Fup5ia

without mechanical air-conditioning with a loss of privacy and natural comfort. Therefore a development of regulatory requirements for traditional neighborhoods in towns is needed for new housing development in order to encourage the use of courtyard types of designs and the use of greenery by following the below actions:

- The builder or designer should choose the right materials and accommodate all needs of the residents and requirements rather than making the economical saving to be his/her main objective.
- Removal of all falling and disused buildings and replacing them with new ones taking into consideration the past heritage of natural ventilation, shading, and greenery usage to achieve healthy and comfortable living environment which are to be inspired by the traditional concept in the continuously changing urban form of traditional features of the towns of Bahrain.

Fig. 12 Type of plants that can be grown in courtyards which have resistance to heat and dust and require less maintenance

References

AlKubaisy, F. (n.d.). There is no architecture without thinking: Inspired and extracted from historical roots and adapted for the future. In *Urban renewal in the Arab world* (p. 202).

Alsammarae, R. (2019). Alousi's cube house in Baghdad. Image courtesy of Tamayouz Excellence Award, Architect Journal. Retrieved from https://bit.ly/2uAef4x

ARCHNET: MIT joined center of Aga Khan trust for culture. n.d.. Retrieved from https://archnet.org/authorities/774/sites/892

The Use of Vegetation in Hot Arid Climates for Sustainable Urban Environments

Jose Manuel Ochoa, Irene Marincic, and Helena Coch

Introduction

The environmental conditions we might have on a summer day sitting under the shade of a tree in a park are not the same as in the middle of a square surrounded by concrete pavement. Such different conditions would result in thermal, acoustic and visual comfort sensations also very different, although both spaces could only be a few metres from each other.

Not always we can say that being under the shade of a tree is pleasant or that being in a concrete square is unpleasant; this will depend on the climatic and environmental conditions of the site.

These environmental conditions can be modified to a certain degree to improve its appearance and functionality, as well as the living conditions for the people who will use the site. Environmental adjustments can be achieved through elements such as tree barriers, plant pergolas or fountains, which will reduce wind, filter solar radiation or moisten the environment.

In hot-dry climates, it is essential to improve the habitability of outdoor spaces and those adjacent to buildings, as these spaces not only benefit the people who use them, but also improve the energy efficiency of buildings by drawing near the weather conditions outside to those required inside.

Our study is located at the Sonoran Desert in North America, which covers part of Arizona, part of California in the United States, and also a significant area of

J. M. Ochoa (✉) · I. Marincic
Departamento de Arquitectura y Diseño, Universidad de Sonora, Hermosillo, Sonora, Mexico
e-mail: josemanuel.ochoa@unison.mx; irene.marincic@unison.mx

H. Coch
Architecture & Energy Research Group, Department of Architectural Technology, School of Architecture, UPC, Barcelona, Spain
e-mail: helena.coch@upc.edu

A. Sayigh, A. Trombadore (eds.), *The Importance of Greenery in Sustainable Buildings*, Innovative Renewable Energy,
https://doi.org/10.1007/978-3-030-68556-0_12

Fig. 1 Landscapes of El Pinacate Biosphere Reserve in the Great Altar Desert that is part of the Sonoran Desert

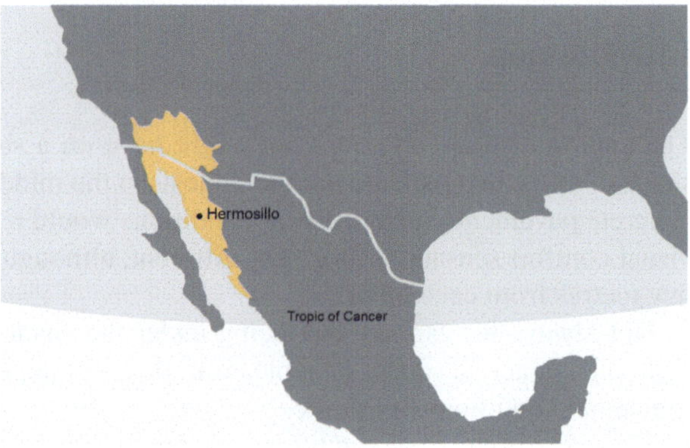

Fig. 2 Location of the city of Hermosillo at 28° 29' North latitude. The yellow colour marks the area covered by the Sonoran Desert (own work based on (Ricketts 1999))

Sonora and Baja California in Mexico. This desert has different regions and subclimates, but in most cases, the climatic conditions are very harsh (Fig. 1).

One of the main cities of the studied region is Hermosillo, the capital of the state of Sonora, with about 800,000 population, which is located at 28° 29' of North latitude, in the northwest of Mexico, at 275 km south from the border to the United States, and 2037 km north from Mexico City (Fig. 2).

The local climate (Climate.OneBuilding 2019) characterises by high rates of solar radiation, clear skies and large temperature fluctuations throughout the year. The air average daily high temperature is over 38 °C for 90 days a year, including most days from early June to early September, with minimum air temperatures of 25–30 °C and maximums around 40–45 °C. Air temperatures can reach, in extreme cases, up to 50 °C (Fig. 3). The average relative humidity is around 15%; however, the arrival of humid air in August brings the average up to 50%.

The hot season extends over 5–6 months a year, and it is necessary to use air conditioning in buildings. The temperature of the earth's surface could reach 70 °C when exposed to solar radiation, although this is affected by the properties of the surface's materials. The winters are mild, with minimum temperatures between 0

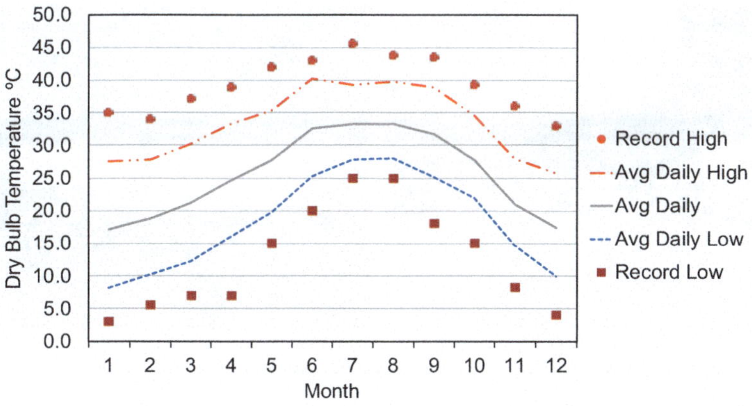

Fig. 3 Monthly average daily dry-bulb temperatures in Hermosillo (own work based on (Climate. OneBuilding 2019))

and 7 °C and a maximum between 25 and 30 °C, so this results in a season with pleasant temperatures.

Hermosillo has almost 300 clear days a year; in spring and summer, the solar radiation can rise to a little more than 1000 W/m^2. Rainfall is scarce, with an annual total of 225 mm. August is the rainiest month of the year with 27 mm. Rainfall is particularly low between April and June. Although storms occur from time to time throughout the year, they are more frequent during the rainy season from July to mid-September; when moist air arrives from the Gulf of California, it can bring strong winds. Winter storms moving inland from the Pacific Ocean sometimes result in significant rainfall, but they are rare.

The average wind speed in Hermosillo has considerable seasonal variations throughout the year. The windiest part of the year lasts 8 months, from November to July, with average wind speeds of more than 3 m/s. The windiest days of the year are in June, with a maximum average wind speed of 4 m/s. The quietest season of the year lasts 4 months, from July to November. The quietest days of the year occur in August, with a maximum average wind speed of 2.5 m/s.

Despite the harsh environmental conditions, the newer areas of the city are not well planned or prepared for the hot-dry desert climate. These neighbourhoods have wide streets with little or no vegetation, large paved areas and single-family homes. Both the low-cost houses and the residential ones are designed with the same criteria of the temperate climate of the central regions of Mexico; therefore, there is an enormous consumption of energy due to air conditioning, during the hot period.

In contrast, the old town bestows narrow streets, adobe houses, small windows, vegetated patios, arcades and light colours, as a sign of regional adaptation (Fig. 4).

During the hot season, the local people have tailored their way of life to these circumstances. In most cases, physical activities and outdoor movements such as walking are reduced to a minimum and rather done during the night or early in the morning. City residents enjoy social activities and family gatherings that can be

Fig. 4 Vegetated square in the old town (right) in contrast with a modern main street (left), both in Hermosillo city

held outdoors all year round, perhaps as a reminder of their rural origins, even though most of the outdoor areas available are improvised and have limited comfort conditions.

As a result of a field study (Ochoa et al. 2006) it was found that, despite the population's dependence on air conditioning, there is a certain level of acclimatization, finding that the average outside temperature in the shade, which people voted most comfortable for spring, is 28.7 °C and for summer 36.2 °C.

The dry air makes the hot temperatures more tolerable but only at the beginning of the summer season; in this period the everyday use of evaporative cooling systems are in public places such as outdoor shopping malls and bar and restaurant terraces. However, when the rainy season begins, the thermal sensation becomes muggy.

Natural ventilation is not a resource that can be used all year round; during spring and autumn the feeling of comfort is improved if the wind speed increases; however, during the summer the wind is usually scorching; thus it does not help the passive cooling or thermal comfort; sometimes it is necessary to protect from the wind at this time.

In rigorous climates, the passive criteria of architectural design and their application must be strictly applied to solve the environmental conditions in order to improve the quality of life of the inhabitants, as well as the energy consumption of the buildings. In the following sections of this chapter, we will discuss the use of vegetation in the outdoor areas of a city with a hot-dry climate and integration into the buildings is an excellent opportunity to improve the climate of the outdoor areas by having better living conditions, as well as better energy efficiency of the buildings at any scale.

Comfort in Extreme Desert Climates

P.O. Fanger (Fanger 1972) defined the sensation of thermal comfort as "the condition of mind that expresses satisfaction with the thermal environment". This pioneering definition would seem to describe the sensation of thermal comfort in a subjective way. However, it is not only the perception or the mood that defines the condition of cold or warm, and comfort or discomfort of a person in a given situation. The personal sensation of cold or warm is part of a cognitive process that integrates many stimuli influenced by physical, physiological and psychological factors, among others.

Expressed in physiological terms, the comfort sensation can be defined as the situation in which the human body makes the least effort to regulate its internal temperature. This situation also involves acclimatization and other factors such as the physical activity that the subject is performing, body size, gender, age and clothing.

Thermal comfort can also be defined in physical terms, as the condition when the heat exchange between the body and the environment is in equilibrium. A positive result of a heat balance means that the human body is gaining energy so the thermal sensation will be warm or hot. If the heat balance is negative, the body is losing energy, and then the thermal sensation will be cool or cold.

The adaptive approach of the thermal comfort, proposed by Nicol and Humphreys (Nicol and Humphreys 2002), combines the thermal characteristics of the environment and the heat transfer exchange with the human body, together with people's behaviour subjective characteristics. This approach takes into account not only the physical interaction between the subject and the environment, but also their long--term psychological and physiological interaction, incorporating the effects of acclimatization, expectations and people's choices, in order to reach more comfortable conditions.

The mentioned authors, as well as Brager, de Dear, Auliciems and others (Nicol and Humphreys 2002), have worked on adaptive models, which take into account people's preferences and expectations.

To address a comfort study based on an adaptive approach is necessary to collect, through a field survey, information on the thermal sensation of people (votes) in different environments and relate them to the thermal variables in those spaces, particularly the dry-bulb temperature. That is, a field survey must be applied to part of the population, asking about their thermal sensation into indoor environments (generally). With the collected data, it is possible to estimate the comfort or neutral temperature Tn, which is the temperature at which the effort of the human body to achieve thermal comfort is minimal. With the data obtained in this type of survey, the comfort range limits can be calculated, which include a group of temperatures in which the people's majority feels comfortable, usually considered as Tn ± 2°C.

It has been demonstrated (Nicol and Humphreys 2002), comparing the results of adaptive models with quantitative ones (based on heat transfer balance between human body and environment), that the model generalizations, in terms of different climates and thermal preferences of the people, are not appropriate to determine the comfort conditions, since the models strongly depend on the environment and the human context. So, it is necessary to determine the thermal comfort conditions for specific climates and analyse population groups. For the application of comfort standards based on quantitative models (such as ISO 7730 or the ASHRAE comfort zone) (International Organization of Standardization 2005a; American Society of Heating, Refrigerating and Air-Conditioning Engineers ASHRAE 2001), it would be necessary to have a broader knowledge, so that standards could be adjusted to specific local conditions.

For the evaluation of thermal comfort in outdoor spaces, the direct application of the criteria based on indoor spaces is not correct. One reason is the large difference between the climatic conditions of indoor and outdoor spaces. The variability of the external climate is broader than that of the interior: the external climate variables present more extensive ranges and change at a faster rate than in the interior, which can be seen in the daily variations (day/night) and the seasonal variations throughout the year. Also, variations of external climatic conditions occur as the subject moves along a space, or when he or she moves to another site with more comfortable conditions.

The acclimatization of local people also impacts their comfort sensation, and the effect includes not only the adaptation to the local climate, but also the use of appropriate clothing, local life patterns and even higher or lower tolerance to specific climatic parameters. Likewise, the tolerance to climatic conditions increases in outer space with respect to an indoor space, since people take into account the lower possibility of climate control outdoors. Therefore, their expectations of thermal comfort outdoors decrease, since they usually do not expect to achieve a thermal comfort sensation comparable to that which can be obtained in artificially acclimatized indoor areas. For these reasons, the evaluation of thermal comfort outdoors should be addressed, considering the factors mentioned above.

Adaptive Thermal Comfort in Social Housing in Urban Areas

To evaluate the comfort of people living in social housing units in a desert climate, we take as a case study the city of Hermosillo, located in the Sonoran Desert. Field surveys have been applied to subjects who live in tract housing units to study different aspects of the dwellings and their occupants. Nine different tract housing groups have been surveyed as a representative sample in the city. The methodology for the surveys' application and environmental measurements has been carried out according to ISO 7730: 2005 (E) (International Organization of Standardization 2005b), ISO 10551: 1995 (E) (International Organization of Standardization 1995) and ISO 7726: 1998 (E) (International Organization of Standardization 1998).

These low-cost houses, which are very numerous in the city, as well as in other cities of Mexico, consist of one floor with a construction area of 33.5–39 m^2, depending on the model, and a lot of area that varies between 117 and 122 m^2. All of them have one bedroom, a living room, kitchen and bathroom. In all cases, they also have an exterior area for parking.

In order to collect the required information, two types of surveys have been carried out: one aimed at getting the physical characteristics of the homes as well as the users' profile and the perception of the housing by them (Romero et al. 2009). The second survey (Gómez-Azpeitia et al. 2014), which was carried out during two periods of the year, one in winter and one in summer, was oriented towards the thermal sensation of the users, while at the same time we monitored the thermal conditions inside the dwellings. We collected information about users, such as age, weight, gender and clothing. The thermal sensation of the dwellers was recorded on a 7-point scale, from cold to hot (ASHRAE scale from −3 to +3). At the same time, the air temperature (dry bulb and wet bulb), black globe temperature, relative humidity and wind speed were monitored. Next, we collected information on the electrical consumption of the surveyed dwellings (Marincic et al. 2009, 2013).

According to the user profile, the dwellers are adapted and acclimated to local conditions. Although they have few years of permanence in the complex of houses, since in general they are of recent construction, the origins of the population are mostly from the same city and in second place from the rest of the state of Sonora. We consider, therefore, that the results correspond to an acclimated population sample.

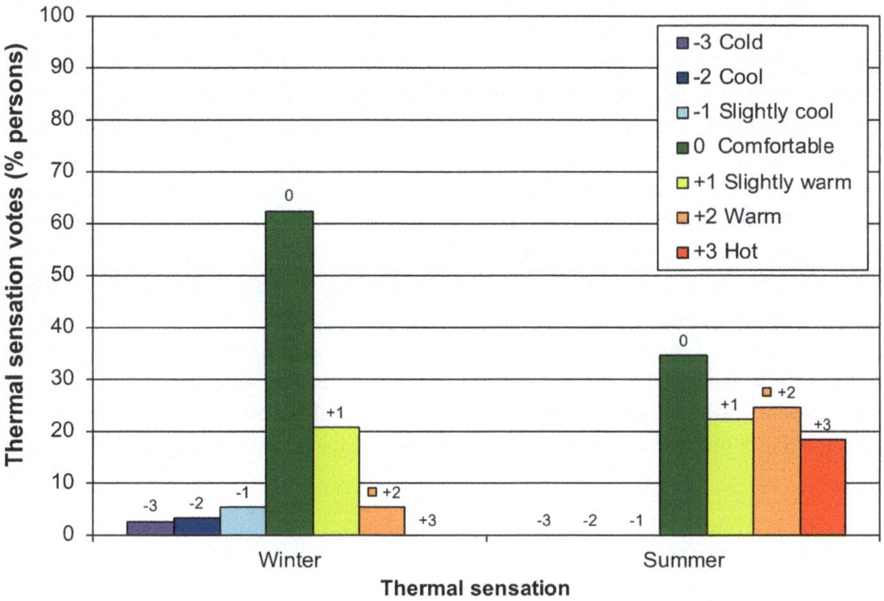

Fig. 5 Thermal sensation response in winter and summer is also given (percentage of people's votes) (Marincic et al. 2013).

Other population groups in Hermosillo are habituated to air-conditioned spaces, either at work, at home or both. This situation is not the case of users of the surveyed housing units, where the use of air conditioning is not part of their daily lives.

In Fig. 5, thermal responses of users are summarized for each period, as a percentage of the total responses for the period. The responses are based on the 7-point scale used by ASHRAE (American Society of Heating, Refrigerating and Air--Conditioning Engineers ASHRAE 2001). Although each period was processed separately, the results are presented together in the same graph

As mentioned in the description of the local climate, thermal conditions in winter are mild, and it would not be complicated to adapt to them. Figure 5 shows that more than 60% of the thermal sensation responses are *comfortable* (0). However, indoor conditions in summer are not comfortable, which is expressed in the percentage of high thermal sensation responses (from+1 to +3) in Fig. 5.

One way to evaluate thermal comfort is through the determination of the neutral temperature (Tn), or comfort temperature, which can be estimated for each particular climate and population. Tn is a representative temperature (obtained through surveys) in which most people feel comfortable. Additionally, a comfort range is established, which is a range of temperatures within which the temperature can vary, and persons still maintain the comfort sensation (Table 1). Knowing specifically comfort ranges for each population group and climate, instead of using directly international standards, obtained for other situations, is useful to propose passive thermal strategies taking into account the real thermal perception of the occupants. Furthermore, in case of using artificial air conditioning, it is possible to establish the actual temperatures at which the air-conditioning equipment thermostat should be set.

As can be seen in the results, the Tn is quite high, since it depends on the mean outdoor temperature of the evaluated period, and the comfort range is quite broad, which is related to the wide temperature oscillations of the desert climate.

As mentioned, the thermal sensation is a subjective response that depends on multiple climatic, physiological, social and psychological factors. Although it is formed by the combination of all these factors, taking into account the user's profile, it is possible to isolate some of them to determine the relative impact of each one on the thermal sensation (Marincic et al. 2012).

The methodology used to determine the neutral temperature and the comfort range is based on the fact that Hermosillo's climate can be considered as asymmetric (Nicol 1993). On this type of climates, the thermal sensation shows a tendency towards only one side of the scale. In this case, the "warm" responses do not have a symmetrical response to the opposite side of the scale (cold). In this case, to estimate the Tn, a method of statistical regression by layers was used (Gómez-Azpeitia

Table 1 Neutral temperature Tn and comfort ranges in Hermosillo (Marincic et al. 2013)

Period	Climate	Lower limit (°C)	Tn (°C)	Upper limit (°C)
Summer	Hot dry	29.7	32.2	34.5
Winter	Temperate	23.5	26.9	31.3

et al. 2014). The method is based on Nicol's proposal for asymmetric climates (Nicol 1993).

In addition to the overall results for this type of population (Table 1), three population characteristics can be studied separately to determine their impact on the comfort sensation: age, body size and gender. The comfort sensation responses have been studied as a function of the indoor temperature for the different cases, thus obtaining the different Tn for each population characteristic (Nicol 1993).

In general terms, it is observed that for a population acclimated to local climatic conditions, and with relatively low comfort expectations, the thermal as well as functional conditions inside the dwellings can be considered as not satisfactory. However, the level of dissatisfaction of the basic needs, such as thermal comfort, is perceived by the user better than expected, taking into account comfort ranges and comfort zones reported in the bibliography for other climates. Of course, this does not imply that there is no need to improve their quality of life.

Thermal Comfort in Outdoor Spaces

Most of the studies on thermal comfort in the world (Fanger 1972; Nicol and Humphreys 2002; Gómez-Azpeitia et al. 2014; Marincic et al. 2012; Nicol 1993) have been done considering the environmental conditions of the interior of buildings. However, for outdoor spaces, this type of study is not frequent, and the way to estimate the comfort sensation is not entirely defined. This situation may be because, since outdoor spaces are generally not artificially acclimatized, their conditions are not directly related to energy consumption. On the other hand, there are no international standards for comfort in outdoor spaces such as ISO 7730 (International Organization of Standardization 2005a) or regulations such as those of ASHRAE (American Society of Heating, Refrigerating and Air-Conditioning Engineers ASHRAE 2001) that are used exclusively for indoors.

In a thermal comfort study for outdoor spaces, surveys have been carried out to examine the comfort sensation of acclimatized local people in the same desert city.

The surveys have been conducted by advanced degree architecture students, who received specific training for the field surveys during spring and summer. The sensation scale that has been used is proposed by Nikolopoulou et al. (Nikolopoulou et al. 2004) for outdoor spaces, which establishes a 5-point scale of thermal sensation or ASV. It was intended that the conditions of the respondents were similar: 51% of people were men and 49% women, and mostly (91%) were sitting and reading, talking or resting, and the rest (9%) strolling. Regarding age, 83% were people between 18 and 25 years old, 6.5% between 26 and 35 years old and 10.5% between 36 and 45 years old, which corresponds to the profile of the university population including students, professors and workers. The surveys were carried out in different outdoor spaces at the campus of the University of Sonora, where people were in transit. In the selection of the different spaces, passive thermal strategies or the lack of them

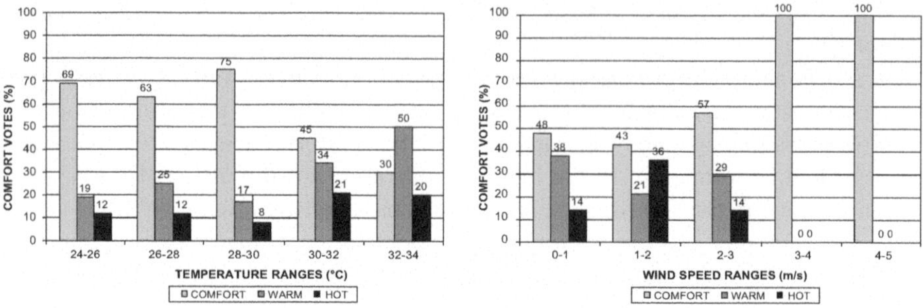

Fig. 6 Thermal sensation responses to different temperature and wind speed ranges (percentage of comfort votes) in spring (Ochoa et al. 2006)

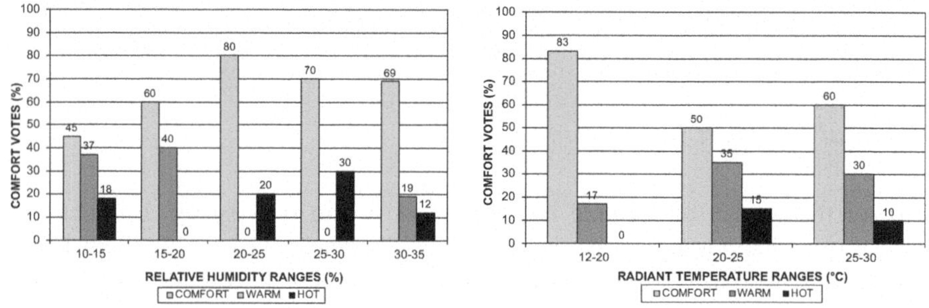

Fig. 7 Thermal sensation responses to different ranges of relative humidity and radiant temperature (percentage of comfort votes) in spring (Ochoa et al. 2006)

were taken into account, which include the use of vegetation to control climatic parameters.

Spring Survey

During the spring surveyed period, the climatic parameters that have the most impact on the sensations of comfort or discomfort of the users corresponded to the air temperature and the wind. In Fig. 6 (left) it can be seen that the temperature considered comfortable by most of the people (63–75%) is between 24 and 30 °C, between 30 and 32 °C 45% of them and between 32 and 34 °C 30%.

Regarding the wind, in Fig. 6 (right) it can be seen that while the speed increases, the percentage of comfortable people also increases, going from 48% for 0–1 m/s to 100% for speeds between 3 and 5 m/s. However, this is only valid for air temperatures below 30 °C.

The relative humidity does not seem to affect the thermal sensation very much; however, as winter and spring are dry seasons, there were days with relative humidity less than 15%. For these conditions, the people's comfort sensation was 45%, while for more humid conditions, 30–35% relative humidity, 69% of people considered the sensation as comfortable (Fig. 7, left).

Radiant temperature (Fig. 7, right) cannot be considered a problem in this period of the year. Measurements have indicated that it was always lower than the ambient temperature. Thus, people's comfort votes were between 50 and 83%, depending on the radiant temperature values.

Summer Survey

Summer conditions are quite different from those of spring, as temperatures vary in a range of 34–40 °C during the time the surveys were conducted. Also, the wind speed was lower (0–3 m/s), and relative humidity increased (30–50%). Due to the high solar radiation of this period, with a range of 895–965 W/m^2, also the radiant temperature increased considerably. Therefore, variations have been observed in the way that climatic variables have affected the thermal sensation of the respondents.

As can be seen in Fig. 8 (left), as the temperature increases, the percentage of people who feel comfortable decreases, with the most unfavourable range being from 38 to 40 °C.

In Fig. 8 (right) it can be seen that with the increase of wind speed the percentage of comfort votes increases. However, during the surveyed period, higher wind speed does not coincide with higher temperatures. Therefore, we do not have evidence that high wind speed contributes to the comfortable sensations with hot wind.

The thermal sensation does not vary much when the relative humidity is between 30 and 45%, which is between 18 and 31% of the comfort surveyed opinions. However, when the humidity is above 45%, the thermal sensation is 100% hot (Fig. 9, left).

Unlike the spring period, in summer, the radiant temperature plays a fundamental role since it is related to the intensity of solar radiation and the ambient temperature, both parameters with high values in summer. In Fig. 9 (right) it can be seen that while radiant temperature between 25 and 30 °C is acceptable for 57% of the

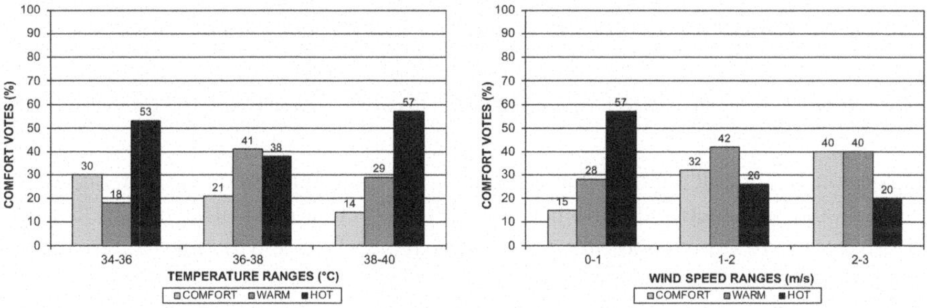

Fig. 8 Thermal sensation responses to different temperature and wind speed ranges (percentage of comfort votes) in summer (Ochoa et al. 2006)

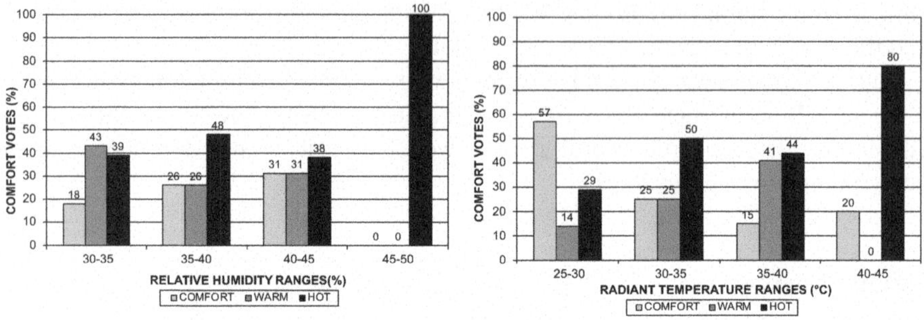

Fig. 9 Thermal sensation responses to different ranges of relative humidity and radiant temperature (percentage of comfort votes) in summer (Ochoa et al. 2006)

sample, for 80% of the people radiant temperature above 40 °C gives rise to a thermal sensation of hot.

The most frequent temperature that people indicated as comfortable in spring was 28.7 °C, and in summer was 36.2 °C. Therefore, it can be affirmed from the first results of the study that there is a certain degree of seasonal acclimatization.

In both seasons, with higher wind speed, higher comfort votes of the people are registered. However, during the summer, when strong winds coincided with temperatures higher than 40 °C, the people's comfort sensation dropped abruptly comfortable to hot, so hot winds do not contribute to a better thermal sensation.

The use of vegetation in outdoor spaces can help to control air temperature and increase humidity; it can also block or channel winds, which in a hot-dry climate can be beneficial for human comfort and energy efficiency of buildings.

As a future stage of this research, it is planned to expand the sample to a higher number of people, and to collect data during different times of the day, for example in the morning, at noon, in the afternoon and in the evening, to have the opportunity to analyse, among other things, what happens when there is no solar radiation. It is also planned to expand the number of study sites in order to generate a model applicable to most cases in the region.

Use of Vegetation for Microclimatic Control in Arid Zones

Characteristics of Plant Species for Microclimate Control

Vegetation has characteristics that are especially useful for modifying the microclimate of outdoor space and buildings. Vegetation is capable of efficiently reflecting, absorbing and filtering solar radiation; it can also deflect and channel wind, as well as moisten and cool the surrounding air.

Table 2 Microclimatic classification of vegetation (Ochoa 2009)

Structural characteristics	Foliage geometry Foliage distribution size of leaves Foliage density
Physiological characteristics	Foliage permanence Environmental adaptation
Microclimatic functional characteristics	Radiation control Wind control Temperature control Moisture control

When we treat vegetation as an element of modification of the microclimate, we must consider specific characteristics of each plant, referring to its structure, and if it is a tree, shrub, ground cover or climber.

Also important are physiological characteristics, such as the permanence of the foliage, that is, whether they are deciduous or evergreen, and their environmental adaptation; these characteristics are fundamental in hot-dry climate areas.

Based on the above, we can define the functional characteristics, which cover the effects on climate parameters, namely solar radiation, long-wave radiation, wind, air temperature and humidity. In Table 2, the main characteristics of a microclimatic classification of vegetation are listed.

The vegetation of hot-dry climates is generally not as lush or exuberant as that of temperate and warm-humid climates. However, it can be observed that it has a rapid growth in the rainy season and during the dry season, it is considerably reduced and often loses its foliage. Because of this, in urban areas, it is common to use plant species from other regions, and although some of them have become acclimatized, they require more irrigation than local species.

For this work, the characteristics of urban vegetation for hot-dry climates will be described below, which indicate the functions that a plant species has within the microclimate control strategies, as well as the type of elements that can form within the spatial configuration of the site.

Thermal Radiation Control

The primary source of thermal radiation in an urban space is the sun, which is also one of the main factors affecting the environmental comfort of people and the thermal performance of buildings.

When the solar radiation crosses the Earth's atmosphere, it can be divided into two components: direct and diffuse radiation.

Direct radiation arrives in a straight line from the sun, whose direction can be predicted quite accurately throughout the year.

Diffuse radiation is the result of the multiple reflections and dispersions through the atmosphere, and it is anisotropic.

As the sun's radiation falls on the various surfaces of the urban landscape, it is reflected and absorbed by them. When the surfaces are heated, they emit long-wave radiation (infrared) depending on their emissivity and surface temperature.The most interesting characteristics of the vegetation for this purpose are (Ochoa 2005):

(a) *Shape and arrangement of foliage*. This determines whether a tree or shrub is appropriate to provide shade for people under its foliage (umbrella form) when the sun is near the zenith or to protect them from the sun when it is at low altitude (distributed along the trunk), near dawn or dusk.
(b) *Maximum expected height and diameter of growth and time taken to reach them.* This information is useful for planning the type of plant species and the layout of the plantation so that each plant has the space to grow and the desired effect.
(c) *Density and transmittance*. The amount of radiation that passes through the foliage depends on these parameters. For deciduous species, it is necessary to know the transmittance both in winter and summer.
(d) *Approximate dates of the foliation of deciduous species*. This information is useful for planning the type of plant species that would be used, according to the annual distribution of the site's climatic parameters.
(e) *Sun exposure*. The resistance to excess or lack of solar radiation will give an indication of the plant species to be planted in each case; a plant that provides shade should resist the sun along the season. However, when several layers or groups are formed, some specimens will always remain in the shade.

Wind Control

The wind is another component of the microclimate that affects people and buildings, which can be significantly modified by vegetation and solid barriers.

The wind is one of the most variable and unpredictable climate parameters, both in speed and direction. In urban environments, it is easily modified by either stable structures such as buildings or permeable barriers with vegetation; some will change direction or reduce speed and others may increase it. The wind is also influenced by the orientation and relative location of the site.There are four actions in which vegetation affects the wind:

Obstruction. It blocks the flow of air in an area.
Deflection. It deflects the wind and slows it down.
Filtration. It reduces the wind speed when passing through a permeable barrier.
Channelling. It changes the direction of the wind, driving it into an area where ventilation is required.

The most interesting vegetation's characteristics for wind control are (Ochoa 2005):

(a) *Shape and arrangement of foliage*. For high barriers, we recommend trees with uniform foliage distributed along the trunk. For lower barriers, shrubs are better.

Mixed barriers, incorporating trees and shrubs, can be configured to have covered the full height of the barrier.

(b) *Wind stress*. The species used must be resistant to wind, to retain their foliage, and should not suffer excessive deformation. A firm type of soil and planting is necessary because the wind can blow them away.

(c) *Sun exposure*. For wind control, the chosen species must resist either sun or shade, as their location will not be associated with the incidence of the sun.

(d) *Time for growth*. In this case, the planting of trees and shrubs that are as mature as possible and of rapid growth should be considered, since, due to the mechanical action of the wind, very young specimens can suffer malformations or even die, before reaching their functional size.

(e) *Foliage character*. In climates with cold winters, evergreens are the most useful when it is necessary to divert the cold winter wind, as they preserve their foliage throughout the year. However, care must be taken in their location to avoid the projection of undesirable shadows. The deciduous trees only work in summer, when they have 100% of their foliage; at this time they are also suitable for use as shade elements.

(f) *Penetrability*. This parameter consists of the capacity of a plant to reduce the wind. It depends on the density of its foliage.

Temperature and Humidity Control

The air inside the foliage of trees and shrubs can be cooled by direct evaporation of water from the soil and by plant transpiration when the relative humidity of the air is less than 50%. However, the cooling effect can be quickly dissipated by the wind, unless walls, tree barriers or buildings confine the site. Another aspect to consider is the scale of the site; the higher the number of trees per unit of area and the higher the area covered by vegetation, the more noticeable the effect on air temperature and humidity.

The most interesting characteristics of the vegetation for the control of air temperature and humidity are (Ochoa 2005):

(a) *Density of foliage*. A high-density foliage allows more leaf area in contact with the air and therefore higher evapotranspiration. Likewise, the higher the density of the foliage, the less the sun's rays will penetrate. The latter also applies to green facades and green roofs, influencing heat transfer into the building.

(b) *Drought tolerance and irrigation requirements*. In arid regions, it is necessary that the vegetation resist drought and have reduced irrigation requirements. On the other hand, most of the vegetation tends to have a surface temperature similar to the ambient temperature, which reduces the emission of long-wave radiation.

Green Element Microclimate Controllers in Hot-Dry Climates

In order to analyse the effect of vegetation on the microclimate in outdoor spaces, we have proposed a classification of green elements that interact in different ways with the buildings and adjacent areas (Ochoa 2009; Ochoa 2005; Ochoa 2017).

These are lined trees, tree group, surface cover and pergolas, which are composed of plants of different types and species.

There is no clear differentiation if an element is used exclusively in streets, squares, parks or gardens, since in each type of space it is possible to have a combination of different elements, in order to achieve the required environmental conditions.

Lined Trees

This element is composed of rows of trees or bushes, or a combination of these, planted following a linear line, either curved or straight, at a more or less regular distance. This configuration is perhaps the most common among the plant components, and is used mainly as road trees planted along the sidewalk in streets for vehicular traffic, or along a road or a pedestrian walk. However, it is also found in other locations such as parks, squares and around buildings. Due to its "wall" shape, it works perfectly as a visual, acoustic and wind barrier, as well as to delimit spaces such as sports courts, swimming pools and access squares to buildings, among others (Fig. 10).

Fig. 10 The vegetation typical of arid zones is not very lush, but it can be used as a visual barrier against the wind (left) or sun (right) (source: own work)

Effects on Microclimate and Habitability

The factor most affected by this type of element is the incidence of solar radiation on people and site surfaces, preventing them from heating up. Also, the vegetation reflects the infrared radiation emitted by the soil and other surfaces, thus preventing radiative cooling at night.

The deciduous species are especially useful in these cases since during the warm season they provide shade and during the cold season, when they lose their leaves, they allow the incidence of the sun's rays in the place.

The wind is also affected in different ways by the tree line, the most obvious effect being to deflect the wind.

Table 3 Schemes of microclimatic effects of trees in line

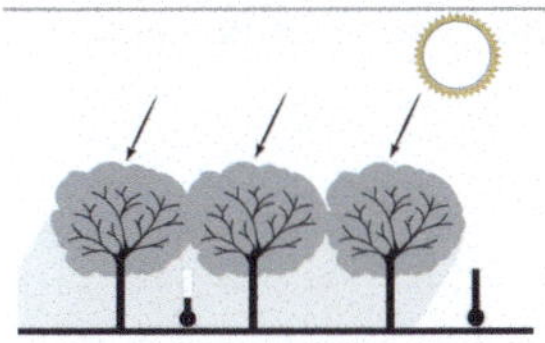	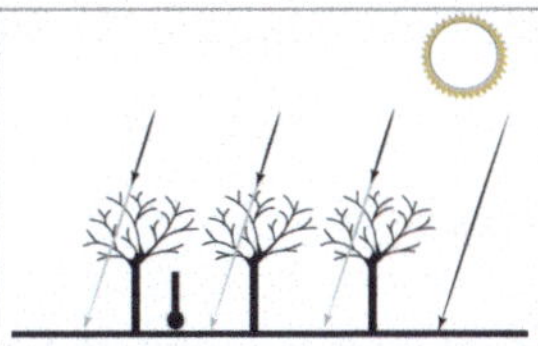
During the day, trees in line provide shade and reduce the radiant temperature of the ground under their foliage, and during the night prevent radiative cooling of the ground surface.	In winter, deciduous trees allow the incidence of solar radiation during the day and radiative cooling at night when the sky is clear.
	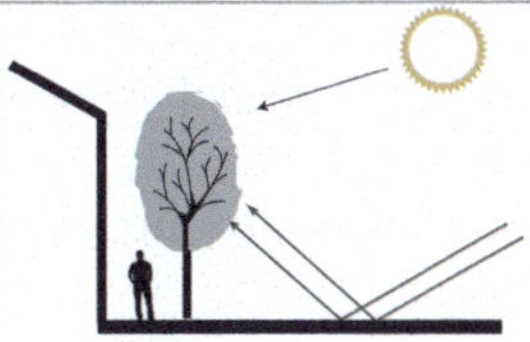
Lined trees combined with shrubs can form dense vertical barriers that protect buildings from reflected radiation from the surrounding ground, as well as from solar radiation near sunrise or sunset.	Barriers of lined trees protect buildings and pedestrians from solar radiation in walkways.
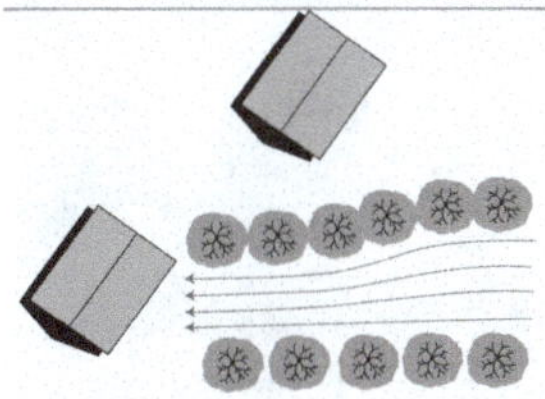	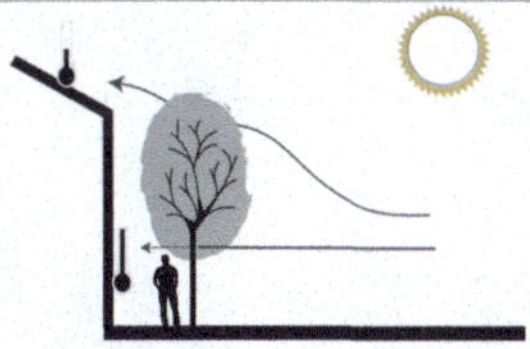
Trees can deflect wind and protect buildings, either during the winter (cold winds) or during the summer (hot winds). Also, they can drive the wind to ventilate specific areas in intermediate seasons.	Barriers of perennial lined trees in walkways protect buildings and pedestrians from convection cooling or heating from wind, depending on the season.

Source: Own work

When the direction of the wind is normal to the tree line, the vegetation deflects it vertically, producing a wind shadow after the barrier, which protects the area next to the barrier. When it is oblique, it deflects it horizontally, redirecting the wind to where it is needed.

Given the physical constitution of the vegetation, it is impossible to create a solid barrier, like a brick wall, "filtering" the wind and slowing its speed.

In both cases (radiation and wind), it is essential to consider evergreen species, which retain their foliage throughout the year. There are positive and negative effects: for example, during the winter a plant barrier can efficiently deflect winds, reducing convection losses from a building; however, if it is badly located it will also obstruct the sun, very necessary at this period. On the other hand, in summer, if it is poorly located, it will obstruct cool breezes, even if they provide shade.

Not only do the microclimatic effects of this element depend on its spatial configuration, but we must also consider the physical properties of the vegetation, such as shade factor, leaf area index, density of foliage, albedo and time of foliation. In Table 3 schemes of microclimatic actions of trees in line and group can be seen.

Group of Trees

This element refers to a group of trees planted in a more or less regular grid; they can be compact groups or have more spaced elements that allow circulation between them. It is found more often in garden squares and urban parks (Fig. 11).

Effects on Microclimate and Habitability

The degree of influence of this element will depend on the density of trees in the considered area, as well as the confinement of the site. As seen in the schemes in Table 4, a group of trees have a more significant effect on solar radiation than a

Fig. 11 A group of trees can reduce the incidence of solar radiation and decrease air and urban surface temperatures (left) mainly if used in combination with ground cover (right) (source: own work)

Table 4 Schemes of microclimatic effects of trees in a group

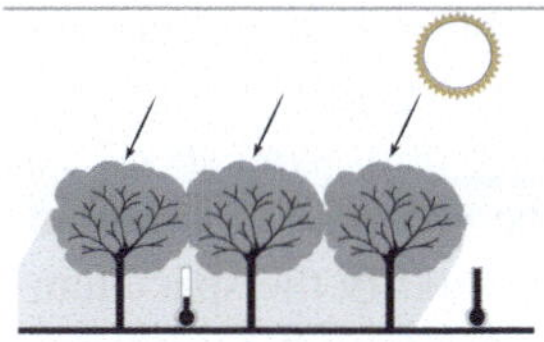 During the day, trees in groups and line provide shade and reduce the radiant temperature of the ground.	 During the night, clusters of trees prevent radiative cooling of the ground surface under their foliage.
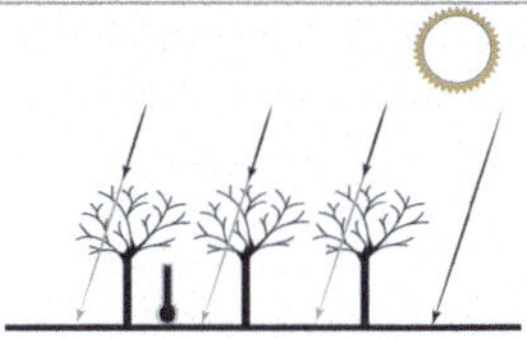 In winter, deciduous trees allow the incidence of solar radiation during the day and radiative cooling at night when the sky is clear.	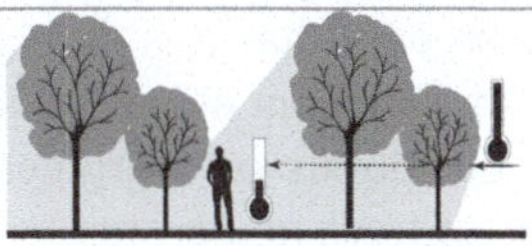 Groups of trees can lower the air temperature under their foliage if they are confined, or in the middle of a large group.
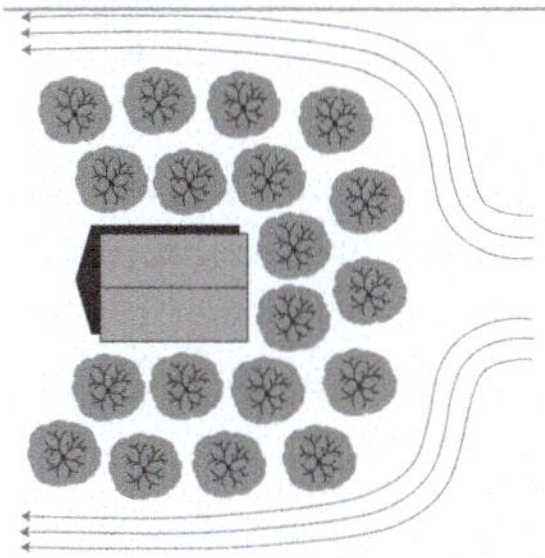 Trees can deflect wind and protect buildings, either during the winter (cold winds) or during the summer (hot winds).	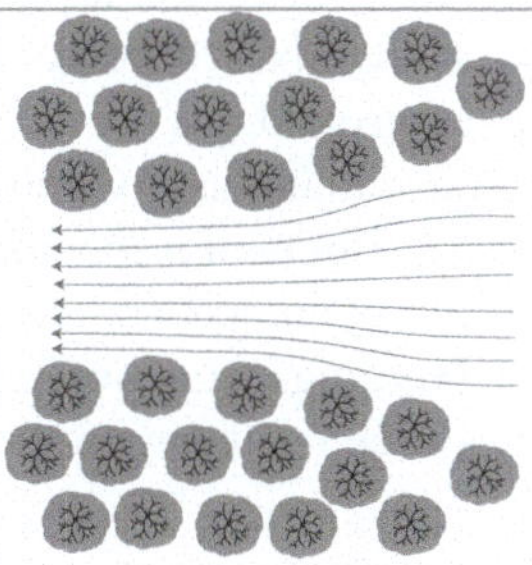 Trees can also drive the wind to ventilate specific areas in intermediate seasons.
 Barriers of trees and shrubs combined are more efficient at deflecting wind.	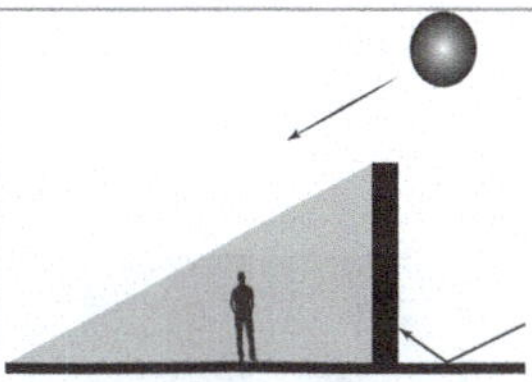 Shrubs can form dense vertical barriers that protect from reflected radiation from the surrounding ground, as well as from solar radiation near sunrise or sunset.

Source: Own work

linear tree array, because it forms a more uniform and larger layer between the sun and people or the ground. The most considerable shadow is obtained in the centre of the group, as there are fewer side reflections.

The area of highest wind protection is in the centre of the group; wind speed can be reduced from 50 to 90%, depending on the density of the group and the permeability of the vegetation (García and Fuentes 1995). We can modify the space temperature in specific cases. In a large and dense group of trees, it is possible to observe differences in temperature with adjacent, urbanized and unvegetated areas. Furthermore, the soil must be well irrigated so that we could increase evapotranspiration. It is better if the earth has a high hygroscopicity so it can absorb moisture. The effectiveness of this element also depends on the density and height of foliage, and if vegetation is deciduous or evergreen. In Table 4, we can see schemes of microclimatic actions of trees in a group.

Vegetation Covers

These elements consist of a vegetation layer attached to a surface, whether vertical, horizontal or inclined. It shall be considered as a surface cover when it has a height of up to 50 cm above the ground.

The surface covers have many uses within the design of exterior spaces; they can be transit areas or serve as limits of a specific area, combining textures, colours and shapes, depending on the plant species used. Likewise, they can be placed on the facades or at roofs of buildings, functioning as thermal insulation, solar radiation control or only as a resource for design or decoration (Fig. 12).

Effects on Microclimate and Habitability

The main effect of a plant layer is to change the physical properties of the substrate on which it is placed.

Fig. 12 Vegetable cover on a green wall shading a resting place (left) and a grass cover in a garden surrounding a building (right) (source: own work)

Table 5 Schemes of microclimatic effects of vegetation covers

 Vegetal covers have during the day a lower radiant temperature than pavements.	 During the night the radiative cooling is lower in the vegetation covers than in the stone pavements.
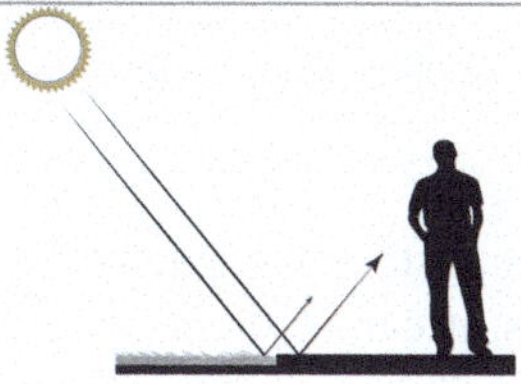 Vegetation reflects less solar radiation than light-coloured pavements.	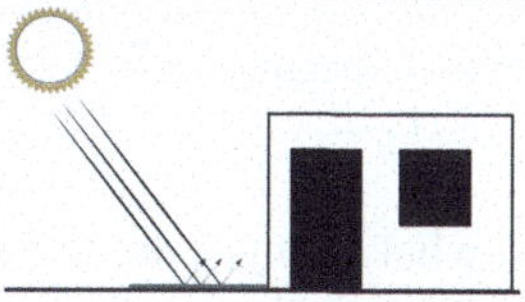 When vegetation ground covers are placed around buildings, the reflection of solar and long-wave radiation is reduced.
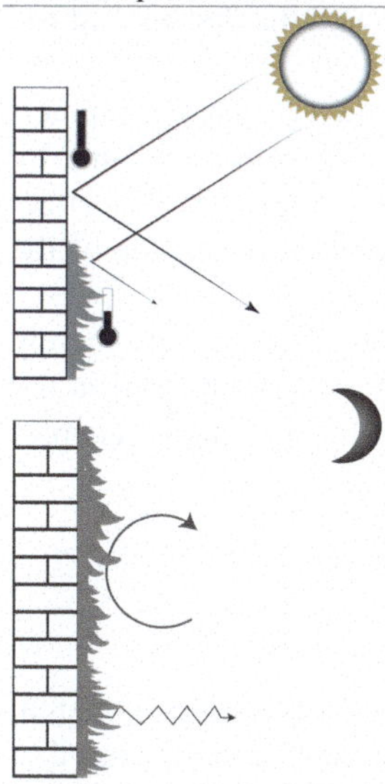 Green walls and roofs also reduce the incidence of solar radiation and wind, as well as radiative cooling. As there are evergreen and deciduous plant covers, their effects would be similar to those of other plant elements.	

Source: Own work

Fig. 13 Examples of pergolas covering a pedestrian walkway in a park (left) and as shading at a resting area (right) (source: own work)

Basically, what changes are the albedo, emissivity, thermal storage capacity and surface roughness, and when it is placed on walls or roofs, it can also change the thermal conductance of the substrate.

The main climatic repercussions, when used as a ground cover, would be the reduction of the reflection and absorption of solar radiation and the decrease of the radiant temperature.

When placed on walls and roofs of buildings, it reduces the effects of wind incidence by avoiding losses and gains of heat by convection, as well as heating by direct contributions of solar radiation, and avoids radiative cooling. Another effect that can be beneficial in buildings is the increase in the thermal resistance of the envelope.

The few disadvantages that surface coverings in buildings can have are merely functional since it is necessary to foresee how the vegetation will be fixed in the building envelope, to avoid humidity and maintenance problems. In Table 5 schemes of microclimatic effects of vegetal covers can be seen.

Pergolas

Pergolas are the only one element, which is not made up exclusively of vegetation. It consists of a structure, which may be fabricated of wood, metal, concrete or other structural material, that serves to support climbing plants.

The only condition is that this structure must be thin enough to obstruct as less as possible the passage of the sun's rays and wind (Fig. 13).

Effects on Microclimate and Habitability

The essential function of the pergola is to provide shade, although varying its inclination, it can also serve as wind, acoustic or visual barrier.

Table 6 Schemes of microclimatic effects of pergolas

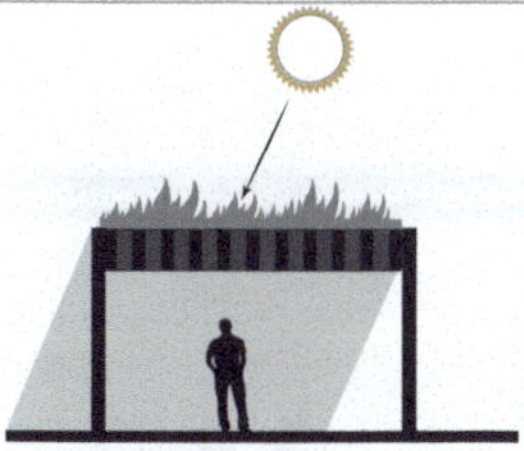 During the day the pergolas provide shade and act as a cold cover, as the vegetation generally does not warm up above air temperature.	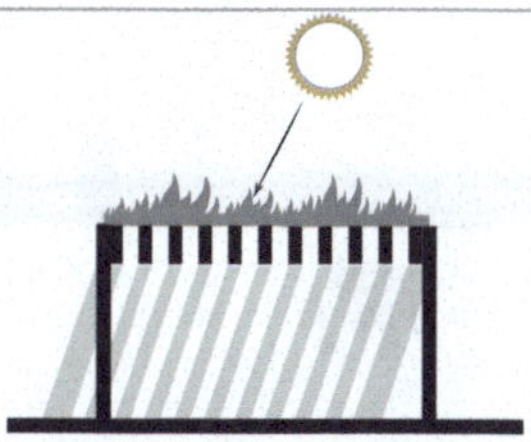 In winter, pergolas with deciduous plants allow the incidence of solar radiation during the day and radiative cooling at night when the sky is clear
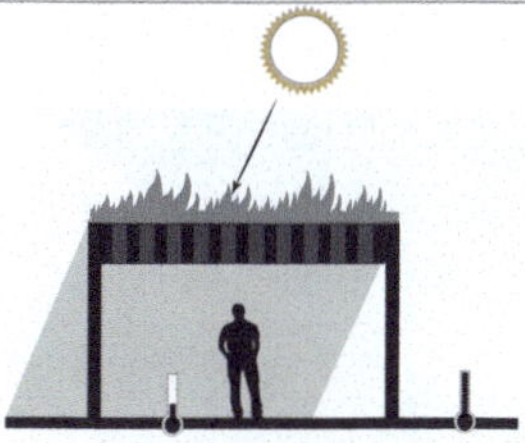 During the day, pergolas prevent radiative heating of the soil surface under their foliage.	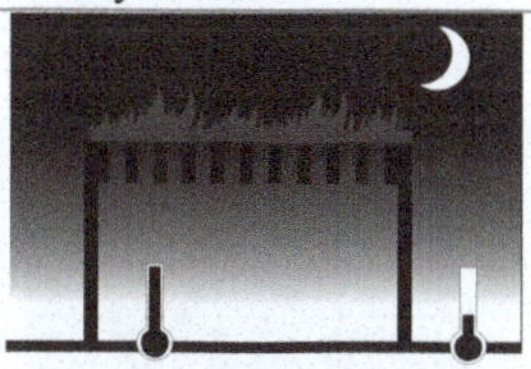 During the night, pergolas prevent radiative cooling of the pavement's surface under their foliage.
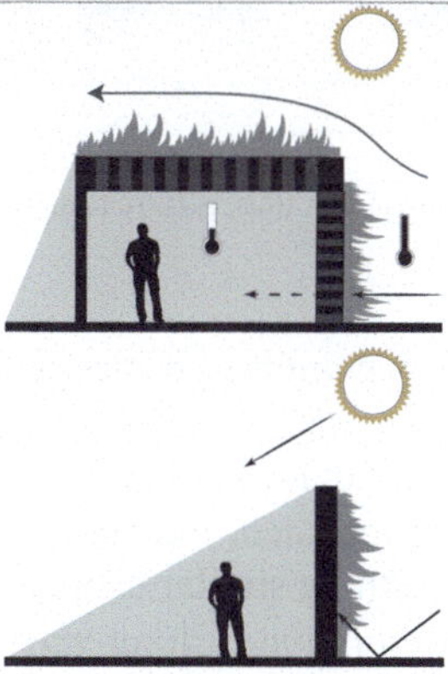 Vertical pergolas can be used as wind and solar radiation barriers at low solar altitudes. They also protect people and buildings from reflections and emissions of long-wave radiation of the surroundings.	

Source: Own work

Depending on its spatial configuration, the environmental impact of the pergola will be diverse.

For example, if it is horizontal, its effect will be to reduce the incidence of solar radiation on the surface below.

If it is vertical, the most noticeable effects will be modifying of wind speed and direction. On a smaller scale and only for very low solar altitudes, it also affects the

Fig. 14 Narrow streets (left) and vegetated courtyards of Ghardaïa, Algeria (right) (source: own work)

Fig. 15 Neighbourhood's park (left) and new residential narrow street (right) with desert vegetation in Hermosillo, Mexico (source: own work)

incidence of solar radiation. Best results can be obtained with combined horizontal and vertical pergolas.

The radiant temperature is another microclimatic parameter that can be affected by pergolas, since the vegetation, in most cases, when well irrigated, maintains a surface temperature similar to the ambient temperature. Something similar happens with the shaded soil under the pergola, which reduces the emission of long-wave radiation on people who are on the site. In Table 6 schemes of microclimatic effects of pergolas can be seen.

Final Thoughts and Conclusions

An outdoor microclimatic space requires specific physical, environmental and functional conditions to be habitable so that humans could carry out their usual activities comfortably and efficiently.

In the case of hot-dry climates like those in the Sonoran Desert, the sense of environmental comfort is intimately related to the search for habitability and looking for how to achieve it. Sometimes it is necessary to make considerable modifications to the physical environment modifying its natural condition. In this way, large cities could be established in the region, such as Tucson and Phoenix in the United States or Mexicali and Hermosillo in Mexico. In these locations, to obtain indoor spaces with comfortable conditions it is necessary to use high-energy-consuming air-conditioning and heating equipment.

However, in outdoor spaces, to reach habitable conditions through mechanical equipment would be unaffordable, and in most cases, effective control of environmental parameters would not be achieved. As we have seen in the previous paragraphs, the use of vegetation is a helpful resource in these cases to control outdoor climatic parameters, although its use goes hand in hand with the consumption of water, which is a scarce asset in desert areas.

In order to achieve habitable conditions, desert cities should be rethought for the future, reconsidering the urban structure, the use of water and energy as well as new lifestyles that could enable us to have more sustainable cities.

According to John Meunier (Meunier 2014), we have much to learn from the old desert cities built before the era of the car and air conditioning, about how to live comfortably in today's desert cities without excessive dependence on non-renewable resources and without putting so much pressure on the environment.

Cities such as Yazd in Iran, Marrakesh in Morocco and Ghardaïa in Algeria (Fig. 14) give us an example of sustainable design with their narrow pedestrian streets shaded by buildings, overhangs and other shading devices, leading to squares and interior courtyards with vegetation. The high density of construction in these cities, in contrast to the large modern cities in the desert, allows the most efficient use of resources.

The vegetation in desert cities' landscapes should behave similar to the ecosystems that can be found in the natural desert. In xeriscape-designed outdoors, grass and other plants that would not naturally survive in a desert climate can be replaced with desert plants, helping to save water and to create an environment more consistent with desert life (Fig. 15).

Acknowledgements We would like to thank María de Lourdes Verdugo, Itzel Martínez and Jaqueline Magaña for their assistance in preparing some images for this work.

References

American Society of Heating, Refrigerating and Air-Conditioning Engineers ASHRAE. (2001). *Handbook fundamentals*. Atlanta: ASHRAE.

Climate.OneBuilding.Org, Repository of free climate data for building performance simulation, Consulted 15 November 2019., http://climate.onebuilding.org/.

Fanger, P. O. (1972). *Thermal comfort*. New York: McGraw Hill.

García, J. R., & Fuentes, V. (1995). *Viento y Arquitectura*. Mexico: Editorial Trillas.

Gómez-Azpeitia, L. G., Bojórquez-Morales, G., Ruiz, R. P., Marincic, I., González, E., Tejeda, A. (2014). Extreme adaptation to extreme environments in hot dry, hot sub-humid and hot

humid climates in Mexico. *Journal of Civil Engineering and Architecture, 8*(8) (serial Nr 81), 929–942, USA. David Publishing Company.
International Organization of Standardization. (1995). ISO 10551:1995 (E) Ergonomics of the thermal environment—assessment of the influence of thermal environment using subjective judgment scales.
International Organization of Standardization. (1998). ISO 7726:1998 (E) Ergonomics of the thermal environment—instruments for measuring physical quantities.
International Organization of Standardization. (2005a). ISO 7730:2005 (E) Ergonomics of the thermal environment—analytical determination and interpretation of thermal comfort using calculation of PMV and PPD indices and local thermal comfort criteria.
International Organization of Standardization. (2005b). ISO 7730:2005 (E) Ergonomics of the thermal environment—analytical determination and interpretation of thermal comfort using calculation of PMV and PPD indices and local thermal comfort criteria.
Marincic, I., Ochoa, J. M., & Alpuche, M. G. (2013). La vivienda económica en Hermosillo: Diagnóstico para mejorar las condiciones de confort térmico y eficientar el uso de la energía. In D. C. Á. Ramírez, S. A. Orozco, & F. C. Canela (Eds.), *Procesos de certificación ambiental de las edificaciones sustentables* (pp. 39–63). Jalisco, Mexico: Centro Universitario de Arte, Arquitectura y Diseño de la Universidad de Guadalajara y Secretaria de Medio Ambiente y Desarrollo Sustentable del Edo. de Jalisco.
Marincic, I., Ochoa, J. M., Alpuche, M. G., & Gómez-Azpeitia, G. (2012). Adaptive thermal comfort for occupants of low-cost dwellings in hot dry climate. *Journal of Civil Engineering and Architecture, 6*(3) (serial Nr 52), 356–363, USA. David Publishing Company.
Marincic, I., Ochoa, J. M., Alpuche, M. G., & Vázquez, E. (2009). Perfil del usuario de la vivienda económica en Hermosillo y patrones de consumo de energía eléctrica. In: XXXIII Semana Nacional de Energía Solar, Guadalajara, Mexico, October 2009, pp. 53–57.
Meunier, J. (2014). Making desert cities. In P. Lusk & A. Simon (Eds.), *Building to endure: Design lessons of arid lands* (pp. 61–88). Albuquerque: The University of New Mexico Press.
Nicol, F. (1993). *Thermal comfort. A handbook for field studies toward an adaptive model*. London: University of East London.
Nicol, J. F., & Humphreys, M. A. (2002). Adaptive thermal comfort and sustainable thermal standards for buildings. *Energy and Buildings, 34*(6), 563–572.
Nikolopoulou, M., Lykoudis, S., & Kikira, M. (2004). Thermal comfort models for urban spaces. In M. Nikolopoulou (Ed.), *Designing open spaces in the urban environment: A bioclimatic approach* (pp. 2–6). Greece: Centre for Renewable Energy Sources.
Ochoa, J. M. (2005). La vegetación como instrumento para el control microclimático (PhD thesis). Universitat Politècnica de Catalunya, Departament de Construccions Arquitectòniques I. http://www.tdx.cat/TDX-0425107-095813. Consulted 15 November 2019.
Ochoa, J. M. (2009). *Ciudad vegetación e impacto climático*. Spain: Erasmus ediciones.
Ochoa, J. M. (2017). La vegetación como elemento de control bioclimáticos en espacios exteriores. In P. Elias & V. Fuentes (Eds.), Estudios de Arquitectura Bioclimática Vol XIII. Mexico: Universidad Autónoma Metropolitana.
Ochoa, J. M., Marincic, I., & Alpuche, M. G. (2006). Confort térmico en espacios exteriores en climas extremos: el clima desértico. In: XXX Semana Nacional de Energía Solar ANES 2006, Veracruz, Mexico, October 2–6, 2006, pp. 115–119.
Ricketts, T. H. (1999). *Terrestrial ecoregions of North America: A conservation assessment*. Washington, D.C: Island Press.
Romero, R., Vázquez, E., Bojórquez, G., Valladares, R., Marincic, I., Gómez, C., Macedo, J., Poujol, F., Estrada, J. (2009). Thermal comfort and occupant perception in dwellings for the low-income sector in hot climates of Mexico. In: 26th conference on passive and low energy architecture PLEA 2009. Quebec City, Canada. June 22–24, 2009.

Conclusions

Greenery can be divided into four types: indoor plants, green walls, green roofs, and green landscaping. Greenery is not a decoration; it is part of architectural design whether inside the buildings, part of the facade, or in the landscape.

Prof. Wim Zeiler quotes in his chapter several added benefits for using greenery in buildings: they have health benefits, absorb CO_2, and improve the acoustics and air quality; it adds to the social cohesion, it will improve the urban heat island (**UHI**) situation, and it can be used as a noise reduction element. Finally, Prof. Zeiler expressed, "Quite a number of studies have shown that biophilic workspaces and interaction with plants may change human behavior, improve productivity and the overall well-being. Evapotranspiration from plants helps lowering the temperature around the planting environment and this can be utilized for air cooling and humidification on city scale as well. In addition to literature overview several recent applications of greenery in projects in the Netherlands are presented to illustrate the architectural added value as well."

Prof. Battisti from Italy in her chapter, "**Green Dreams: Regenerating Cities Through Nature**," in addition to showing several examples of buildings with well-designed greenery parts, concluded, "The recovery of a direct relationship with the earth and the environment represents one of the few social levers to transform the misery of the present into the richness of a possible future."

Prof. Adli and Dr. Abd El Aziz from Egypt in chapter "**Greening the Urban Environment: An Integrated Approach to Planning Sustainable Cities: The Case of Greater Cairo"** deal with not only buildings but also the whole city that must have many green zones and areas which enhance social and creativity incentives in making people happy. They expressed this greater need for greenery stating: Green strategies should capitalize on the great potential for all inhabitants, private sector, and investors to change the current status of green spaces in the GCR. Trained professionals and leadership with clear mechanisms for gathering and updating physical social, economic, and environmental data could cause a colossal uplift in the quality of urban plans. To conclude, further endeavors are essential to create a

A. Sayigh, A. Trombadore (eds.), *The Importance of Greenery in Sustainable Buildings*, Innovative Renewable Energy,
https://doi.org/10.1007/978-3-030-68556-0

comprehensive and pertinent vision for greenways in the GCR with huge prospects for enhancing the quality of life for millions of residents and visitors alike.

In chapter "**Urban Vegetation and Microclimatic Comfort in Warm Climates**," Gustavo Cantuaria (Brazil) and Manuel Correia Guedes (Portugal) limit their analysis to dry hot climate of some regions in Brazil. They analyzed the importance of water and the lack of it resulting in 40,000 children dying each day due to diseases related to lack of water. Planting trees around buildings creating shading is very important in reducing incoming heat through the facades. This was followed by classification of various green plants used to reduce air-conditioning loads in buildings. In arid zones where water is scarce, planting trees will reduce the use of water for cooling; however, they must be suited to the specific region and not any type of trees. The authors found that "Although there is no way to accurately simulate the self-regulating temperature of vegetation, we know it exists and has a very important role in the microclimate and its energy mechanisms. By analysing and comparing different tree microclimates, it is possible to predict certain tendencies. The mango tree has proved to be the equivalent of a natural air conditioner and humidifier. It is the best alternative in creating thermally more comfortable microclimates, and more bearable outside spaces."

Prof. Carolina Ganem Karlen, Argentina, in chapter "**From Vernacular to Sustainable Contemporary Architecture: Urban Green and Patios**" implies that greenery is nature and bringing nature to our habitats means we shall be gaining in beneficial physiological and psychological aspects of our lives. Greenery in buildings has always had multiple functions, from symbolic, aesthetic, or ornamental to production and regulation of the microclimate. The integration of greenery with the constructed spaces is of particular importance today. It is a solution to environmental discomfort and pollution load of urban spaces. Most of her work was related to the oasis city of Mendoza, Argentina. The concept of using greenery in building dates back as far as 3500 years ago to the Hanging Gardens of Babylon in Iraq.

In her view Professor Karlen posits: "Greenery systems can be used as insulation, block solar radiation, and reduce the wall and roof surface temperature resulting in a reduction in heat transfer through building envelopes, which can be as high as 80–90%. The greenery systems have a significant impact especially in hot and dry climates since they were able to block high solar radiation and produce the cooling effect from the evapotranspiration process. To enrich all the described benefits regarding the use of greenery to improve thermal behaviour and energy efficiency of buildings, it seems significant to conclude this chapter by highlighting how much greenery improves comfort and appeal in sustainable buildings."

Dr. Lucia Nelli, Italy, in her chapter deals mostly with rooftop greenhouses: "**Rooftop Greenhouses: Smart and Inclusive Design"** highlights food growing by having different urban rooftop farming forms. "The rooftop gardens produce food for hospitals, markets, universities, research centres and communities giving great contribution to; sustainability, social networking, social inclusion, community engagement and education."

The author stressed that rooftop farming is a complex system which requires several disciplines such as qualitative research, geographic information systems, life cycle assessment, and life cycle costing.

In the chapter written by Prof. Aboulnaga and his associate Ms. Fouad, Egypt, "**Urban Green Coverage: Importance of Green Roofs and Urban Farming Policies in Enhancing Livability in Buildings and Cities: Global and Regional Outlook**, a review study on the policies and laws regulating the design and implementation of green roofs internationally and regionally is presented. The global examples show a very rich contribution and awareness of the importance and sustainable usage of green roofs, walls, and patios.

The authors conclude that many countries in Europe, North America, MENA countries, and Asia have set up specific policies and encouragement of the use of green roofs. They provide several tables representing various cities globally where these policies have been implemented.

The following chapter by Dr. Trombadore and associates, Italy, **Food in the City: Enabling Integrated Solutions for Urban Agriculture and Food Production in Buildings**, highlights the need to supply food to the city by utilizing the roof space by creating circular urban agriculture projects. The authors then designed a Living Tower in Amsterdam. The Living Tower provides a wide range of social, environmental, and economic benefits. The authors concluded, "How to produce food in a healthy way and involve consumers into the production process is fundamental to making them realize that another food system is possible, shifting the paradigm from horizontal, intensive, soil-disruptive crops, to integrated, vertical and sustainable crops.

This is crucial if we want to develop more integrated Vertical Farming projects in our neighborhoods, promoting the image of an architectonic environment as a living organism, being part of the new food production chain, sponsoring zero kilometer products, increasing buildings' sustainable comfort while reducing their energy consumption."

Dr. Seyedehmamak Salavatian, Iran, concentrated in her chapter on Iran, "**The Role of Natural Factors in Courtyard Houses of Hot-Arid Climate of Iran**." In the traditional courtyard houses of Iran one-third of the space is allocated to greenery producing thermal comfort and a pleasant living environment. This cultural adaptation is common in the Middle East especially during the summer months where the temperature can reach 50 °C; the courtyard can act as a cool living space.

The author gives several dimensions to the shape and size of greenery in the courtyards in Iran and concludes, "In most cases a longitudinal pond along the main axis and in the middle of courtyard is created surrounded by greenery zones. The zone can be a large-sized flower bed in the center of the courtyard and the pond is adjacent to it in front of the main façade. However some courtyards are arranged differently mostly due to the physical alterations through time or because of specific social or religious functions."

A chapter was written by Dr. Trombadore and associates, Italy, on "**Adaptive Design of Green Facades and Vertical Farm: Examples of Technological Integration of Microalgae for Energy Production in Resilient Architecture**."

The chapter deals with growing algae and microalgae on the buildings' facades to absorb CO2 emission and generate more oxygen. The chapter outlines the process of growing algae on the walls and harvesting the crops. They identify different algae suited for different climate zones or locality and the period needed to get the maximum crop.

In chapter "**Significance of Courtyard House Design in the Arab World,**" Dr. Falah AlKubaisy, Bahrain, concentrated on the Arab countries and their courtyard housing design. It is a very extensive analysis of how the environment dictates the shape and the architecture in major crowded cities of the Middle East.

As a representative of the Arab countries the chapter deals with the Kingdom of Bahrain building development, which could be applicable to any Arab country.

From thousands of years till the 1950s the courtyard-type houses have dominated the building design in the Middle East. This phenomenon was driven by two significant factors: firstly, the internal privacy of the family activities, and secondly, to reduce the harsh hot weather and dust effect which occurred most of the year in the region.

Chapter "**Use of Vegetation for Microclimatic Control in Arid Zones**" by Prof. J. M. Ochoa; Prof. I. Marincic, Mexico; and Prof. Helena COCH, Spain, discussed the structure characteristics of greenery in buildings.

The chapter listed the greenery influence on building as structural characteristics, physiological characteristics, and microclimatic characteristics of greenery in buildings. The first part deals with the plants used, the second part deals with the environment, while the last part deals with the climate parameters.

The mayor of the city of Utrecht, the fourth largest city in the Netherlands, introduced a law that "no roofs unused"; they should have either plants or mosses or solar systems to reinvigorate biodiversity and create a happy environment. His aim is to green all buildings in the city. The municipality gives grants of up to €20,000 to any householder to green up his or her building. Now it is the greenest city in Holland and all its electricity is generated by solar energy. The city also encourages the use of bicycles for local transport. 316 bus stops are topped with yellow flowering sedum plants, and all its buses are electrically powered from Dutch wind turbines (Ref. The Guardian, 28 March 2020, page 38).

I close my conclusion with a paragraph from the ICON magazine issue 199, spring 2020, page 55. "The Earth can be more productive by leaning into wildest tendencies of its residents; that future generations will need to interact differently with the land in order for the Earth to survive."

If you have a piece of land next to your building, creating a permaculture garden is possible which utilizes some of the less desirable plants alongside other crops producing fresh food as well as a pleasing environment.

All large cities have very many high-rise buildings, most of which have abundant rooftop spaces which can be utilized for rooftop farming, thus providing additional food supply, increased employment, and improved social cohesion.

Ali Sayigh 2020

Index

A. Sayigh, A. Trombadore (eds.), *The Importance of Greenery in Sustainable Buildings*, Innovative Renewable Energy,
https://doi.org/10.1007/978-3-030-68556-0

G

The manufacturer's authorised representative in the EU is Springer Nature Customer Service Centre GmbH, Europaplatz 3, 69115 Heidelberg, Germany. If you have any concerns regarding our products, please contact ProductSafety@springernature.com

Printed and bound by CPI Group (UK) Ltd, Croydon, CR0 4YY
30/09/2026
02225691-0009